Petroleum and Gas Field Processing

Second Edition

CHEMICAL INDUSTRIES

A Series of Reference Books and Textbooks

Founding Editor

HEINZ HEINEMANN
Berkeley, California

Series Editor

JAMES G. SPEIGHT
CD & W, Inc.
Laramie, Wyoming

MOST RECENTLY PUBLISHED

Petroleum and Gas Field Processing, Second Edition, Hussein K. Abdel-Aal, Mohamed A. Aggour, and Mohamed A. Fahim

Handbook of Refinery Desulfurization, Nour Shafik El-Gendy and James G. Speight

Refining Used Lubricating Oils, James Speight and Douglas I. Exall

The Chemistry and Technology of Petroleum, Fifth Edition, James G. Speight

Educating Scientists and Engineers for Academic and Non-Academic Career Success, James Speight

Transport Phenomena Fundamentals, Third Edition, Joel Plawsky

Synthetics, Mineral Oils, and Bio-Based Lubricants: Chemistry and Technology, Second Edition, Leslie R. Rudnick

Modeling of Processes and Reactors for Upgrading of Heavy Petroleum, Jorge Ancheyta

Synthetics, Mineral Oils, and Bio-Based Lubricants: Chemistry and Technology, Second Edition, Leslie R. Rudnick

Fundamentals of Automatic Process Control, Uttam Ray Chaudhuri and Utpal Ray Chaudhuri

The Chemistry and Technology of Coal, Third Edition, James G. Speight

Practical Handbook on Biodiesel Production and Properties, Mushtaq Ahmad, Mir Ajab Khan, Muhammad Zafar, and Shazia Sultana

Introduction to Process Control, Second Edition, Jose A. Romagnoli and Ahmet Palazoglu

Fundamentals of Petroleum and Petrochemical Engineering, Uttam Ray Chaudhuri

Advances in Fluid Catalytic Cracking: Testing, Characterization, and Environmental Regulations, edited by Mario L. Occelli

Advances in Fischer-Tropsch Synthesis, Catalysts, and Catalysis, edited by Burton H. Davis and Mario L. Occelli

Petroleum and Gas Field Processing

Second Edition

Hussein K. Abdel-Aal
Mohamed A. Aggour
Mohamed A. Fahim

CRC Press is an imprint of the
Taylor & Francis Group, an **informa** business

CRC Press
Taylor & Francis Group
6000 Broken Sound Parkway NW, Suite 300
Boca Raton, FL 33487-2742

Printed on acid-free paper
Version Date: 20150817

International Standard Book Number-13: 978-1-4822-5592-8 (Paperback)

Visit the Taylor & Francis Web site at
http://www.taylorandfrancis.com

and the CRC Press Web site at
http://www.crcpress.com

Contents

Section II Separation of Produced Fluids

Section III Crude Oil Treatment

Section V Surface Production Facilities

Preface

As oil exploration and production (E&P) activities over the world continue to grow, we academicians and engineering professionals must fully grasp the technology and processes involved within their function to proactively support oil and gas field operations. A competent understanding of technology and various processes that drive E&P also provides an overall appreciation of the role played by surface production operations.

In addition, the very high cost and risk involved in E&P demands the reevaluation of all operations encountered in handling the oil–gas mixture from well head all the way to quality petroleum oil and gas products. Many oil production processes present a significant challenge to the oil and gas field processing facilities and equipment design. The optimization of the sequential operations of handling the oil–gas mixture can be a major factor in increasing oil and gas production rates and reducing operating costs.

Fully revised and updated to reflect major changes over the past 10 years or so, this second edition offers thorough coverage of every sector in the field processing of produced crude oil along with its associated gas. We intend to go forward to continue building on our foundation success attained in the first edition. Our mission remains unchanged: to deliver an expanded and updated volume that covers the principles and procedures related to the processing of reservoir fluids for the separation, handling, treatment, and production of quality petroleum oil and gas products.

Adding new information about field facilities to this edition, this book contains 18 chapters (compared to 13 in the first edition) grouped into five sections. Section I presents a broad background that covers the production of oil and gas (Chapter 1), a preview of principal field processing operations (Chapter 2), composition and types of crude oil and products (Chapter 3), composition and characterization of natural gas (Chapter 4), and the role of economics in oil and gas field operations (Chapter 5). Section II handles two-phase and three-phase gas–oil separation (Chapters 6 and 7, respectively). Crude oil treatment is detailed in Section III, which covers emulsion treatment and dehydration of crude oil (Chapter 8), desalting (Chapter 9), stabilization and sweetening of crude oil (Chapter 10), and other treatment options (Chapter 11). Gas handling and treatment, covered in Section IV, includes sour gas treatment, gas dehydration, and separation and production of natural gas liquids (NGLs), Chapters 12, 13, and 14, respectively. Section V, on the other hand, is devoted to surface production facilities, a new addition to the second edition. It includes four chapters covering the topics of produced water management and disposal (Chapter 15), field storage tanks, vapor recovery units (VRUs) and tank blanketing (Chapter 16), oil field chemicals (Chapter 17), and piping and pumps (Chapter 18).

This second edition takes advantage of recent publications with immense knowledge in the area of surface petroleum operations by the inclusion of new subjects, in particular natural gas, economics and profitability, oil field chemicals, and piping and pumps. These additions contribute new features to our book, especially when it comes to the dollar sign in an economic study.

The concept of unit operation, which is discussed in the "Introduction" of this book, is presented in many surface operations in this new edition. Unit operations are identified in crude oil field treatment and natural gas processing. Distillation, mixing, and absorption are typical examples.

An all-inclusive guide to surface petroleum operations, the text provides a comprehensive and visionary approach to solve problems encountered in field processing of oil and gas. It contains examples and case studies from a variety of oil field operations. Example step-by-step exercises are worked out. This book is arranged so that it can be used both as a text and as a reference. As a textbook, it would fit nicely for courses on surface petroleum operations taught in many schools all over the world. It would be suitable for use in a one- or two-semester course for students majoring in petroleum engineering, chemical engineering, and allied engineering. On the other hand, it would be invaluable for experts, engineers, and practicing professionals working in the petroleum industry.

The authors are indebted to the many oil organizations and individuals who have provided information and comments on the subject materials presented in this edition. As far as the production and the publication of this edition, we feel a deep sense of gratitude to Barbara Glunn, Robert Sims, and Kari Budyk of Taylor & Francis, and Adel Rosario of MTC.

Hussein K. Abdel-Aal
Mohamed A. Aggour
Mohamed A. Fahim

Authors

Prof. Hussein K. Abdel-Aal is an emeritus professor of chemical engineering and petroleum refining at NRC, Cairo, Egypt, and KFUPM, Dhahran, Saudi Arabia. He worked in the oil industry (1956–1960) as a process engineer in Suez oil refineries before working on his graduate studies in the United States.

From 1971 to 1988, Prof. Abdel-Aal was with the Department of Chemical Engineering, KFUPM, Dhahran, where he was the head from 1972 to 1974. He was a visiting professor with the Chemical Engineering Department at Texas A&M in 1980–1981. In 1985–1988, Prof. Abdel-Aal assumed the responsibilities of the head of the solar energy department in NRC, Cairo.

Prof. Abdel-Aal has contributed to more than 90 technical papers and is the main author of the text book *Petroleum and Gas Field Processing* (Marcel Dekker Inc., 2003); he is also the editor of *Petroleum Economics & Engineering, Third Edition* (CRC Press, 2014).

He is a fellow and a founding member of the board of directors of the International Association of Hydrogen Energy, Miami, Florida. He is on the honorary editorial board of *International Journal of Hydrogen Energy.*

Dr. Mohamed A. Aggour is a professor and former chairman of the petroleum engineering program of Texas A&M University at Qatar. Prior to this, he was a professor of petroleum engineering at the Petroleum Institute, Abu Dhabi, and King Fahd University of Petroleum and Minerals, Saudi Arabia.

Dr. Aggour was the leader of the Production Technology Research Group of Esso Resources Canada and a staff production engineer at the East Texas Division of Exxon Company. He has more than 48 years of combined academic, industry, and research experience. He has numerous publications and three patents.

He received 19 departmental and six university-level Distinguished Teaching Awards, and three Distinguished Research Awards. He was the recipient of the 2012 Best Petroleum Engineering Faculty from the Institute of Academic Excellence and the 2013 Society of Petroleum Engineering Regional Award for Distinguished Achievement by the Petroleum Engineering Faculty.

Dr. Mohamed A. Fahim was a professor and the chairman of the Chemical Engineering Department for 40 years at the University of Kuwait and the University of United Arab Emirates.

Dr. Fahim published more than 150 papers in the field of petroleum refining and gas processing.

Prof. Fahim is the main author of *Fundamentals of Petroleum Refining* (Elsevier, 2010) and a coauthor of *Petroleum and Gas Field Processing* (Marcel Dekker Inc., 2003).

Introduction

This new edition of *Petroleum and Gas Field Processing* comes at a time when oil producers are taking a close look at the economy of oil field operations to improve the ultimate recovery and to maximize the yield of oil and gas obtained during the surface processing operations.

Oil field operations in general encompass three main phases, as shown in the following block diagram.

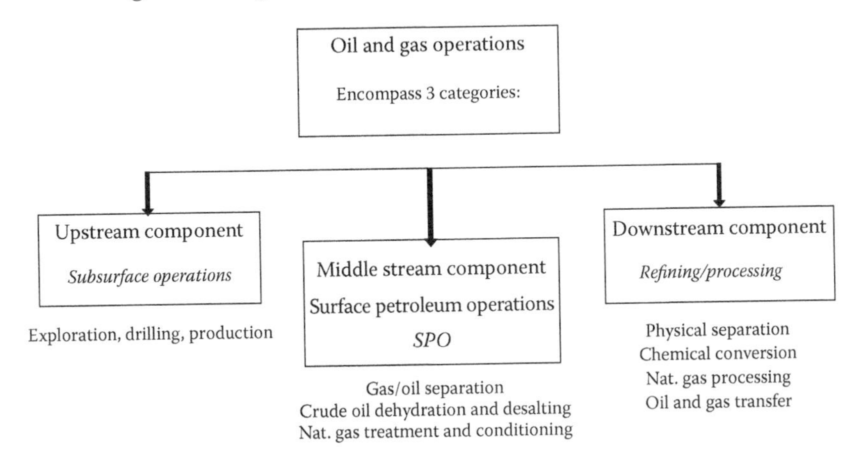

This is illustrated further by the following flow diagram:

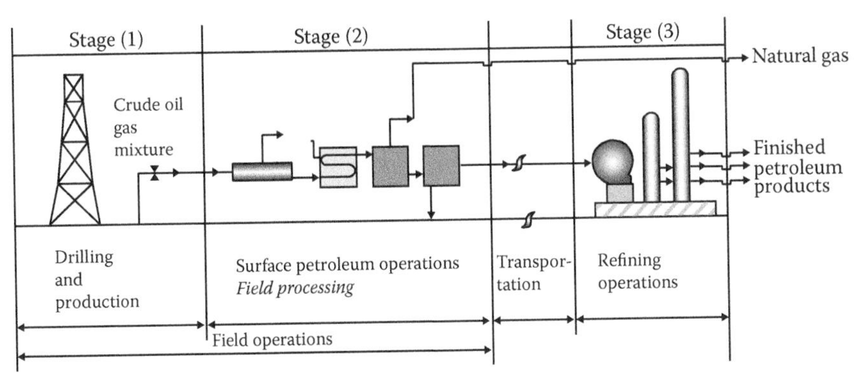

Petroleum and gas field processing operations, referred to as surface petroleum operations (SPO), cover the myriad procedures required to handle the crude oil mixture between the well head and the delivery points for refining operations and other usages.

Crude oil is far from being one homogeneous substance. Its physical characteristics differ depending on where the crude oil is found, and those variations determine its usage and price as well. As the U.S. Energy Information Administration (EIA) puts it succinctly, "not all crude is created equal" (U.S. Energy Information Administration, 2015). It is worth mentioning that more than 40,000 oil fields are scattered around the globe, on land and offshore. The largest are the Ghawar Field in Saudi Arabia and the Burgan Field in Kuwait, with more than 60 billion barrels (9.5×10^9 m^3) estimated in each. Most oil fields are much smaller. According to the EIA, as of 2003 the United States alone had more than 30,000 oil fields. More than half the United States' oil reserves are located in its 100 largest fields. According to a new EIA report, these massive fields account for 20.6 billion of the 36.5 billion barrels of oil, or 56 percent of the total ("List of Oil Fields," 2015).

Many oil production processes present a significant challenge to oil and gas field processing facilities. This applies to the design and operations of the processing equipment. A typical example is the fact that the nature of crude oil emulsions changes continuously as the producing field depletes. Therefore, conditions change as well.

In this new edition, we attempt to introduce the concept of unit operations used by chemical engineers to provide the readers with tools for creating a *surface* process system that will economically separate and treat oil–gas mixtures as they exit the wellhead into quality salable oil and gas products.

What Is a Unit Operation?

A unit operation represents a basic physical operation in a chemical or petroleum process plant. Examples are distillation, absorption, fluid flow, and heat and mass transfer. Fundamentals pertaining to a given unit operation are the same regardless of its industrial applications. This is how pioneers came up with the term *unit operation*.

Unit operations deal mainly with the transfer and the change of both materials and energy primarily by physical means, arranged as needed by a given petroleum or chemical industry. The following is a partial list of some important unit operations:

- Fluid flow—Deals with the principles governing the flow and transportation of fluids.

- Heat transfer—Deals with the principles underlying the heat transfer by different modes.
- Distillation, absorption, extraction, and drying—Known as diffusional mass transfer unit operations. Separation of petroleum hydrocarbons, by these unit operations, is accomplished by the transfer of molecules from one phase to the other by diffusion. Typical examples are distillation and absorption.

The significance of introducing the unit operation concept in understanding the processing surface operations in an oil field will be apparent to readers when it is realized that most of these surface operations are *physical operations*, or *nonreacting processes*. They deal mainly with the transfer and the transformation of *energy*, and the transfer, separation, and conditioning (treating) of *materials* by physical means. Three modes of transfer that take place in oil field processing operations are recognized as follows:

1. Momentum transfer (gained by fluid flow)
2. Heat transfer of oil using heat exchangers and furnaces
3. Mass transfer in distillation columns, absorbers, and others that lead to enrichment and separation (transfer is due to the diffusion of the molecules that separates the light from the heavy)

These three modes of transfer are usually covered under the topic of transport phenomena.

Examples of some common unit operations that take place during oil and gas field processing are listed next along with their specific applications.

Unit Operation	Application
Equilibrium flashing	Gas–oil separation
Distillation/stripping	Crude oil stabilization/sweetening
Absorption	Treatment of natural gas
Fluid flow	Most of field operations
Heat transfer	Most of field operations

The proposed treatment of the subject matter of petroleum and gas field processing will follow a chronological sequence of field operations in transient. Each is described with the unit operation concept when applicable as the oil–gas mixture proceeds from the wellhead until the crude oil is finally separated, treated, and stored ready for shipping or refining. The same is true for the associated gas, passing through its journey until it is finally a quality sale gas.

Process System

A process system is a collection of equipment that affects the required separation or treatment through physical methods or chemical changes. For example, by means of a properly designed processing system, crude oil desalting is accomplished by intimate mixing of the crude oil with dilution water.

By applying the unit operation concept, we are able to identify the process system that handles crude oil–gas mixtures all the way through as shown next.

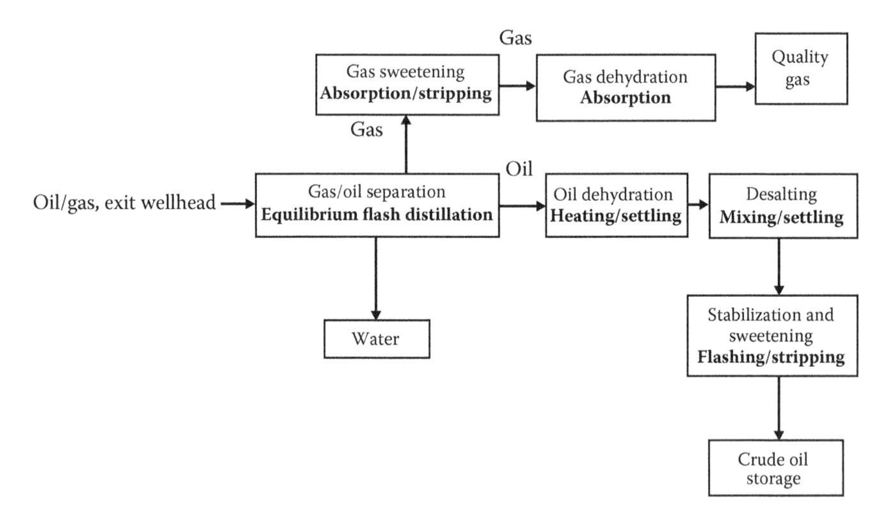

Following the order of events using this flow sheet, one can determine the function of every piece of equipment involved in an oil field process design, highlighted in bold. The immediate next step is the determination of the size and the type of equipment needed to carry out the physical changes, using the unit operations principle underlying this process. Remember that the fundamentals pertaining to a given unit operation are the same regardless of its application, hence unit operation.

Section I

Background

In this section, oil and gas from formation to production is highlighted first (Chapter 1). A preview of the principal field surface processing operations is given next (Chapter 2), followed by an elaborate presentation on the composition, and types of crude oil and its products (Chapter 3), and the composition and characterization of natural gas and its products (Chapter 4). An important part in this background, introduced in this new edition, is the focus on the role of economics in oil and gas field operations as presented in Chapter 5.

1

Oil and Gas from Formation to Production

1.1 Introduction

The main goal of our book is to present field surface operations and their facilities for handling and processing the produced oil, gas, and water. This is an important aspect in the overall planning and development of the field and must be considered and integrated into the early stages of planning, economic evaluation, and development of the field. One has to be aware that the digital oil field (DOF) and its defining principles have matured in recent years to permeate standard business practice in the petroleum industry. Integrated operations (IO) is an evolutionary term that refers to work processes and ways of performing hydrocarbon exploration and production, facilitated by information and communication technology. The impact and value of IO, or its potential, is far reaching and profound.

Before considering the surface production operations and facilities, however, extensive work and studies are first made to characterize and evaluate the reservoir, determine the production strategy for the life of the field, design the well completions that are compatible with the production strategy, and design the well-drilling programs.

Petroleum engineering students would, normally, have covered the subject matters related to the reservoir, well completion, drilling, and subsurface production methods before taking a course on surface handling and treatment of the produced fluids. Non-petroleum-engineering students taking this course, however, would be lacking such important and useful knowledge and background. This chapter, therefore, provides brief background information for the non-petroleum-engineering students to appreciate all operations related to the production of oil and gas.

A brief description of how oil and gas are formed and accumulated underground is first presented. An overview of the life cycle of oil and gas fields from exploration to abandonment is then presented. The exploration activities used in finding (discovering) petroleum reservoirs are then highlighted followed by descriptions of the various types of petroleum reservoir according to their geologic and production classifications. Finally, an overview of the field development work, including the reservoir, drilling, and production engineering aspects of the development, are summarized.

1.2 Formation and Accumulation of Oil and Gas

1.2.1 Formation of Oil and Gas

Several theories have been proposed to explain the formation and origin of oil and gas (petroleum); these can be classified as the organic theory of petroleum origin and the inorganic theory of origin. The organic theory provides the explanation most accepted by scientists and geologists.

It is believed, and there is evidence, that ancient seas covered much of the present land area millions of years ago. The Arabian Gulf and the Gulf of Mexico, for example, are parts of such ancient seas. Over the years, rivers flowing to these seas carried large volumes of mud and sedimentary materials into the sea. The mud and sedimentary materials were distributed and deposited layer upon layer over the sea floor. The buildup of thousands of feet of mud and sediment layers caused the sea floors to slowly sink and be squeezed. This eventually became the sedimentary rocks (the sandstone and shale, and the carbonates) where petroleum is found today.

The very large amount of small plant and animal life, which came into the sea with river mud and sedimentary materials, and the much larger amount of small marine life remains already on the sea floors constituted the source of petroleum. These small organisms died and were buried by the depositing silt and thus were protected from ordinary decay. Over many years, pressure, temperature, bacteria, and other reactions caused these dead organisms to change into oil and gas. The gas was formed under the higher temperature conditions, whereas the oil was formed under the lower temperature conditions. The rocks where oil and gas were formed are known as the source rock.

1.2.2 Accumulation of Oil and Gas

The oil, gas, and salt water occupied the pore spaces between the grains of the sandstones, or the pore spaces, cracks, and vugs of the limestones and dolomites. Whenever these rocks were sealed by a layer of impermeable rock, the cap rock, the petroleum accumulating within the pore spaces of the source rock, was trapped and formed the petroleum reservoir. However, when such conditions of trapping the petroleum within the source rocks did not exist, oil gas moved (migrated), under the effects of pressure and gravity, from the source rock until it was trapped in another capped (sealed) rock.

Gas, oil, and water segregate within the trap rocks, because of the differences in density. Gas, when existing, occupied the upper part of the trap and water occupied the bottom part of the trap, with the oil between the gas and water. Complete displacement of water by gas or oil never occurred. Some salt water stayed with the gas and/or oil within the pore spaces and as a film covering the surfaces of the rock grains; this water is known as the connate water, and it may occupy from 10% up to 50% of the pore volume.

The geologic structure in which petroleum has been trapped and has accumulated, whether it was the source rock or the rock to which petroleum has migrated, is called the petroleum reservoir.

In summary, the formation of a petroleum reservoir involves first the accumulation of the remains of land and sea life and their burial in the mud and sedimentary materials of ancient seas. This is followed by the decomposition of these remains under conditions that recombine the hydrogen and carbon to form the petroleum mixtures. Finally, the formed petroleum is either trapped within the porous source rock when a cap rock exists or it migrates from the source rock to another capped (sealed) structure.

1.3 Life Cycle of Oil and Gas Fields

Oil and gas field development starts with the exploration activities. As described later, geological and geophysical surveys and studies are used to determine the location where a hydrocarbon reservoir may potentially exist. The results of such studies merely provide information about the potential location of the reservoir, its area, depth, and some characteristics such as faults and fractures. Based on the information available, a location (normally at the center of the potential reservoir) is selected to drill the first exploration well, called *wild cat*. The design of this well is based on experience since no data are yet available for proper design of the well. As this well is being drilled, rock and fluid properties data for all penetrated formations are collected and analyzed. More attention is given, and more data are collected and analyzed through the target depth of the potential reservoir. If hydrocarbons (oil and/or gas) are found, the well is tested to determine the production potential; otherwise, the well is considered a *dry well* and is abandoned.

If the wild cat is successful, more exploration wells will be drilled and tested. The number and locations of these wells are determined to provide as much information as possible about the reservoir volume, the amount of hydrocarbons in place, and the production potential of the discovery. Preliminary reservoir simulation studies coupled with economic evaluations are made at this stage to determine whether the discovery is commercially viable. Once a decision is made to develop the field, extensive simulation studies will be conducted to examine various development and production strategies with the objective of determining the optimum development and production plan, which yield the maximum recovery and best economics.

Following this, well completion designs will be made with the objective of having wells work for the entire life of the field, providing maximum recovery in the most economic and safe manner. Based on the completion designs, the well drilling designs and programs will be developed. At this stage and based on the production forecast, the surface facilities for separation and treatment of produced fluids are designed. Procurement of materials and equipment is planned and made to secure their availability on time for actual field development and production.

Drilling operations then start according to schedule. Each drilled well is tested and evaluated, and the drilling program could be modified based on the data collected. To accelerate revenue, all or part of the surface production and processing facility should be on location to produce wells as they are drilled and completed.

Production data (production rates, pressures, temperatures, gas–oil ratio, and water cut, if any) are collected for a period of time and then compared against the forecasted (predicted) data from reservoir simulations. Normally, no match would be obtained. Then a process called history matching is performed where the reservoir simulations are modified (by changing the data used in developing the original simulations that have the least certainty) until the simulation data match the actual production data. The modified simulations are then used to forecast future production. Again after a period of time, the actual production data are compared against the recent simulation data. Again, no match would be obtained. The history matching process is repeated and this would probably continue until the end of the life of the field.

It should be noted that several operations, such as pressure maintenance, improved/enhanced recovery, and artificial lift, might be implemented during the production life of the field. When no more hydrocarbons can be economically produced, the field is abandoned. To do so, wells have to be killed, filled with layers of cement and sand, and the surface casing capped. This process is governed by either company or government regulations.

1.4 Finding Oil and Gas: Exploration

As explained in the previous sections, oil and gas exist in reservoirs located thousands of feet below the earth's surface and ocean floors. These reservoirs would exist only in certain locations depending on the geologic history of the earth. Therefore, determining the location of petroleum reservoirs is a very difficult task and probably is the most challenging aspect of the petroleum industry. Finding or discovering a petroleum reservoir involves three major activities: geologic surveying, geophysical surveying, and exploratory drilling activities. The following sections provide a brief background on each of these activities.

1.4.1 Geologic Survey

Geologic surveying is the oldest and first used tool for determining potential locations of underground petroleum reservoirs. It involves examination of the surface geology, formation outcrops, and surface rock samples. The collected information is used in conjunction with geologic theories to determine whether petroleum reservoirs could be present underground at the surveyed location. The results of the geologic survey are not conclusive and only offer a possibility of finding petroleum reservoirs. The rate of success of finding petroleum reservoirs using geologic surveys alone has been historically low. Currently, geologic surveys are used together with other geophysical surveys to provide higher rates of success in finding petroleum reservoirs.

1.4.2 Geophysical Surveys

There are mainly four types of geophysical surveys used in the industry: gravity survey, magnetic survey, seismic survey, and remote sensing.

The gravity survey is the least expensive method of locating a possible petroleum reservoir. It involves the use of an instrument, a gravimeter, which picks up a reflection of the density of the subsurface rock. For example, because salt is less dense than rocks, the gravimeter can detect the presence of salt domes, which would indicate the presence of an anticline structure. Such a structure is a candidate for possible accumulation of oil and gas.

The magnetic survey involves measurement of the magnetic pull, which is affected by the type and depth of the subsurface rocks. The magnetic survey can be used to determine the existence and depth of subsurface volcanic formations, or basement rocks, which contain high concentrations of magnetite. Such information is utilized to identify the presence of sedimentary formations above the basement rocks.

The seismic survey involves sending strong pressure (sound) waves through the earth and receiving the reflected waves off the various surfaces of the subsurface rock layers. The sound waves are generated either by using huge land vibrators or using explosives. The very large amount of data collected, which include the waves' travel times and characteristics, are analyzed to provide definitions of the subsurface geological structures and to determine the locations of traps that are suitable for petroleum accumulation.

This type of survey is the most important and most accurate of all of the geophysical surveys. Significant technological developments in the field of seismic surveying have been achieved in recent years. Improvements in the data collection, manipulation, analysis, and interpretation have increased the significance and accuracy of seismic surveying. Further, the development of three-dimensional (3D) seismic surveying technology has made it possible to provide 3D descriptions of the subsurface geologic structures.

Remote sensing is a modern technique that involves using infrared, heat-sensitive color photography to detect the presence of underground mineral deposits, water, faults, and other structural features. The sensing device, normally on a satellite, feeds the signals into special computers that produces maps of the subsurface structures.

1.4.3 Exploratory Drilling

The data collected from the geologic and geophysical surveys are used to formulate probable definitions and realizations of the geologic structure that may contain oil or gas. However, we still have to determine whether petroleum exists in these geologic traps, and if it does exist, would it be available in such a quantity that makes the development of the oil/gas field economical? The only way to provide a definite answer is to drill and test exploratory wells.

The exploratory well, known as the wildcat well, is drilled in a location determined by the geologists and geophysicists. The well is drilled with insufficient data available about the nature of the various rock layers that will be drilled or the fluids and pressures that may exist in the various formations. Therefore, the well completion and the drilling program are usually overdesigned to ensure safety of the operation. This first well, therefore, does not represent the optimum design and would probably cost much more than the rest of the wells that will be drilled in the field.

As this exploratory well is drilled, samples of the rock cuttings are collected and examined for their composition and fluid content. The data are used to identify the type of formation versus depth and to check on the presence of hydrocarbon materials within the rock. Cores of the formations are also obtained, preserved, and sent to specialized laboratories for analysis. Whenever a petroleum-bearing formation is drilled, the well is tested while placed on controlled production. After the well has been drilled, and sometimes at various intervals during drilling, various logs are taken. There are several logging tools or techniques (electric logs, radioactivity logs, and acoustic logs) that are used to gather information about the drilled formations. These tools are lowered into the well on a wireline (electric cable) and, as they are lowered, the measured signals are transmitted to the surface and recorded on computers. The signals collected are interpreted and produced in the form of rock and fluid properties versus depth.

The exploratory well will provide important data on rock and fluid properties, type and saturation of fluids, initial reservoir pressure, reservoir productivity, and so forth. These are essential and important data and information that are needed for the development of the field. In most situations, however, the data provided by the exploratory well will not be sufficient. Additional wells might need to be drilled to provide a better definition of the size and characteristics of the new reservoir. Of course, not every exploratory well will result in a discovery. Exploratory wells may result in hitting dry holes or they may prove the reservoir to be an uneconomical development.

1.5 Types of Petroleum Reservoirs

Petroleum reservoirs are generally classified according to their geologic structure and their production (drive) mechanism.

1.5.1 Geologic Classification of Petroleum Reservoirs

Petroleum reservoirs exist in many different sizes and shapes of geologic structures. It is usually convenient to classify the reservoirs according to the conditions of their formation as follows.

- Dome-shaped and anticline reservoirs—These reservoirs are formed by the folding of the rock layers as shown in Figure 1.1. The dome is circular in outline, and the anticline is long and narrow. Oil or gas moved or migrated upward through the porous strata where it was trapped by the sealing cap rock and the shape of the structure.
- Faulted reservoirs—These reservoirs are formed by shearing and offsetting of the strata (faulting), as shown in Figure 1.2. The movement of the nonporous rock opposite the porous formation containing the oil/gas creates the sealing. The tilt of the petroleum-bearing rock and the faulting trap the oil/gas in the reservoir.
- Salt-dome reservoirs—This type of reservoir structure, which takes the shape of a dome, was formed due to the upward movement of a large, impermeable salt dome that deformed and lifted the overlying layers of rock. As shown in Figure 1.3, petroleum is trapped between the cap rock and an underlying impermeable rock layer, or between two impermeable layers of rock and the salt dome.

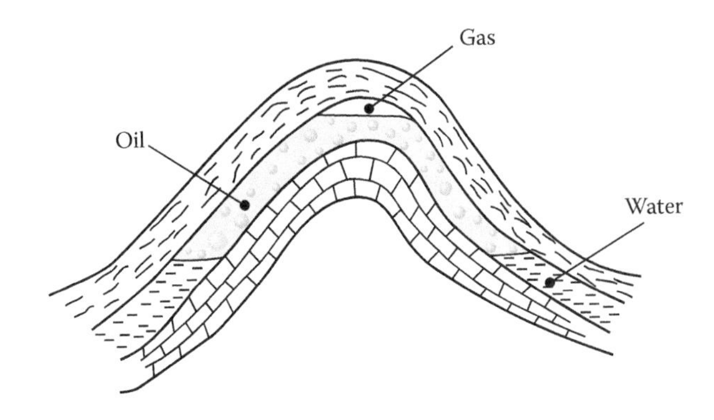

FIGURE 1.1
A reservoir formed by folding of rock layers.

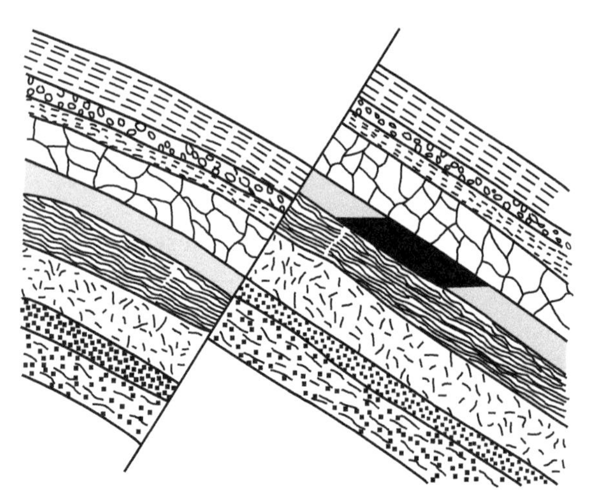

FIGURE 1.2
A cross section of a faulted reservoir.

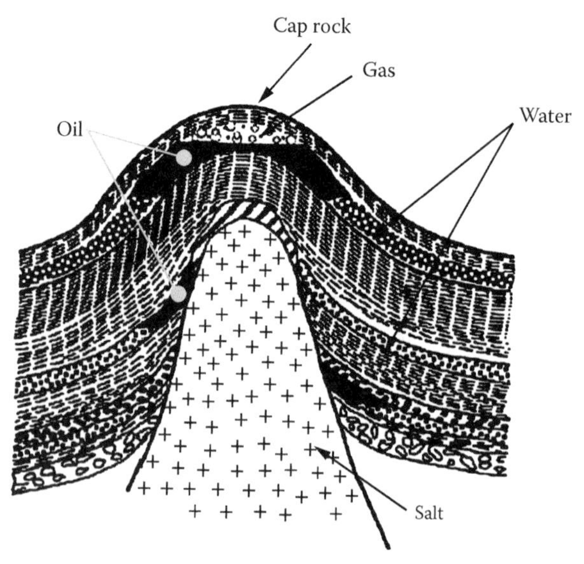

FIGURE 1.3
Section in a salt dome structure.

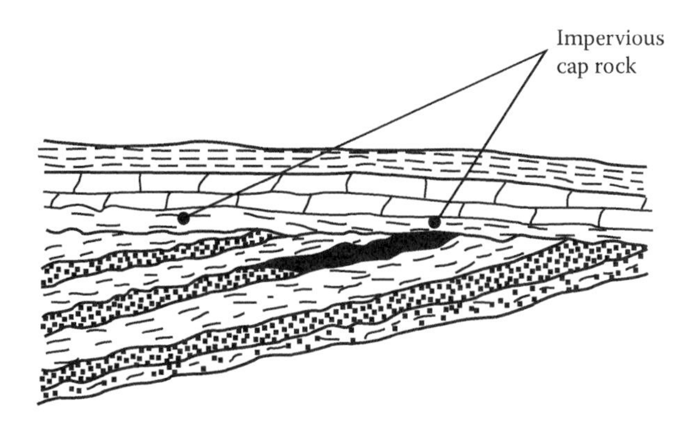

FIGURE 1.4
A reservoir formed by unconformity.

- Unconformities—This type of reservoir structure, shown in Figure 1.4, was formed as a result of an unconformity where the impermeable cap rock was laid down across the cutoff surfaces of the lower beds.
- Lense-type reservoirs—In this type of reservoir, the petroleum-bearing porous formation is sealed by the surrounding, nonporous formation. Irregular deposition of sediments and shale at the time the formation was laid down is the probable cause for this abrupt change in formation porosity. An example of this type of reservoir is shown in Figure 1.5.
- Combination reservoirs—In this case, combinations of folding, faulting, abrupt changes in porosity, or other conditions create the trap from this common type of reservoir.

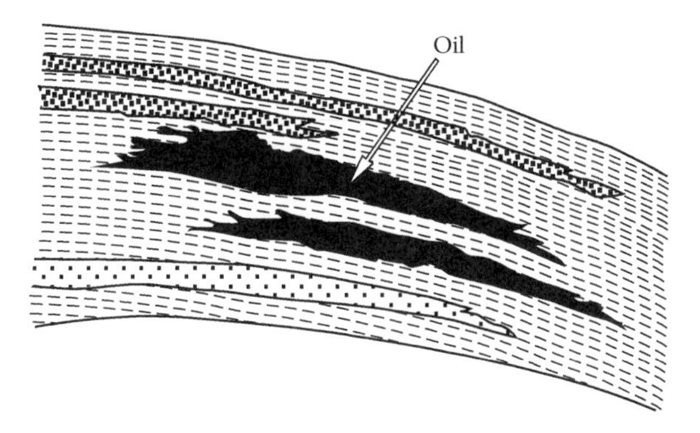

FIGURE 1.5
An example of a sandstone lense-type reservoir.

1.5.2 Reservoir Drive Mechanisms

At the time oil was forming and accumulating in the reservoir, the pressure energy of the associated gas and water was also stored. When a well is drilled through the reservoir and the pressure in the well is made to be lower than the pressure in the oil formation, it is that energy of the gas, or the water, or both that would displace the oil from the formation into the well and lift it up to the surface. Therefore, another way of classifying petroleum reservoirs, which is of interest to reservoir and production engineers, is to characterize the reservoir according to the production (drive) mechanism responsible for displacing the oil from the formation into the wellbore and up to the surface. There are three main drive mechanisms:

1. Solution gas drive reservoirs—Depending on the reservoir pressure and temperature, the oil in the reservoir would have varying amounts of gas dissolved within the oil (solution gas). Solution gas would evolve out of the oil only if the pressure is lowered below a certain value, known as the bubble point pressure, which is a property of the oil. When a well is drilled through the reservoir and the pressure conditions are controlled to create a pressure that is lower than the bubble point pressure, the liberated gas expands and drives the oil out of the formation and assists in lifting it to the surface. Reservoirs with the energy of the escaping and expanding dissolved gas as the only source of energy are called solution gas drive reservoirs. This drive mechanism is the least effective of all drive mechanisms; it generally yields recoveries between 15% and 25% of the oil in the reservoir.
2. Gas cap drive reservoirs—Many reservoirs have free gas existing as a gas cap above the oil. The formation of this gas cap is due to the presence of a larger amount of gas than can be dissolved in the oil at the pressure and temperature of the reservoir. The excess gas is segregated by gravity to occupy the top portion of the reservoir.

   In such reservoirs, the oil is produced by the expansion of the gas in the gas cap, which pushes the oil downward and fills the pore spaces formerly occupied by the produced oil. In most cases, however, solution gas is also contributing to the drive of the oil out of the formation. Under favorable conditions, some of the solution gas may move upward into the gas cap and, thus, enlarge the gas cap and conserves its energy. Reservoirs produced by the expansion of the gas cap are known as gas cap drive reservoirs. This drive is more efficient than the solution gas drive and could yield recoveries between 25% and 50% of the original oil in the reservoir.
3. Water drive reservoirs—Many other reservoirs exist as huge, continuous, porous formations with the oil/gas occupying only a small

portion of the formation. In such cases, the vast formation below the oil/gas is saturated with salt water at very high pressure. When oil/gas is produced by lowering the pressure in the well opposite the petroleum formation, the salt water expands and moves upward, pushing the oil/gas out of the formation and occupying the pore spaces vacated by the produced oil/gas. The movement of the water to displace the oil/gas retards the decline in oil, or gas pressure, and conserves the expansive energy of the hydrocarbons.

Reservoirs produced by the expansion and movement of the salt water below the oil/gas are known as water drive reservoirs. This is the most efficient drive mechanism; it could yield recoveries up to 50% of the original oil.

1.6 Development of Oil and Gas Fields

The very large volume of information and data collected from the various geologic and geophysical surveys and the exploratory wells are used to construct various types of maps. Contour maps are lines drawn at regular intervals of depth to show the geologic structure relative to reference points called the correlation markers. Isopach maps illustrate the variations in thickness between the correlation markers. Other important maps such as porosity maps, permeability maps, and maps showing variations in rock characteristics and structural arrangements are also produced. With all data and formation maps available, conceptual models describing the details of the structure and the location of the oil and gas within the structure are prepared.

The data available at this stage will be sufficient to estimate the petroleum reserves and decide and plan for the development of the field for commercial operation.

The development of petroleum fields involves the collective and integrated efforts and experience of many disciplines. Geologists and geophysicists are needed, as described in Section 1.3, to define, describe, and characterize the reservoir. Reservoir engineers set the strategy for producing the petroleum reserves and managing the reservoir for the life of the field. Production and completion engineers design the well completions and production facilities to handle the varying production methods and conditions, and drilling engineers design the well drilling programs based on well completion design. In the past, each group used to work separately and deliver its product to the next group. That is, when geologists and geophysicists finish their work, they deliver the product to the reservoir engineering group. Then, reservoir engineering would deliver the results of their work to production engineering, and so on. In almost all cases, it was necessary for each group to go back to the previous group for discussion, clarification, or request additional work. This has been realized as a very inefficient operation. In recent years, most major companies have adopted what is known as the multidisciplinary team approach for field developments. In this approach, a team consisting of engineers and scientists covering all needed disciplines is formed. The team members work together as one group throughout the field development stage. Of course, other specialists such as computer scientists, planners, cost engineers, economists, and so forth work closely with the team or may become an integral part of the team. Experience has shown that this field development approach is very efficient; more and more companies are moving in this direction.

The following sections provide brief descriptions of the roles and functions of drilling, reservoir, and production engineering.

1.7 Drilling Engineering and Operations

Following the preparation stage of field development (i.e., setting the production strategy, determining the locations of the wells in the field, and designing the well completions), the drilling-related activities begin. The drilling program is first designed. Then, plans are prepared and executed to acquire the required equipment and materials. The drilling sites in the field are then prepared for the equipment and materials to be moved in, and the drilling operations begin. Depending on the organization of activities within the oil company, drilling engineers may only be responsible for drilling and casing of the well, and production engineers will be responsible for completion of the well. Alternatively, drilling engineers may be responsible for drilling and completion of the wells.

The drilling program consists of three main stages: (1) drilling the hole to the target depth, (2) setting the various casings, and (3) cementing the casing.

1.7.1 Drilling the Well

Well drilling has gone through major developments of drilling methods to reach the modern method of rotary drilling. In this method, a drilling bit is attached to the bottom end of a string of pipe joints known as the drilling string. The drilling string is rotated at the surface, causing rotation of the drilling bit. The rotation of the bit and the weight applied on it through the drilling string causes the crushing and cutting of the rock into small pieces (cuttings). To remove the cuttings from the hole, a special fluid, called the drilling fluid or the drilling mud, is pumped down through the drilling string, where it exists through nozzles in the bit as jets of fluid. This fluid cleans the bit from the cuttings and carries the cuttings to the surface through the annular space between the drilling string and the wall of the hole. At the surface, the mud is screened to remove the cuttings and is circulated back into the drilling string. The drilling operation is performed using huge and complex equipment known as the drilling rig. This is briefly described next.

1.7.1.1 The Drilling Rig

Figure 1.6 shows a schematic of a rotary drilling rig. It consists of two main sections: the substructure (bottom section) and the derrick (top section). The substructure, which ranges from 15 to 30 ft in height, is basically a rigid

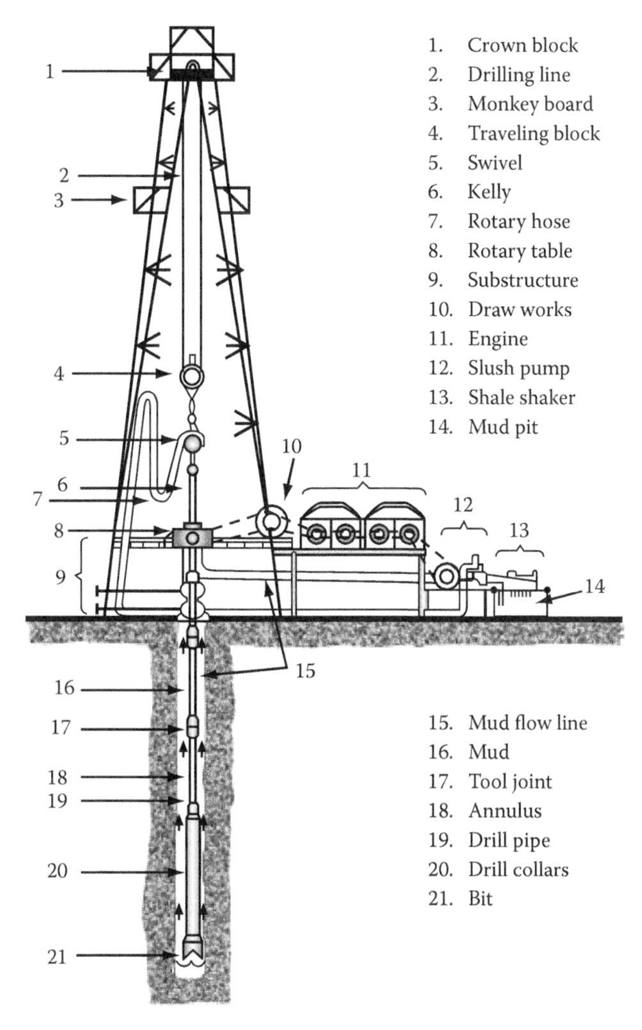

FIGURE 1.6
A schematic of a rotary drilling rig.

platform that supports the derrick. The rig is composed of several systems and components. The major systems are as follows:

- Power system—It consists of diesel-engine-driven electric generators to supply electric power to the various systems and components of the rig. About 85% of the power generated is consumed by the drilling-mud circulating system.
- Hoisting system—The function of this system is to lower and raise the drilling string into and out the well. The main components of the system are the crown block, the drawworks, the traveling block, the hook, the swivel, the elevator, the drilling line, and the deadline anchor.
- Rotating system—The system consists mainly of a motor-driven rotating table that is used to rotate a pipe with a square or hexagonal cross section, called the kelly, while allowing it to slide through. The kelly is suspended by the hoisting system and is connected at its bottom to the drill pipe and the bit.
- Circulating system—The system consists mainly of drilling-mud storage tanks, high-pressure pumps, circulating hoses and pipes, and a shale shaker. Its function is to circulate the mud through the well to bring the cuttings to the surface.
- Drilling bit—The drilling bit is the device that does the actual drilling by crushing and cutting the rock as it rotates with some force applied by the drilling string. There are different types and shapes used for different types of rock.

1.7.1.2 Drilling Fluid (Mud)

The drilling fluid is a very important element of the drilling operation. Its importance stems from the many essential functions it serves. Some of these functions are as follows:

- Transporting the cuttings from the bottom of the hole to the surface
- Cooling of the bit and lubrication of the drill string
- Exerting hydrostatic pressure to overbalance the pressure of the formation and thus prevent flow of formation fluids into the well
- Supporting the walls of the hole to prevent it from caving in
- Enhancing drilling by its jetting action through the bit nozzles

The drilling fluid can be prepared in different formulations to provide the desired properties (density, viscosity, and filtration) under the bottom hole conditions. The basic drilling fluid consists of water and clay (water-base mud). Other materials and chemicals are also added to control the properties of the fluid. Other fluids such as foam and air have also been used in drilling operations.

1.7.2 Casing the Well

The casing is a steel pipe that is placed in the drilled hole (well) to support the wall of the hole and prevent it from collapsing. When cemented to the wall, it seals the subsurface formation layers and prevents communications between the various layers.

Normally, four strings of casing of different diameters are installed in the well at various depths that are specified by the geologist. These are the conductor, the surface casing, the intermediate casing, and the production casing. The conductor has the largest diameter and shortest length of the four casing strings; the production casing has the smallest diameter and longest casing.

Casings of various outside diameters are available in different grades and weights. The grade refers to the type of casing steel alloy and its minimum yield strength. Commonly available grades are H-40, J-55, N-80, C-75, L-80, and P-105. The letter (H, J, etc.) identifies the type of alloy and heat treatment; the number (40, 55, etc.) refers to the minimum yield strength in thousands of pounds per square inch (psi). For a given outside diameter and grade, casings are available in different weights (i.e., various inside diameters) expressed in pounds per linear foot of casing. The weight and grade of the casing specify its resistance to various loads such as burst, collapse, and tension loads. In designing casing strings, weight and grade must be selected such that the casing string will not fail under all loads to which it will be subjected during drilling, setting casing, and production.

To set a casing string, the drilling operation is stopped when the desired depth is reached and the drill string and the pit are pulled out of the hole. The casing string is then lowered into the hole, joint by joint, using the hoisting system of the rig, until the total length of casing is in the hole. A round, smooth object called the guide shoe is attached to the bottom of the first joint of casing to ease and guide the movement of the casing into the hole. The casing is then cemented to the wall and drilling operation is resumed until the target depth for the next casing string is reached. Normally, before setting the production casing, the petroleum formation is logged and evaluated. The casing will be set only if the logging results indicate the presence of a productive formation. Otherwise, the well will be abandoned.

1.7.3 Cementing the Casing

To cement the casing, the annulus between the casing and the wall of the hole must be filled with cement. To achieve this, the required volume of the cement slurry (prepared on location) is pumped through the casing. A special rubber plug is normally inserted ahead of the cement to separate it from the mud and prevent any contamination of the cement with mud. Another plug is inserted after pumping the specified volume of cement. This is followed by pumping a fluid (normally mud) to displace the cement. When the first plug reaches the bottom, pumping pressure is increased to rupture the plug and allow the flow of cement behind the casing. When the top plug reaches the bottom, the cement must have filled the annulus to the surface. The cement is then left undisturbed until it sets and acquires enough compressive strength before resuming the drilling operation for the next casing string.

Once the casing is cemented, it becomes permanently fixed into the hole. It is very important to have a good cement bond between the casing and the wall of the hole. For this purpose, a special log (cement bond log) is conducted to check the integrity of the cementing operation. Failure to have a good cementing job will necessitate expensive remedial cementing operations.

1.8 Reservoir Engineering Role and Functions

Reservoir engineers play a major role in field development and operation. Some of the major functions of reservoir engineering are discussed next.

1.8.1 Estimation of Reserves

Estimation of oil and gas reserves in a discovered reservoir is one of the most important factors in evaluating the discovery and deciding on its viability for commercial development. To determine the volume of oil or gas present in the reservoir, the bulk volume of the reservoir (V_b) is first determined using the available reservoir description data. The volume of fluids in the pore spaces of the reservoir rock is then calculated by multiplying the bulk volume by the rock porosity (φ); this is also known as the pore volume of the rock. Porosity is a property of the rock defined as the ratio of the volume of the pore spaces within the rock to the bulk volume of the rock. The pore volume is normally occupied by oil (or gas) and water. The fraction of the pore volume occupied by water is known as the water saturation (S_w). The porosity and initial water saturation are determined from the logs and core samples obtained from the exploratory wells. Therefore, the initial volume of oil (V_0) at the reservoir conditions is determined by

$$V_0 = V_b\varphi(1 - S_w).$$

This volume of oil is called the initial oil in place (IOIP) or the original oil in place (OOIP). This method of estimating the OOIP is known as the volumetric method. Other methods (the material balance method and decline curve method) also exist for estimating the OOIP.

It is impossible to recover all of the OOIP; certain forces within the reservoir rock prevent the movement of some oil from the rock to the well. The fraction of the OOIP that could be recovered is called the recovery factor (E_r), and the total recoverable volume of oil ($E_r V_0$) is called the proven reserves.

The proven reserve for a reservoir changes with time as a result of three factors. First, the volume of oil in place decreases as oil is produced from the reservoir. Second, as more oil is produced, more reservoir data become available, which could change or modify the initial estimate of the OOIP. Third, new developments and improvements in recovery and production methods may increase the possible recovery factor. Therefore, the proven reserve of any field is continuously updated.

1.8.2 Well Location, Spacing, and Production Rates

Another important function of reservoir engineering is to determine the optimum locations of the wells to be drilled and the production rate from each well for the most effective depletion of the reservoir. Where the wells should be drilled depends largely on the structural shape of the reservoir and the reservoir drive mechanism. As a general rule, no wells should be drilled into locations of the reservoir where water or gas is expected to invade that part of the reservoir early after starting production. Figures 1.7 to 1.10 illustrate the preferred locations of wells for various reservoir structures and drive mechanisms.

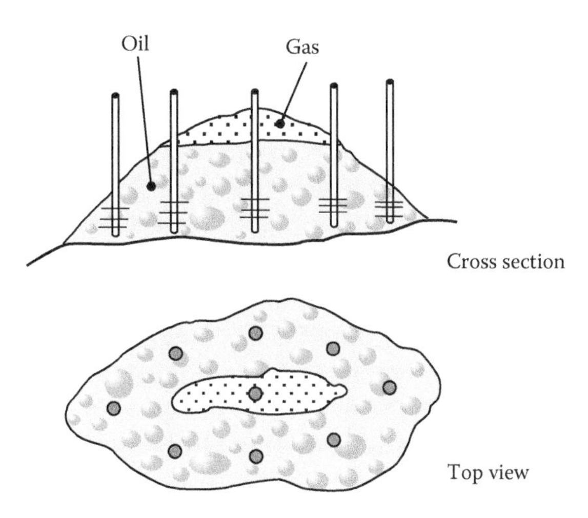

FIGURE 1.7
Gas cap drive reservoir; completion near bottom of oil zone.

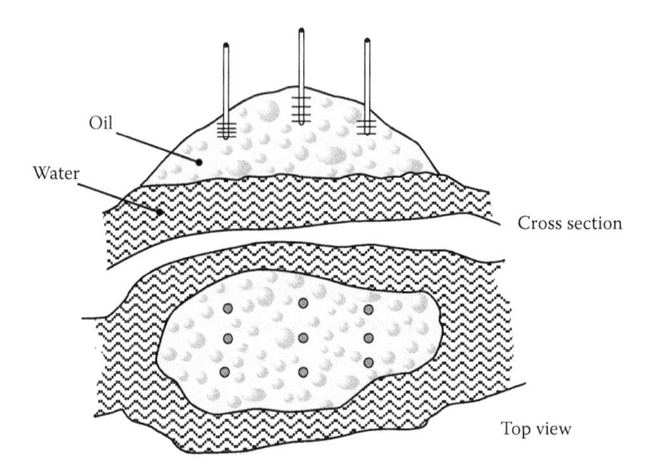

FIGURE 1.8
Bottom water drive reservoir; completion near top of oil zone.

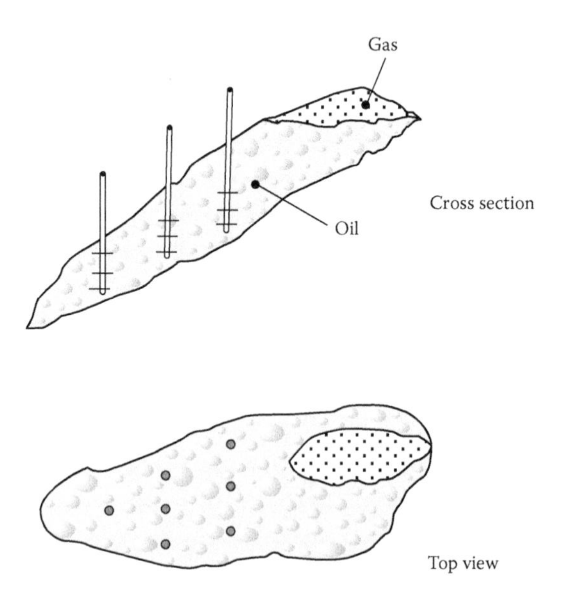

FIGURE 1.9
Gas cap drive, high angle of dip; wells located at bottom of structure.

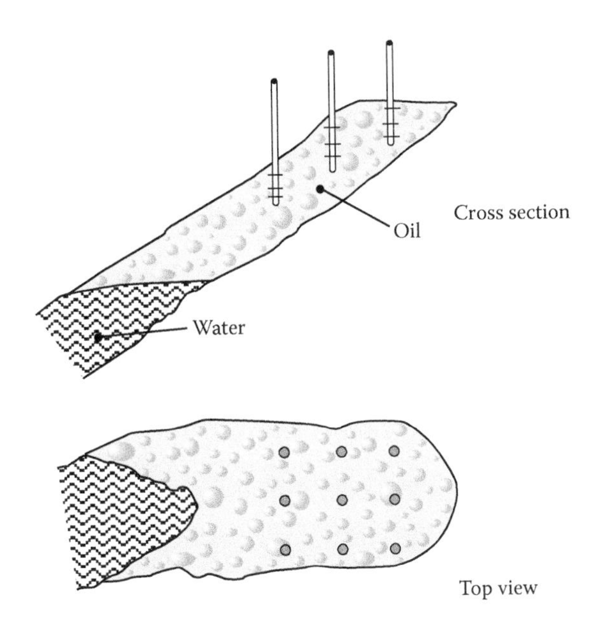

FIGURE 1.10
Water drive, high angle of dip; wells located near the upper section of the structure.

The spaces between wells and the production rate from each well are determined with the objective of recovering the maximum volume of petroleum in the most economical way. This is not an easy task and requires extensive reservoir simulation studies. Such studies continue after the development stage and, as more data become available, the studies may result in having to drill additional wells (infill wells) for higher recovery.

1.8.3 Reservoir Simulation

The complexity of the reservoir structures and the large number of nonlinear equations governing the flow of fluids in the reservoir make it impossible to obtain analytical solutions to the fluid flow problem. Reservoir simulators implementing various numerical techniques are constructed for solving such a problem. Running a simulator requires inputting many parameters that are related to the reservoir rock and fluids. The majority of these parameters may either be known with various degrees of certainty or may have to be assumed. Therefore, the accuracy of the simulation results depends, first, on the accuracy of the input data.

Simulation studies are first conducted during the development stage to predict and evaluate the performance of the reservoir for the life of the field. Once production starts, data are collected and the actual reservoir performance is checked against the simulation-predicted performance. In almost all cases, the actual and predicted results do not match. Therefore, reservoir (simulation) engineers modify the simulation by changing some or all of the input data using either newly acquired data or making new assumptions, or both. This is done until a match between the simulation results and the actual results, obtained from the producing field, is obtained. This process is known as history matching. The modified simulator is then used to predict the field performance and direct the production strategy for better management of the reservoir. The whole process is repeated on a continuous basis for every specified period of production.

1.8.4 Reservoir Management

The aspects of reservoir management are numerous and involve extensive and complicated studies. In simple terms, however, the role of reservoir management is to specify specific producing strategies that will control the movements of the various fluids within the reservoir in order to achieve maximum recovery of the hydrocarbon materials while minimizing the production of undesired fluids. The production strategies may include assignment of production rates to individual wells in the field, shutting off certain wells for a specific duration, drilling of new wells at specific locations, and implementing specific pressure maintenance or improved recovery operations. The importance of reservoir management has been recognized in recent years. This led to the formation of specialized groups and departments for reservoir management in most major oil companies.

1.8.5 Improved Petroleum Recovery

The natural reservoir drive mechanisms described in Section 1.5.2 normally result in low recovery factors. In fact, if reservoirs were produced solely by the natural driving forces, very poor recoveries would be obtained. The specific rock and fluids properties and the forces that control the movements of the fluids within the reservoir are responsible for such poor recoveries. Petroleum recovery by these natural drive mechanisms is called primary recovery.

In order to achieve recoveries higher than the primary recovery, we must intervene into the reservoir to artificially control, or alter, the natural driving forces and the rock and fluids properties. Reservoir engineers have developed, and continued research is being conducted for further developments, refinement, and improvement, and various techniques to achieve higher than primary recoveries. Some techniques involve supplementing the natural driving force by injecting high-pressure fluids into the reservoir. Other techniques aim at changing the fluids or rock properties to enhance the mobility of the petroleum fluids and suppress the forces that hinder their movements. All such techniques are known as enhanced recovery methods or improved recovery methods. A brief description of some of the improved recovery methods follows.

1.8.5.1 Pressure Maintenance

As oil is produced, the reservoir pressure declines at a rate that depends on the reservoir drive mechanism, the strength of that drive, and the amount of oil produced. Solution gas drive reservoirs experience the highest rate of pressure decline, followed by the gas cap drive reservoirs, with the water drive reservoirs being the least affected. The decrease in reservoir pressure reduces the ability of the formation to produce oil. The loss of productivity becomes very severe if the reservoir pressure drops below the bubble point pressure.

One way of maintaining a high reservoir pressure to maintain productivity and increase recovery is to inject a fluid into the reservoir at such quantities and pressure that it will keep the reservoir pressure at the desired high level. Depending on the type of reservoir, pressure maintenance may be achieved by either the injection of water through wells drilled at the periphery of the reservoir or the injection of gas at the top of the reservoir.

1.8.5.2 Water Flooding

In water flooding, injection wells are drilled between the oil producing wells in a specific regular pattern. For example, for the 5-spot flooding pattern, one injection well is drilled between and at equal distances from four producing wells. Water is injected into the injection well to drive (push) the oil toward the four producers. Simulations of the process are usually made before implementation to determine the optimum well spacing, or pattern, and the injection rate and pressure to achieve efficient displacement of oil by the injected water. Water-flooding operations are always associated with a high water cut in the produced fluid. The process continues until the oil production rate becomes too low for economic operation.

Since the water viscosity is lower than the oil viscosity, water may finger through the oil and reach the producing well prematurely, bypassing a significant amount of oil that would be very difficult to recover. Engineers have to be careful in determining the injection rate for the specific reservoir and rock and fluid properties to avoid water fingering. Water flooding could be improved if the water viscosity were altered to become higher than the oil viscosity. This would create more favorable mobility conditions, make the oil–water interface more stable, and avoid fingering. This is achieved by adding certain polymers to the injection water. This is known as polymer flooding.

1.8.5.3 Chemical Recovery

Some liquids have a higher affinity than others to adhere to solid surfaces. When two liquids are in contact with a common solid surface, one of the two liquids (the one with the higher adhesion affinity) will spread over the surface at the expense of the other. The solid surface is then identified as wettable by that liquid. Most reservoir rocks are wetted by water than by oil and are, therefore, identified as water-wet rocks. Such conditions are favorable for displacing oil by water. There are situations, however, where the reservoir rocks are oil-wet. In these cases, water would not be able to displace the oil. Reservoir engineers and surface chemistry scientists have developed methods in which chemicals are used to change the wettability of the rock for effective water-flooding operations. These methods are known as the chemical recovery methods. The two most common chemical recovery methods are surfactant flooding and caustic flooding.

In surfactant flooding, a slug of water–surfactant solution is first injected through the injection well into the formation. This is followed by injection of ordinary water as in regular water-flooding operations. The surfactant ahead of the floodwater causes changes in the interfacial tension and mobilizes the oil that would otherwise adhere to the surface of the rock. Thus, the displacement of the oil by the floodwater becomes possible. Again, adding polymers to the water to create favorable mobility conditions could increase the flooding effectiveness. In this case, the method can be called surfactant–polymer flooding.

Caustic flooding is essentially a surfactant flooding, with the surfactant being generated within the reservoir rather than being injected. The method is applicable in situations where the reservoir oil contains high concentrations of natural acids, which can react with alkaline to produce surfactants. The most common approach is to inject a slug of caustic soda (NaOH) solution ahead of the floodwater. The alkaline reacts with the acids in the oil to in situ produce surfactants. Then, the process is converted into a surfactant-flooding process. This method is less expensive than the regular surfactant-flooding process.

1.8.5.4 Miscible Recovery

In miscible recovery, a slug of a substance that is miscible in the reservoir oil is injected into the reservoir at pressures high enough to achieve good miscibility. This is then followed by water injection. The process has been used with carbon dioxide, rich natural gas, nitrogen, flue gases, and light hydrocarbon liquids as the miscible fluid. Miscible flooding could achieve very high recovery factors.

1.8.5.5 Thermal Recovery

Heavy oil reservoirs present a unique production problem. The high viscosity of the oil makes it difficult, and in some cases impossible, to produce the oil, even with the aforementioned improved recovery methods. The best method to mobilize the oil is to heat the formation to reduce the oil viscosity. When heating is used in recovering the oil, the recovery method is called thermal recovery. The three most common thermal recovery methods are steam stimulation, steam flooding, and in situ combustion.

In the steam stimulation method steam is injected into the producing well for a specified period of time (normally more than a month); then, the well is shut off for another period of time (normally a few days). The injected steam heats up the surrounding formation, causing significant reduction in oil viscosity. The well is then put on production for a period of time until the oil flow declines. The process is then repeated through the same cycle of injection, shutting off, and production. This process is also known as the Hugh and Pugh method.

Steam flooding is similar to the water-flooding process, except that steam is used instead of water. The steam is injected into an injection well to reduce the oil viscosity while the condensed steam (hot water) displaces the oil toward the producing wells.

In the in situ combustion process, air is injected into the formation through an injection well under conditions that initiate ignition of the oil within the nearby formation. The combustion zone creates a front of distilled oil, steam, and gases. Continued air injection drives the combustion front toward the producing wells. The combination of heating and displacement by the steam, gases, and condensed liquids enhances the recovery of the oil.

1.9 Production Engineering: Role and Functions

Production engineers have probably the most important role in both the development and operating stages of the field. They are responsible for making the development and production strategies prepared by the reservoir engineers a reality. Production engineers are responsible for designing and installing the well completions that are capable of producing the desired volumes of oil/gas with the prescribed methods of production. They are also responsible for maintaining the wells at their best producing conditions throughout the life of the field. These two major responsibilities are classified as subsurface production engineering. Still, production engineers are responsible for designing, installing, operating, and maintaining all surface production facilities starting from the flow lines at the wellhead and ending with the delivery of oil and gas to the end user. This is classified as surface production engineering, which is the main theme of this book. Both the subsurface and surface production engineering aspects of the field are tied together and considered as one production system. The main objective of that system is to obtain maximum recovery in the most economical and safe manner.

In the following sections, the operations related to subsurface production engineering are briefly described.

1.9.1 Well Completion Design

The well completion is the subsurface mechanical configuration of the well that provides the passage for the produced fluids from the face of the formation to the wellhead at the surface. The well completion design is needed by the drilling engineers to properly design, plan, and execute the drilling of the well that is compatible with the well completion.

1.9.1.1 Types of Well Completion

There are three major types of well completion: Open hole completion, cased hole (perforated) completion, and liner completion.

For open hole completions the well is drilled down to a depth that is just above the target petroleum formation. The production casing is then lowered into the well and cemented. The target formation is then drilled and is left uncased (open). Depending on the production rate and the properties of the produced fluids, the well may be produced through the production casing or through production tubing placed above the producing formation with a packer that provides a seal between the tubing and casing. One of the functions of the packer is to protect the casing from the produced fluids. Figure 1.11 illustrates this type of well completion.

For cased hole (perforated) completions the well is drilled all the way through the producing zone and the production casing is lowered and cemented. The casing is then perforated across the producing zone to establish communication between the formation and the well, as illustrated in Figure 1.12. Again, depending on the producing conditions, production could be either through the casing or through a tubing.

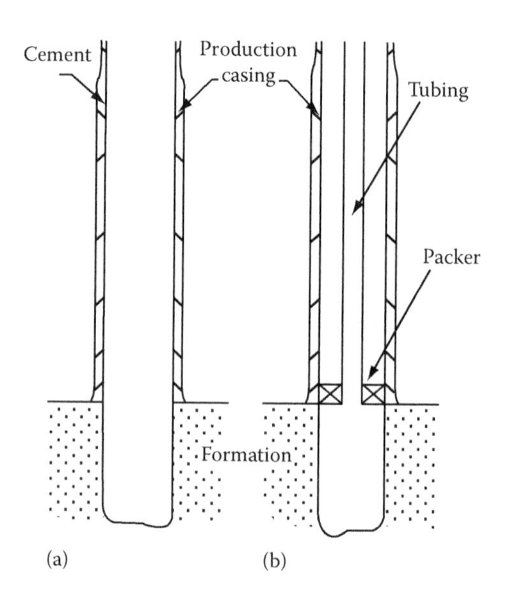

FIGURE 1.11
Open hole completion: (a) production through casing and (b) with tubing and packer.

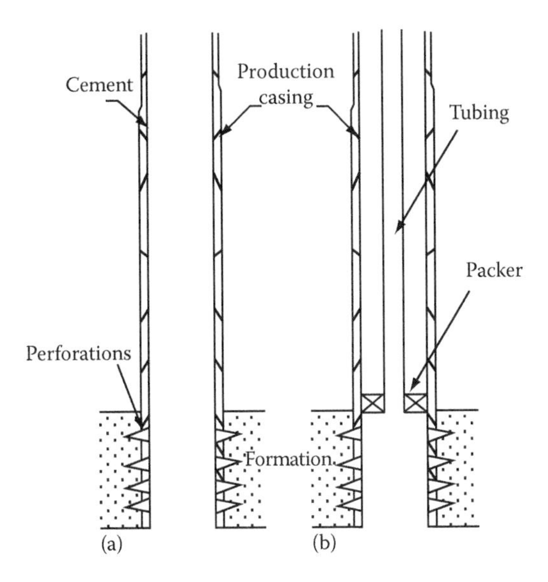

FIGURE 1.12
Cased hole completion: (a) without tubing and (b) with tubing packer.

As illustrated in Figure 1.13, in liner completion the production casing is set and cemented above the petroleum formation similar to an open hole completion. A liner (basically a smaller diameter casing) is then set and cemented across the producing formation. The liner is then perforated to establish communication between the well and the formation. In some cases,

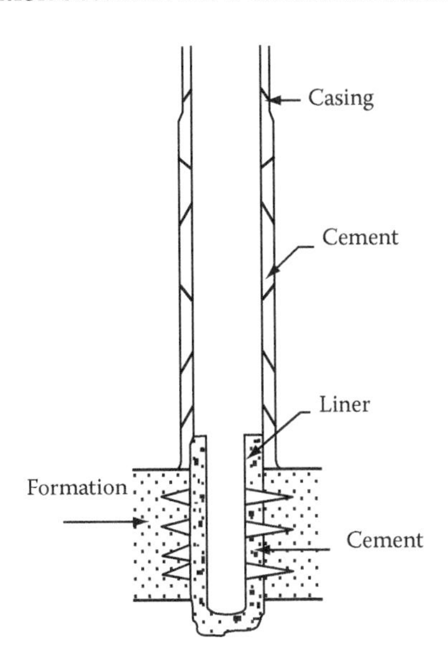

FIGURE 1.13
Liner completion.

an already perforated (or slotted) liner may be set across the formation without cementing.

1.9.1.2 Tubing–Casing Configurations

There are several different tubing–casing configurations. The selection of a particular configuration is normally governed by the characteristics of the reservoir and cost. The type of reservoir and drive mechanism, the rock and fluid properties, and the need for artificial lift and improved recovery are among the technical factors that influence the selection of well completion and configuration. Tubing–casing configurations can be grouped under two main categories: the conventional and the tubingless completions.

A conventional completion is the completion with a production tubing inside the production casing. A packer may or may not be used and the production could be through the tubing, the annulus, or both. Depending on the number of petroleum formations to be produced through the well, we may have either single or multiple completion. Figure 1.14 illustrates a multiple (triple) conventional completion.

The tubingless completion is a special type of completion where a relatively small-diameter production casing is used to produce the well without the need for production tubing. Such completions are low-cost completions and are used for small and short-life fields. Again, we may have multiple tubingless completions or single, as illustrated in Figure 1.15.

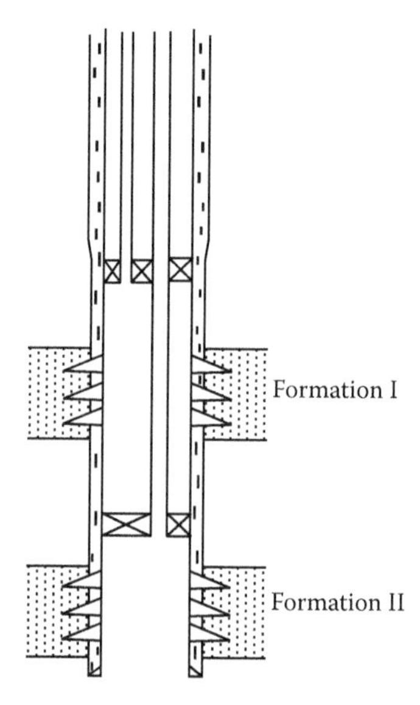

FIGURE 1.14
Multiple conventional completion.

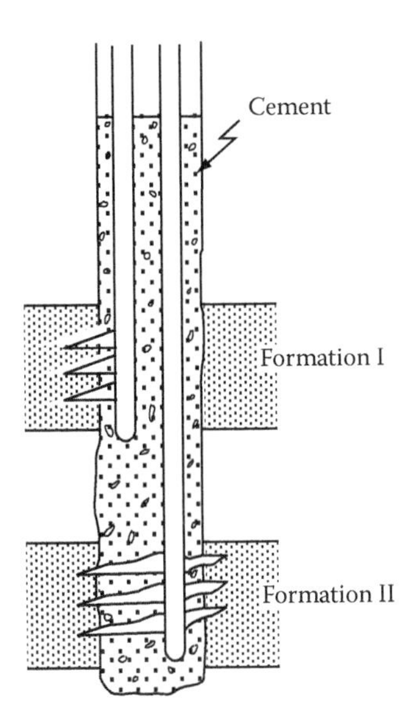

FIGURE 1.15
Multiple tubingless completion.

1.9.2 Tubing Design

Tubing design involves the determination of the size, the grade, and weight of the production tubing. The size of the production tubing is controlled by the production rate, the types of flowing fluid, and the pressures at the bottom of the well and at the surface. The grade and weight of the tubing reflect its strength and are determined through analysis of the various loads that act on the tubing under all expected conditions.

1.9.2.1 Determining Tubing Size

Determining the size of production tubing is the starting point for completion and drilling design, as it directly affects the sizes of all casing strings and, accordingly, the design and cost of the drilling program. The tubing size must be selected to handle the desired production rates under the varying producing conditions for the life of the well. To properly determine the tubing size, the whole production system (from the formation to the surface separator) must be considered. The ability of the formation to produce fluids by the natural drive and improved recovery methods from the start of production until depletion need to be considered. The flow of produced fluids through the production tubular, the wellhead restrictions, and the surface flow line over the life of the well need to be analyzed taking into account possible means of artificially lifting the fluids.

The flow from the formation to the bottom of the well (bottom hole) is governed by what is known as the inflow performance relationship (IPR) of the well, whereas the flow from the bottom hole to the surface is represented by the outflow performance relationship (OPR).

The IPR is the relationship between the flow rate (q) and the flowing bottom-hole pressure (P_{wf}). The relationship is linear for reservoirs producing at pressures above the bubble point pressure (i.e., when P_{wf} is greater than or equal to the bubble point pressure). Otherwise, the relationship takes the shape of a curve, as illustrated in Figure 1.16. When the IPR is linear, it can be represented with what is known as the productivity index (PI), which is the nverse of the slope of the IPR. The PI is basically the production rate per unit drawdown (the difference in pressures between the average reservoir static pressure, P_R, and P_{wf}). The IPR depends on the reservoir rock and fluid characteristics and changes with time, or cumulative production, as illustrated in Figure 1.17. Methods exist for determination of the IPR and for predicting future IPRs.

Outflow performance involves fluid flow through the production tubular, the wellhead, and the surface flow line. In general, analyzing fluid flow involves the determination of the pressure drop across each segment of the flow system. This is a very complex problem, as it involves the simultaneous flow of oil, gas, and water (multiphase flow), which makes the pressure drop dependent on many variables, some of which are interdependent. There is no analytical solution to this problem. Instead, empirical correlations and mechanistic models have been developed and used for predicting the pressure drop in multiphase flow. Computer programs based on such correlations and models are now available for the determination of pressure drops in vertical, inclined, and horizontal pipes.

Determination of the tubing and flow line sizes is a complex process involving the determination and prediction of future well productivity, analysis of multiphase flow under varying production conditions, and economic analysis. However, a simplified approach is summarized in the following steps and illustrated in Figure 1.18:

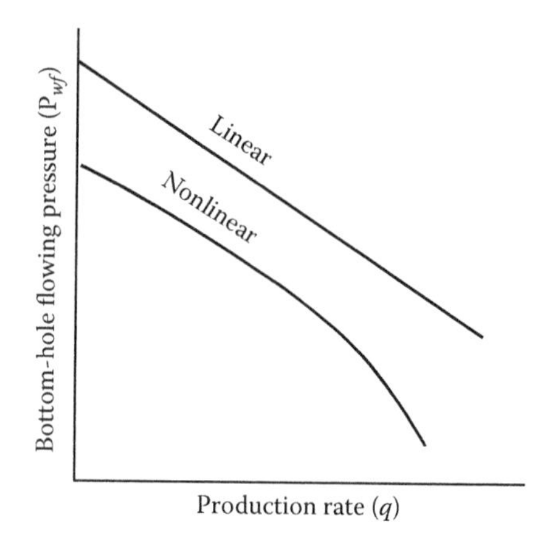

FIGURE 1.16
Inflow performance relations.

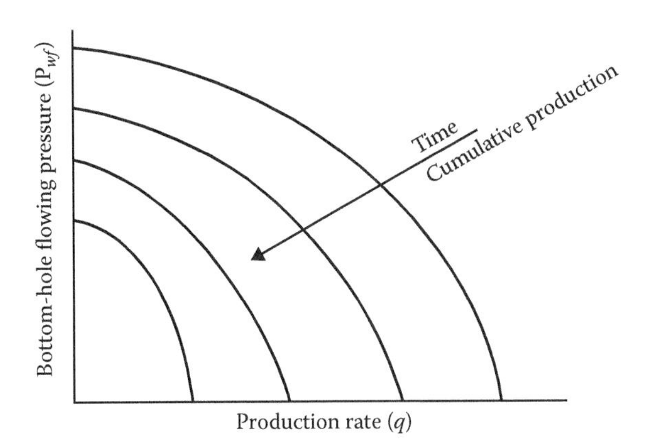

FIGURE 1.17
Dependence of the IPR on time, or cumulative production.

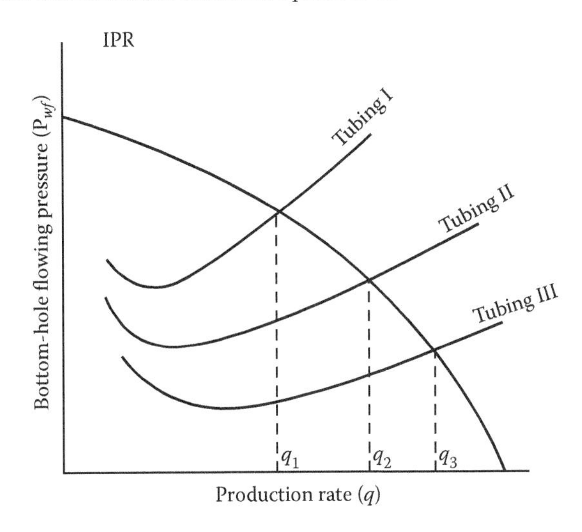

FIGURE 1.18
IPR and outflow performance for different tubings.

1. Determine (predict) the present and future IPR. Plot the results as P_{wf} versus q.
2. Selecting tubing and flow line diameters and starting with a specified value for the surface separator pressure, determine the flowing bottom-hole pressure for a specified production rate, water cut (WC), and gas–oil ratio (GOR), using available multiphase flow correlations or models.
3. Step 2 is repeated for different values of production rates (q) with WC and GOR being fixed. Thus, a relation between q and P_{wf} is established and presented on the same plot of the IPR.

4. Steps 2 and 3 are repeated for various combinations of tubing and flow line sizes.
5. Steps 2–4 are repeated for various expected values of WC and GOR, and for various producing conditions (natural flow and artificial lift).
6. On the P_{wf}–q plot, the intersection of the OPR curve for a particular tubing–flow line combination and producing conditions with the IPR curve of any specific time represents the maximum possible production rate and corresponding P_{wf}. If the OPR curve does not intersect with a particular IPR curve, then the well will not flow when such an IPR and producing conditions are reached.
7. If possible, select the tubing–flow line combination whose OPR curves intersects with present and all corresponding future IPR curves and provides values for q that are consistent with the planned production strategy.
8. When more than one tubing–flow line combination is possible, select the one that provides the best economics.

1.9.2.2 Determining Tubing Grade and Weight

Once the tubing size has been determined, the next step is to determine the grade and weight of the tubing. Similar to casing, tubing grade refers to the type of steel alloy and its minimum yield strength. Tubings are available in the same grades as casing (i.e., H-40, J-55, N-80, C-75, L-80, and P-105). Other high-strength grades that are resistance to sulfide stress cracking are also available.

To determine the tubing grade and weight, the maximum collapse and burst loads that act on the tubing are first calculated and multiplied by a afety factor. These values are then used to make an initial selection of tubing grade and weight that provides sufficient collapse and burst resistance. With the selected weight, the tension load is calculated and multiplied by a safety factor. This is then compared to the tensile strength of the selected tubing. The selected tubing is accepted if its tensile strength matches or exceeds the tensile load. Otherwise, another grade and weight are selected, and the calculations are repeated until a final selection is made.

1.9.3 Completion and Workover Operations

Completion and workover operations are basically similar. However, they are given different associations based on when the operation is performed. Completion operations are any and all operations performed on the well to get it ready for production. Workover operations, however, refer to such operations that are performed to resolve specific problems that are found after production has started. Some of the important and common operations are briefly described in the following subsections.

1.9.3.1 Perforating Operation

For cased hole completions, perforations are made through the casing and cement and into the formation to establish communication between the formation and the wellbore. It is essential to have clean perforations with relatively large diameters and deep penetrations to achieve high well productivity. Further, perforating should be done only through the clean and productive zones within the formation, as determined from the formation evaluation logs. Therefore, extreme care is taken in locating the perforating gun at the right locations. The selection of the type of perforating gun, explosive charges, and completion fluid and the control of the pressure in the well at the time of perforating are very important elements in achieving effective and productive perforations.

Perforations are made by detonating specially shaped explosive charges. The shaped charge consists of a body called the case; a linear that is made of a powder alloy of lead, copper, and tungsten; the explosive material that is contained between the case and the liner; and a detonating cord. Figure 1.19 shows a schematic cross section of a shaped charge. Upon explosion, the case expands and ruptures and the liner collapses into a carrot-shaped jet consisting of lead, copper, and tungsten particles. The jet travels at very high velocity and impacts upon the casing with an extremely high pressure. The high energy of the jet causes the jet to penetrate through the casing, cement, and formation, thus creating the perforation.

Perforating guns are classified as hollow steel carrier, semiexpendable, and fully expendable guns (Figure 1.20). Hollow steel carrier guns are made of steel cylinders that carry the explosive charges and needed accessories. They come in small diameters that can go through the production tubing and in large diameters that can go only through the production casing. They are rigid, can withstand high pressure and temperature conditions, and have the advantage

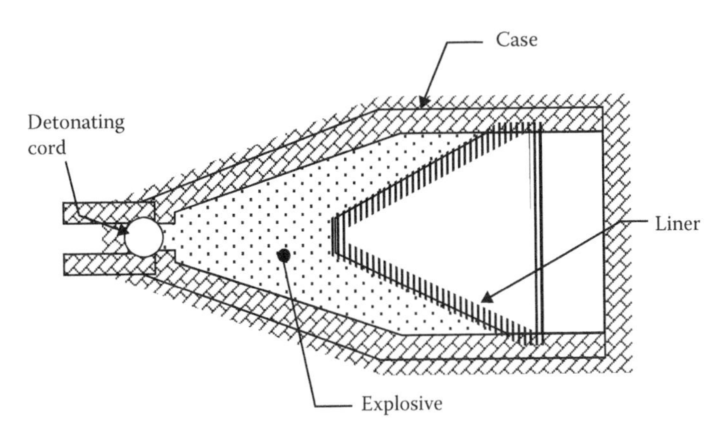

FIGURE 1.19
Shaped charge (cutaway).

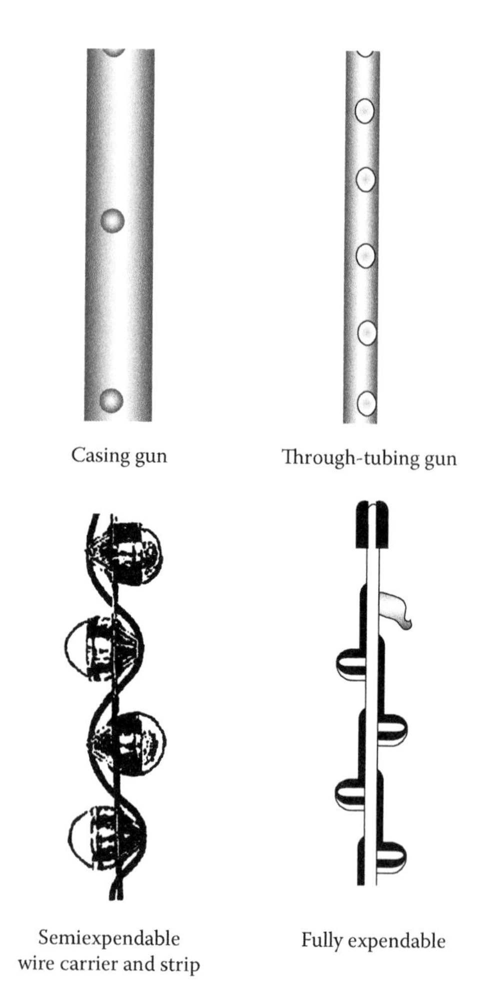

FIGURE 1.20
Types of perforating guns.

of containing all of the debris that is left after perforating. However, because of their rigidity, they would not go through sharp bends. For semiexpendable guns, the charges are fixed to a flexible metal strip. After detonation, most of the debris is expelled into the well, but the metal strip, any unfired charges, and the electronic detonator are retrieved. The flexibility of the guns allows movement through bends and highly deviated holes. Fully expendable guns have advantages similar to those of semiexpendable guns. However, because, after detonation, the gun breaks completely into pieces that are left in the hole, there is no way to know whether all charges have been fired.

To obtain clean and relatively undamaged perforations, the perforating operation should be conducted with the bottom-hole pressure much less than the reservoir pressure. Upon perforating, the differential pressure causes fluids to surge from the formation into the well; this effectively cleans the perforations from any debris. This method of perforating is known as underbalance perforating. To perforate underbalance, the production tubing must be installed in the well and the perforating mode is, therefore, identified as the through-tubing perforating mode.

In the early days, large through-casing hollow steel carrier guns were always needed to obtain large-diameter perforations with deep penetration. This required that the tubing be out of the well, and in order to have control over the well, perforating had to be conducted with the bottom-hole pressure being higher than the reservoir pressure by a safe margin. This mode of perforating overbalance did not offer an effective means for cleaning the perforations of the debris. Another mode of perforating that offers the opportunity to use large casing guns and perforating under-balance is known as tubing-conveyed perforating. Recent technological developments, however, have made it possible to obtain such desired perforations using small-size charges on a semiexpendable gun.

1.9.3.2 Well Stimulation Operations

Well stimulation operations could be classified into two main categories: matrix acidizing and formation fracturing. The objective of matrix acidizing is to remove near-wellbore formation damage that might have been caused by drilling or other workover operations in order to restore or enhance well productivity. Formation fracturing, however, is conducted on formations having very low permeability in order to increase well productivity.

For carbonate formation, matrix acidizing is achieved by injecting hydrochloric acid (HCl) into the formation at low pressure. As acid is injected, it preferentially flows through more permeable passages, reacting with and dissolving the carbonate. This creates irregular highly conductive channels, known as wormholes, for easier flow of fluid from the formation to the well.

For sandstone matrix acidizing, a solution of hydrofluoric and hydrochloric acids (known as mud acid) is injected into the formation to dissolve clay and, to some extent, silica. This removes the damage to restore the near-wellbore permeability.

Fracturing is performed mostly on wells completed in very tight (very low permeability) formations. In carbonate formations, a fracturing fluid followed by HCl is injected at pressures exceeding the formation fracturing pressure. The fracturing fluid initiates a fracture and the acid reacts with the carbonate walls of the fracture, leaving the walls as rough surface. Therefore, when the pressure is reduced, the fracture will not close and will provide a very conductive passage for fluids to flow from the formation into the well. This operation is known as acid fracturing. Fracturing tight sandstone formations is known as hydraulic fracturing. As with carbonate fracturing, a fracturing fluid is injected at high pressure to initiate the fracture. The fracture is kept open by filling it with highly permeable, high-compressive-strength sand known as proppant.

1.9.3.3 Sand Control

When wells are completed in unconsolidated or weakly consolidated formations, sand is likely to be produced with the fluids. Sand production is a very serious problem. Produced sand erodes subsurface and surface equipment, necessitating very costly frequent replacement. Sand also settles in the bottom of the well and in surface processing facilities. This requires periodic shutdowns to clean the well and facilities.

Methods are available to control or prevent sand production. These are classified as mechanical retention methods and chemical, or plastic, consolidation methods. The simplest method of mechanical retention involves the installation of screens opposite the producing zone. Screens, however, are subject to erosion and corrosion and would need to be frequently replaced. Therefore, screens are used only as a temporary solution to sand production problems. The best and most commonly used method of mechanical sand control is known as gravel packing. In simple terms, the method involves the installation of a screen having a smaller diameter than the casing diameter opposite the producing zone. The annular space between the casing and screen is packed with specially sized gravel (sand). Formation sand will bridge against the gravel, which bridges against the screen. This prevents formation sand from flowing with the fluids into the well.

Chemical or plastic consolidation involves the injection of the polymer and catalyst solution into the formation to coat the sand grains around the wellbore with the polymer. The polymer is then cured to harden and bonds the sand grains together and thus consolidates the sand around the wellbore.

Each sand control method has its limitations and care should be taken to select the most suitable method for the situation in hand. All methods of sand control result in loss of productivity; this, however, is acceptable in comparison to the problems associated with sand production.

1.9.3.4 Remedial Cementing

Remedial cementing refers to any cementing operation performed on the well after placing it in production. These are basically repair jobs executed to resolve specific problems in the well. The most common applications of remedial cementing are the following:

- Control of GOR and WOR—Squeezing cement into the perforations producing gas or water can stop excessive gas or water production with the oil and thus reduce an undesired high GOR or water–oil ratio (WOR).

- Channel repair—A channel is a void or crack that developed during the primary cementing of the production casing string. When the channel behind the casing communicates two or more zones, unwanted fluids (such as water) may flow through the channel into the formation and be produced with the oil. Alternatively, the oil may flow through the channel and be lost into another lower-pressure formation. A channel is repaired by creating a few perforations at the top and bottom of the channel and squeezing cement into the channel to seal it off.
- Recompletion—When one well is used to produce from multiple reservoirs or multiple zones with one reservoir or zone produced at a time, the depleted zone must be plugged off before producing the next zone. Cement is squeezed into the old perforations to seal the depleted zone completely; then, the new zone is perforated.
- Casing leak repair—Depending on the nature of the leak, cement may be used to seal parts of the casing, leaking unwanted fluids into the well or allowing produced oil to leak into other formations.
- Setting liners—Cementing liners to convert an open hole completion into a perforated liner completion.
- Plug and abandonment—For abandoning depleted wells, cement is used to squeeze and seal off all perforations and to set several cement plugs in the well.

1.9.4 Producing the Well

Producing the well means bringing the fluids that flowed from the formation into the borehole from the bottom of the well to the surface. Fluids need to be brought to the surface at the desired rate and with sufficient pressure to flow through the surface-treating facilities. Some reservoirs possess such a high pressure that it can produce the desired rates at high bottom-hole pressures that could push the fluids to the surface at the desired wellhead pressure. This mode of production is known as natural flow. Reservoirs with high initial pressures and with strong pressure support (i.e., from bottom water) can be produced under natural flow for extended periods of time.

When the reservoir pressure declines, pressure support may be provided by injecting water or gas into the reservoir to maintain natural flow at desired rates and surface pressures. In some cases, however, pressure support may not be sufficient to maintain desired natural flow. This usually occurs due to increased water production, which increases the hydrostatic head and friction losses in the tubing. Therefore, the fluids may reach the surface at lower than desired pressures or may not even reach the surface. In such cases, external means of lifting the fluids to the surface will be needed. These means are known as the artificial lift methods. Production engineers are responsible for selecting, designing, installing, and operating artificial lifting facilities.

1.9.4.1 Artificial Lift Methods

The objective of artificial lift is to create low bottom-hole pressure to allow high rates of production from the formation into the bottom of the well and artificially lift the fluids to the surface with the desired surface pressure. The main methods of artificial lift are sucker rod pumping, hydraulic pumping, electric submersible pump, and gas lift.

Sucker rod pumping (SRP) is the oldest and most common method of artificial lift used in the industry. The method is, however, limited to vertical and straight wells with relatively low productivity. The pumping facility consists mainly of surface equipment that provide reciprocating up-and-down movement to a rod string (the sucker rod) that is connected to the bottom-hole pump (Figure 1.21). The pump may be thought of as a cylinder-plunger system with one-way valves. Produced fluids enter the cylinder above the plunger during the down stroke and are lifted to the surface during the up stroke.

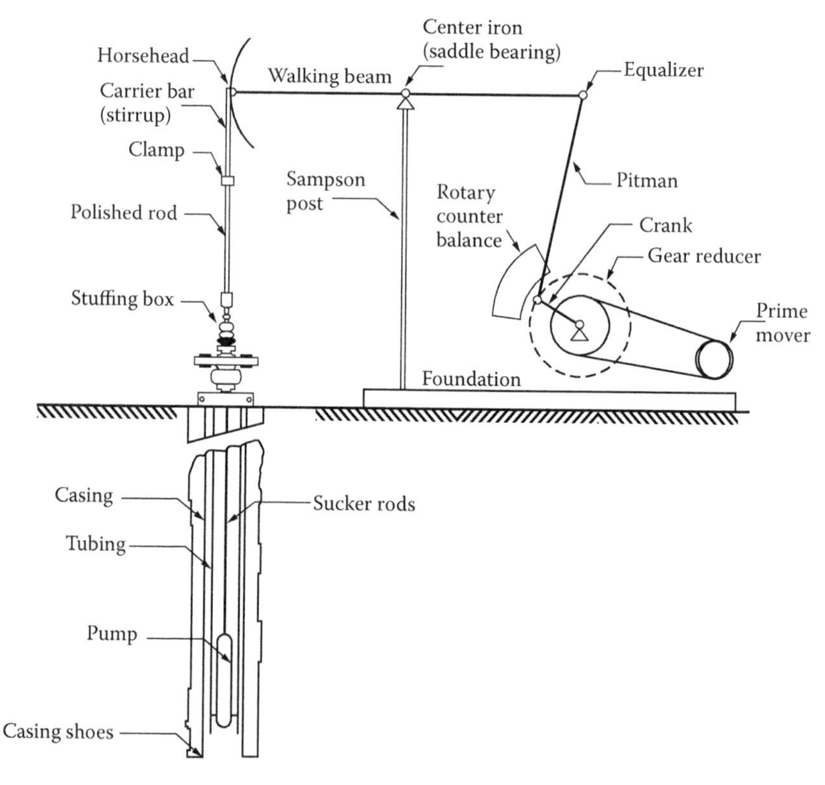

FIGURE 1.21
Schematic of sucker rod pumping.

In the hydraulic pumping (HP) method a pump located at the bottom of the well is powered by high-pressure fluid. The function of the pump is to increase the produced fluid pressure to lift it to the surface with the desired wellhead pressure. Figure 1.22 shows a schematic of hydraulic pumping facility. The surface facility provides the high-pressure hydraulic fluid that powers the pump. Hydraulic fluid may flow through a closed circuit and does not come in contact with the produced fluids. Oil may also be used as the hydraulic fluid, allowing its mixing with the produced fluids. The major limitations of this method are the need for expensive centralized hydraulic power units and expensive clean hydraulic fluid.

The electric submersible pump (ESP) method uses centrifugal pumps powered by electric motors. The pump and motor are both located at the bottom of the well and electric power is provided to the motor from the surface through a special cable (Figure 1.23). As with the HP, the function of the ESP is to increase the pressure of the fluid so that it can move and reach the surface with high pressure. The ESP is capable of producing very high rates with a high surface pressure.

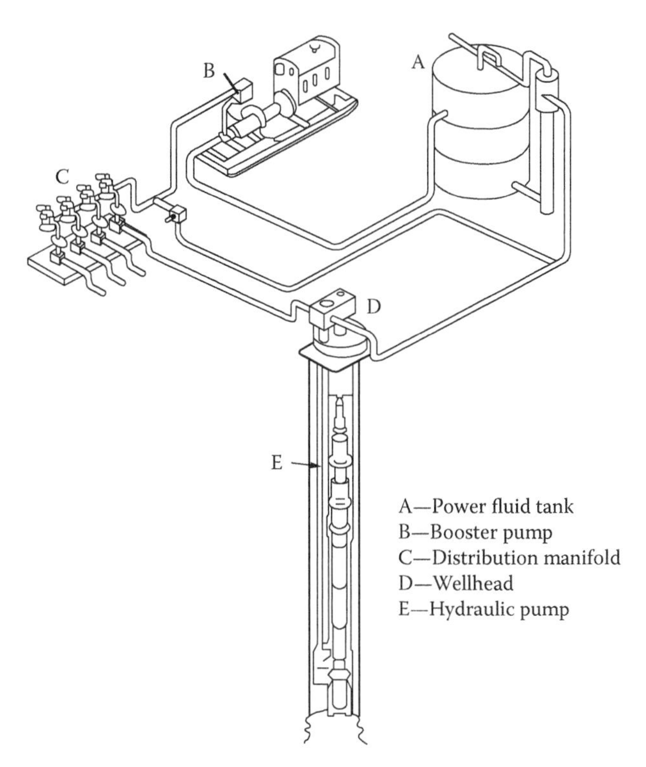

FIGURE 1.22
Schematic of hydraulic pumping installation.

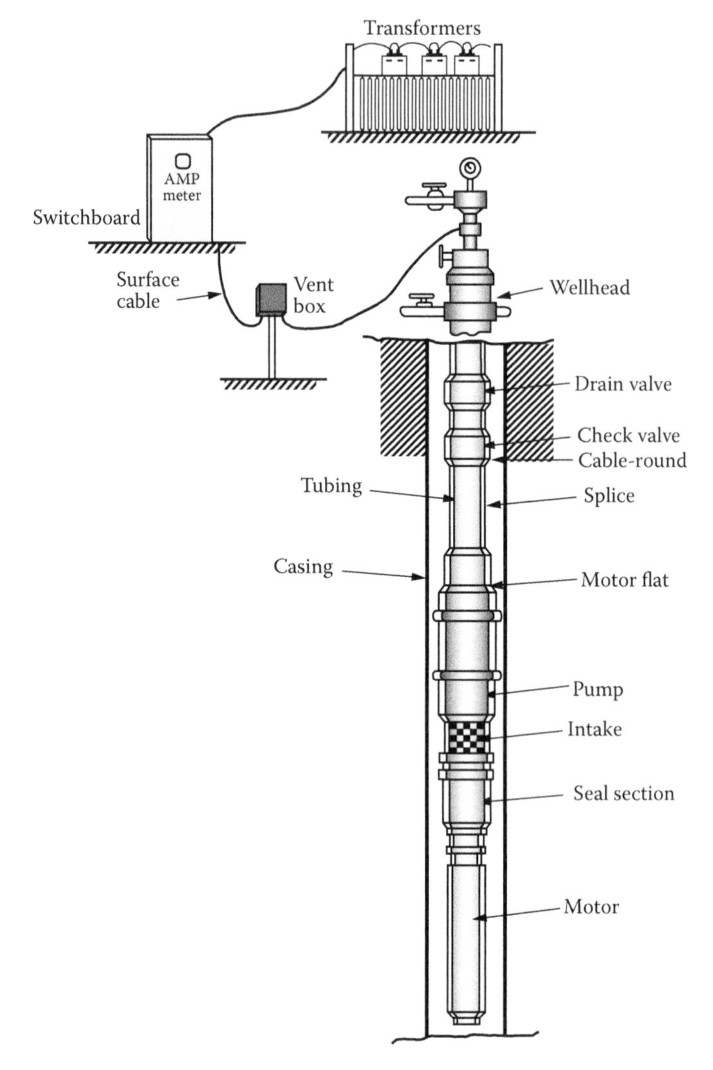

FIGURE 1.23
Typical standard complete pumping system. (Courtesy of Bryon Jackson, Centrilift.)

Although many different types of gas lift (GL) installation exist, the lifting concept is the same. High-pressure gas is injected into the annular space between the casing and tubing, and enters the tubing through special gas-lift valves (Figure 1.24). The gas mixes with fluids in the tubing, reducing its density and, consequently, reducing the hydrostatic head imposed by the fluid at the bottom of the well. The reduction in bottom-hole pressure causes more production of fluids from the formation. The reduced hydrostatic head reduces the pressure losses in the tubing and thus enables the fluid to reach the surface with relatively high pressure.

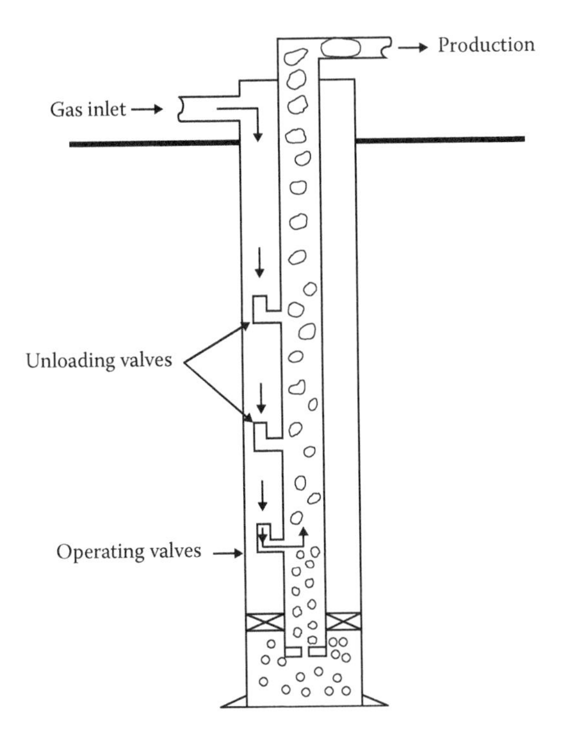

FIGURE 1.24
Schematic of a gas lift.

REVIEW QUESTIONS

1. Briefly describe the organic theory of petroleum formation.
2. Define the petroleum reservoir.
3. What are the necessary conditions that lead to the accumulation of petroleum to form a petroleum reservoir?
4. What is meant by the *source rock*?
5. Name and briefly describe, with illustrations, the different geologic types of petroleum reservoir.
6. What is meant by *reservoir drive mechanism*?
7. Describe the different types of reservoir drive mechanisms.
8. Name and briefly describe the techniques used for petroleum exploration.
9. What are the functions of drilling fluids?
10. What is the purpose of the well casting?
11. Briefly describe the role and functions of reservoir engineering in petroleum field development and operation.

12. Show, using illustrations and giving reasons, where you would locate and complete wells for an anticline gas cap drive reservoir.
13. Show, using illustrations and giving reasons, where you would locate and complete wells for an anticline water drive reservoir.
14. What are the roles of reservoir simulation?
15. What is meant by *improved recovery*?
16. Describe the various methods of improved recovery.
17. Briefly describe the role and functions of production engineering in petroleum field development and operation.
18. What are the main types of well completion?
19. Define the *productivity index* of the well.
20. What is the purpose of *perforating* a well?
21. Explain the function of each of the following *workover operations*:
    a. Stimulation
    b. Sand control
    c. Remedial cementing
22. Describe the difference between *artificial lift* and improved recovery.
23. What are the different types of artificial lift methods?

2

Principal Field Processing Operations: A Preview

This chapter should serve as a preview of the contents of the text and as a way of giving a bird's-eye view of the scope of activities encompassed. Field processing of produced crude oil–gas mixture aims to separate the well stream into quality oil and gas saleable products in order to recover the maximum amount of each at minimum cost. Our objective in this chapter is to offer a basic understanding and to present a concise description for every processing surface unit from wellhead to finished quality stream products. This chapter would prove to be of value to pave the way for the readers. As readers navigate through this chapter, they can maneuver smoothly from stage to stage as presented in Figure 2.1.

2.1 Gas–Oil Separation

The first step in processing the well stream is to separate the crude oil, natural gas, and water phases into separate streams. A gas–oil separator is a vessel that does this job. Gas–oil separators can be horizontal, vertical, or spherical.

Oil-field separators can be classified into two types based on the number of phases to separate:

1. Two-phase separators, which are used to separate gas from oil in oil fields, or gas from water for gas fields
2. Three-phase separators, which are used to separate the gas from the liquid phase, and water from oil

The liquid (oil, emulsion) leaves at the bottom through a level-control or dump valve. The gas leaves the vessel at the top, passing through a mist extractor to remove the small liquid droplets in the gas.

Separators can be categorized according to their operating pressure. Low-pressure units handle pressures of 10 to 180 psi (69 to 1241 kPa). Medium-pressure separators operate from 230 to 700 psi (1586 to 4826 kPa). High-pressure units handle pressures of 975 to 1500 psi (6722 to 10,342 kPa).

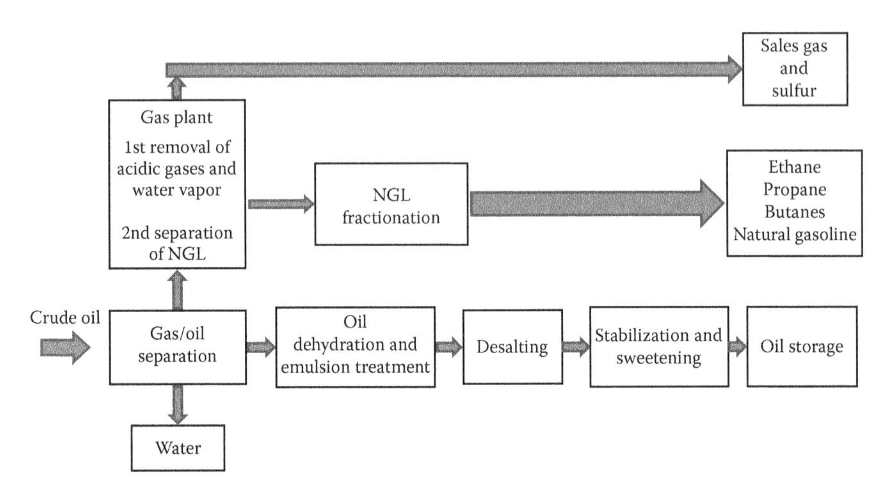

FIGURE 2.1
Field processing of well field effluents.

Gravity segregation is the main force that accomplishes the separation, which means the heaviest fluid settles to the bottom and the lightest fluid rises to the top. The degree of separation between gas and liquid inside the separator depends on the following factors: separator operating pressure, the residence time of the fluid mixture, and the type of flow of the fluid (turbulent flow allows more gas bubbles to escape than laminar flow).

2.2 Oil Dehydration and Emulsion Treatment

Once crude oil is separated, it undergoes further treatment steps. An important aspect during oil field development is the design and operation of wet crude handling facilities.

One has to be aware that not all the water is removed from crude oil by gravity during the first stage of gas–oil separation. Separated crude may contain up to 15% water, which may exist in an emulsified form. The objective of the dehydration step is a dual function: to ensure that the remaining free water is totally removed from the bulk of oil and to apply whatever tools necessary to break the oil emulsion. In general, free water removed in the separator is limited to water droplets of 500 μm and larger. Oil stream leaving the separator would normally contain free water droplets that are smaller in size, in addition to the water emulsified in oil. The full picture of water produced with oil could be visualized as shown in Figure 2.2.

Produced crude oil contains sediment and produced water (BS&W), salt, and other impurities. These are readily removed from the crude oil through this stage. Produced water containing the solids and impurities is discharged

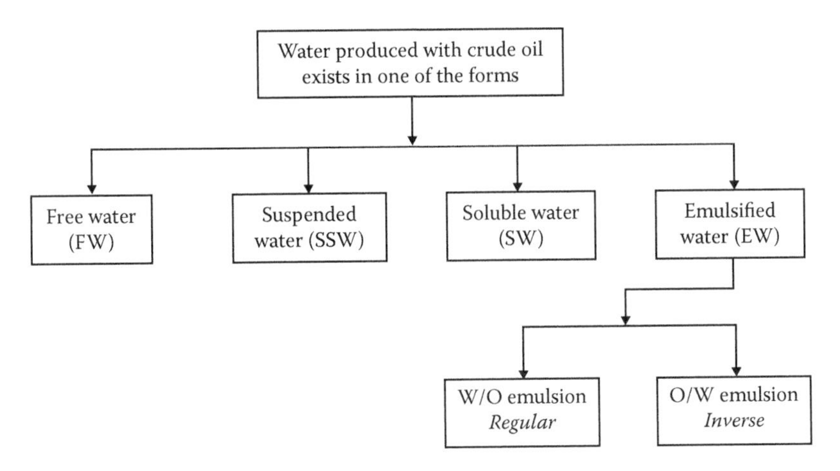

FIGURE 2.2
Forms of saline water produced with crude oil.

to the effluent water treatment system. Clean, dehydrated oil flows from the top of the vessel. Depending on the salt specifications, a combination dehydrator followed by a desalter may be required.

A dehydration system, in general, comprises various types of equipment according to the type of treatment: water removal or emulsion breaking. Most common are the following:

- Free water knockout drum (FWKO)
- Wash tank
- Gunbarrel
- Flow treater
- Chemical injector
- Electrostatic dehydrator

It is very common to use more than one dehydrating aid, particularly for emulsion breaking. Examples are the heater-treater and chem-electric dehydrator.

The role played by adding chemicals to break emulsions should not be overlooked. These chemicals act as de-emulsifiers—once absorbed on the water–oil interface, they will rupture the stabilizing film causing emulsions.

2.3 Desalting

The removal of salt from crude oil is recommended for refinery feed stocks if the salt content exceeds 20 PTB (pounds of salt, expressed as equivalent sodium chloride, per thousand barrels of oil). Salt in crude oil, in most cases, is found dissolved in the remnant water (brine) within the oil. It presents serious corrosion and scaling problems and must be removed.

Electrostatic desalting, whether employed for oil field production dehydration and desalting or at oil refineries, is used to facilitate the removal of inorganic chlorides and water-soluble contaminants from crude oil. In refinery applications, the removal of these water-soluble compounds takes place to prevent corrosion damage to downstream distillation processes.

Salt content in crude oil (PTB) is a function of two parameters: the amount of remnant water in oil (R) and the salinity of remnant water (S). To put it in a mathematical form, we say:

$$PTB = f(R, S)$$

The electrostatic desalting process implies two important consecutive actions:

1. Wash water injection in order to increase the population density of small water droplets suspended in the crude oil (water of dilution)
2. Creating a uniform droplet size distribution by imparting mechanical shearing and dispersion of the dispersed aqueous phase (electrostatic coalescer)

2.4 Stabilization and Sweetening

Once degased, dehydrated, and desalted, crude oil should be pumped to gathering facilities for storage. However, stabilization and sweetening are a must in the presence of hydrogen sulfide (H_2S). H_2S gas is frequently contained in the crude oil as it comes from the wells. It not only has a vile odor, it is also poisonous. It can kill a person if inhaled. It is also corrosive in humid atmosphere forming sulfuric acid. Pipeline specifications require removal of acid gases (carbon dioxide, CO_2) along with H_2S.

The stabilization process, basically a form of partial distillation, is a dual job process. It sweetens *sour* crude oil (removes the hydrogen sulfide and carbon dioxide gases) and reduces vapor pressure, thereby making the crude safe for shipment in tankers. Vapor pressure is exerted by light hydrocarbons, such as methane, ethane, propane, and butane. As the pressure on the crude is lowered, light hydrocarbons vaporize and escape from the bulk of the oil. If a sufficient amount of these light hydrocarbons is removed, the vapor pressure becomes satisfactory for shipment at approximately atmospheric pressure.

2.5 Storage Tanks

Production, refining, and distribution of petroleum products require many different types and sizes of storage tanks. Small bolted or welded tanks might be ideal for production fields, whereas larger, welded storage tanks are used in distribution terminals and refineries throughout the world. Product operating conditions, storage capacities, and specific design issues can affect the tank selection process.

Storage tanks are available in many shapes: vertical and horizontal cylindrical; open top and closed top; flat bottom, cone bottom, slope bottom, and dish bottom. Large tanks tend to be vertical cylindrical or have rounded corners transition from vertical sidewall to bottom. This allows it to withstand hydraulic hydrostatically induced pressure of contained liquid. Most container tanks for handling liquids during transportation are designed to handle varying degrees of pressure.

Most *farm petroleum* storage tanks are aboveground, single-walled, horizontal storage tanks. Such tanks should have secondary containment systems to retain any leakage.

2.5.1 Types of Storage Tanks

Basically there are eight types of tanks used to store liquids:

1. Fixed roof tanks
2. External floating roof tanks
3. Internal floating roof tanks
4. Domed external floating roof tanks
5. Horizontal tanks
6. Pressure tanks
7. Variable vapor space tanks
8. Liquefied natural gas (LNG) tanks

The first 4 tank types are cylindrical in shape with the axis oriented perpendicular to the subgrade. These tanks are almost exclusively aboveground. Horizontal tanks can be used above and below ground. Pressure tanks often are horizontally oriented and spherically shaped to maintain structural integrity at high pressures. They are located aboveground. Variable vapor space tanks can be cylindrical or spherical in shape.

2.6 Gas Sweetening

Having finished crude oil treatment, we now turn to the treatment and processing of natural gas. The actual practice of processing natural gas to pipeline dry gas quality levels can be quite complex but usually involves three main processes to remove the various impurities:

1. Sulfur and carbon dioxide removal (gas sweetening)
2. Water removal (gas dehydration)
3. Separation of natural gas liquids (NGLs)

It should be pointed out that sweetening of natural gas almost always precedes dehydration and other gas plant processes before the separation of NGLs. Dehydration, on the other hand, is usually required for pipeline specifications. It is a necessary step in the recovery of NGLs from natural gas.

Sour natural gas composition can vary over a wide concentration of H_2S and CO_2. It varies from parts per million to about 50 volume percent. Most important are H_2S and CO_2 gases. Gas sweetening is a must for the following reasons: the corrosiveness of both gases in the presence of water; and the toxicity of H_2S gas and a heating value of no less than 980 Btu/SCF.

Some of the desirable characteristics of a sweetening solvent are:

- Required removal of H_2S and other sulfur compounds must be achieved.
- Reactions between solvent and acid gases must be reversible to prevent solvent degradation.
- Solvent must be thermally stable.
- The acid gas pickup per unit of solvent circulated must be high.
- The solvent should be noncorrosive.
- The solvent should not foam in the contactor or still.
- Selective removal of acid gases is desirable.
- The solvent should be cheap and readily available.

Amine gas treating, also known as gas sweetening and acid gas removal, refers to a group of processes that use aqueous solutions of various alkylamines (commonly referred to simply as amines) to remove H_2S and CO_2 from gases. They are known as *regenerative chemical solvents*. It is a common unit process used in refineries, and it is also used in petrochemical plants, natural gas processing plants, and other industries.

2.6.1 Process Description

A typical amine gas treating process includes an absorber unit and a regenerator unit as well as accessory equipment. In the absorber, the amine solution flowing countercurrent to the raw feed gas absorbs H_2S and CO_2 to produce a gas product free of H_2S and CO_2 and an amine solution rich in the absorbed acid gases. Rich amine is then routed into the regenerator producing lean amine to be recycled back to the absorber. The stripped overhead gas from the regenerator is concentrated H_2S and CO_2.

2.7 Gas Dehydration

Glycol dehydration is a liquid desiccant system for the removal of water from natural gas and natural gas liquids. It is the most common and economical means of water removal from these streams. Triethylene glycol (TEG) is used to remove water from the natural gas stream in order to meet the pipeline quality standards. This process is required to prevent hydrates formation at low temperatures or corrosion problems due to the presence of carbon dioxide or hydrogen sulfide (regularly found in natural gas). Dehydration, or water vapor removal, is accomplished by reducing the inlet water dew point (temperature at which vapor begins to condense into a liquid) to the outlet dew point temperature, which will contain a specified amount of water.

2.7.1 Process Description

The wet gas is brought into contact with dry glycol in an absorber. Water vapor is absorbed in the glycol and consequently its dew point reduces. The wet rich glycol then flows from the absorber to a regeneration system in which the entrained gas is separated and fractionated in a column and reboiler. The heating allows boiling off the absorbed water vapor and the water dry lean glycol is cooled (via heat exchange) and pumped back to the absorber.

2.8 Recovery and Separation of Natural Gas Liquids

Although some of the needed processing of natural gas can be accomplished at or near the wellhead (field processing), the complete processing of natural gas takes place at a processing plant, usually located in a natural gas producing region.

NGLs usually consist of the hydrocarbons ethane and heavier (C_2^+) components. In order to recover and to separate NGLs from a bulk of a gas stream, a change in phase is to be induced. In other words, a new phase has to be developed for separation to take place.

Two distinctive operations are in practice for the separation of NGL constituents, dependent on the use of either *energy* or *mass* as a separating agent.

1. Energy separating agent (ESA)—Removing heat by refrigeration will allow heavier components to condense, hence a liquid phase is formed. Production of NGLs at low temperature is practiced in many gas processing plants. For example, to recover, say, C_2, C_3, and C_4 from a gas stream, demethanization by refrigeration is done.
2. Mass separating agent (MSA)—To separate NGLs a new phase is developed by using a liquid (solvent), MSA, to be introduced in contact with the gas stream, that is, absorption. This solvent is selective to absorb the NGL components.

2.9 Fractionation of Natural Gas Liquids

Once NGLs have been separated from a natural gas stream, they are further separated into their component parts, or fractions, using the distillation or fractionation process. This process can take place either in the field or at a terminal location hooked to a petrochemical complex. NGL components are defined as ethane, propane, butane, and pentanes plus (natural gasoline).

Fractionation in gas plants has many common goals. As presented in Figure 2.1, it is aimed at producing on-specification products and making sources available for different hydrocarbons. Fractionation is basically a distillation process leading to fractions or cuts of hydrocarbons. Examples of cuts or fractions are C_3/C_4, known as LPG, and C_5^+, known as natural gasoline. Liquid fractionation towers are used to separate and remove NGLs. They can be controlled to produce pure vapor-phase products from the overhead by optimizing the following factors:

- Inlet flow rate
- Reflux flow rate
- Rebolier temperature
- Reflux temperature
- Column pressure

2.10 Surface Production Facilities

Section V of the text includes a new section that encompasses topics that would serve as facilities for surface production operations. For convenience they are grouped in this section under the heading of Surface Production Facilities. This section is a new addition to the second edition of the book. It includes four chapters that cover the following topics:

- Produced Water Management and Disposal (Chapter 15)
- Field Storage Tanks, Vapor Recovery System (VRS), and Tank Blanketing (Chapter 16)
- Oil Field Chemicals (OFC) (Chapter 17)
- Piping and Pumps (Chapter 18)

These facilities are best recognized as the *four pillars* in surface field facilities.

3

Composition, Types of Crude Oil, and Oil Products

Crude oil is far from being one homogeneous substance. Its physical characteristics differ depending on where in the world it is produced, and those variations determine its usage and price. This is due to the fact that oils from different geographical locations will naturally have their own unique properties. These oils vary dramatically from one another when it comes to their viscosity, volatility, and toxicity.

In its natural, unrefined state, crude oil ranges in density and consistency, from very thin, lightweight, and volatile fluidity to an extremely thick, semi-solid, heavyweight oil. There is also a tremendous gradation in the color that the oil extracted from the ground exhibits, ranging from a light, golden yellow to the very deepest, darkest black imaginable.

In this chapter, the composition of crude oil is presented using the chemical approach and by applying the physical methods traditionally used. Classification and characterization of crude oils based on correlation indexes and crude assays are explained. Different types of well-known crude oils as well as benchmarks are included. Also, main crude oil products are briefly described as produced by the backbone distillation operations.

3.1 Introduction: Facts about Crude Oil

Crude oil, commonly known as petroleum, is a liquid found within the earth comprised of hydrocarbons, organic compounds, and small amounts of metal. Although hydrocarbons are usually the primary component of crude oil, their composition can vary from 50% to 97% depending on the type of crude oil and how it is extracted. Organic compounds like nitrogen, oxygen, and sulfur typically make up between 6% and 10% of crude oil, whereas metals such as copper, nickel, vanadium, and iron account for less than 1% of the total composition.

Figure 3.1 presents comprehensive information about crude oil. It includes types of crude oil, identification parameters of crude oil, elemental composition

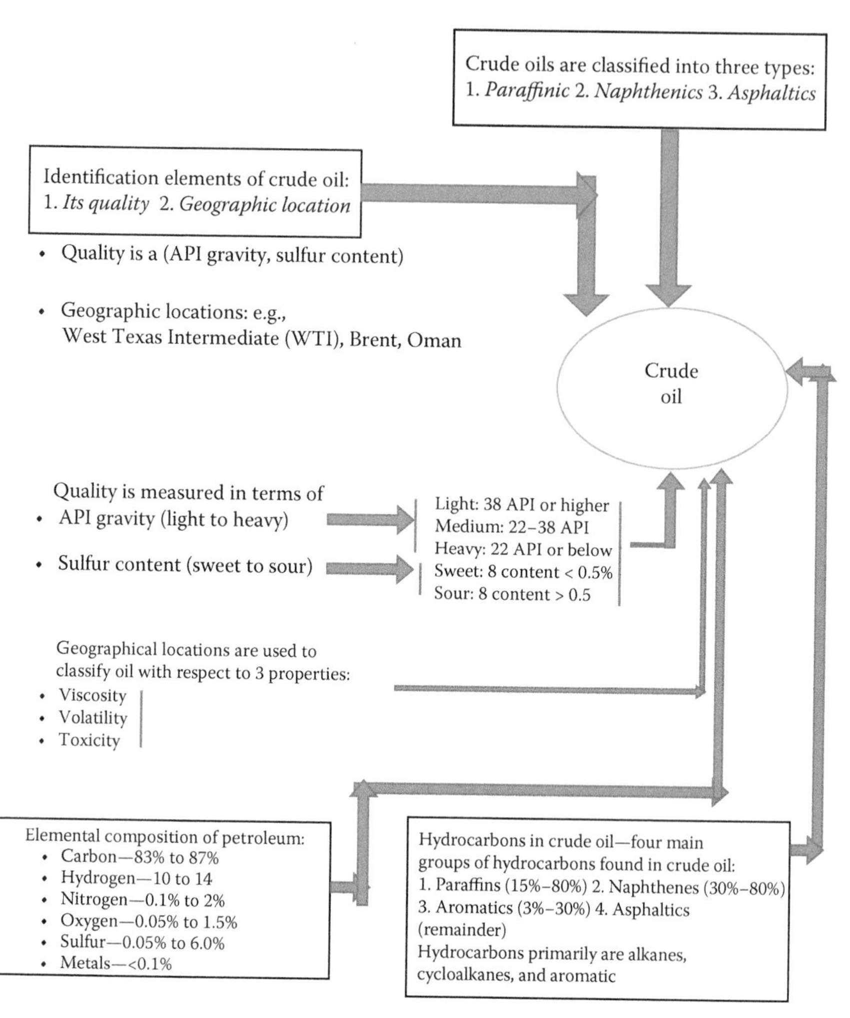

FIGURE 3.1
Facts about crude oil.

of crude oil, the four main groups of hydrocarbons found in crude oil, and how the quality of crude oil is measured by two parameters:

1. The API gravity and classified as light (38 API or higher), medium (22–38 API), or heavy (22 API or below)
2. The sulfur content as sweet (S content <0.5%) or sour (S content >0.5)

This classification is further exemplified by some typical crude oils as shown in Table 3.1.

TABLE 3.1

Examples of Typical Crude Oils

Classification	API Range	Examples (Crude Name, API)
Light	>33	Saudi Super Light, 39.5 Nigerian light, 36 North Sea Brent, 37
Medium	28–33	Kuwait, 31 Venezuela Light, 30
Heavy	<28	Saudi Heavy, 28 Venezuela Heavy, 24

3.2 Crude Oil Composition

Crude oil is essentially a mixture of many different hydrocarbons, all of varying lengths and complexities. There are two broad approaches to studying and quantifying the composition of crude oil: the chemical approach (analysis) and the physical methods. Chemical composition describes and identifies the individual chemical compounds isolated from crude oils over the years. However, no crude oil has ever been completely separated into its individual components, although many components can be identified. A total of 141 compounds were identified in a sample of Oklahoma crude that account for 44% of the total crude volume. Physical representation, on the other hand, involves considering the crude oil and its products as mixtures of hydrocarbons and describing physical laboratory tests or methods for characterizing their quality.

3.2.1 Chemical Approach

Nearly all petroleum deposits are made up of a mixture of chemical compounds that consist of hydrogen and carbon, known as hydrocarbons, with varying amounts of nonhydrocarbons containing S, N_2, O_2, and other some metals. The composition of crude oil by elements is approximated as shown in Figure 3.1. It could be further stated that these hydrocarbon compounds making up oils are grouped chemically into different series of compounds described by the following characteristics:

- Each series consists of compounds similar in their molecular structure and properties (e.g., the alkanes or paraffin series).
- Within a given series, there exists a wide spectrum of compounds that range from extremely light or simple hydrocarbon to a heavy or complex one. As an example, CH_4 stands for the former group and $C_{40}H_{82}$ for the latter in the paraffinic series.

3.2.1.1 First Hydrocarbon Series

The major constituents of most crude oils and its products are hydrocarbon compounds, which are made up of hydrogen and carbon only. These compounds belong to one of the following subclasses:

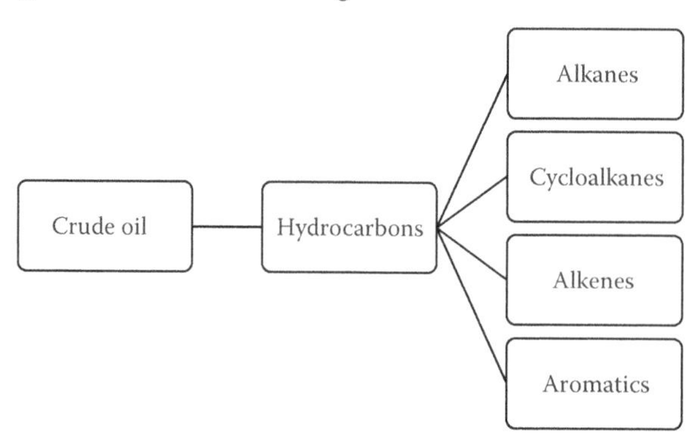

- Alkanes or paraffins—Alkanes are saturated compounds having the general formula C_nH_{2n+2}. Alkanes are relatively nonreactive compounds in comparison to other series. They may either be straight-chain or branched compounds; the latter are more valuable than the former, because they are useful for the production of high-octane gasoline.
- Cycloalkanes or cycloparaffins (naphthenes)—Cycloalkanes and bicycloalkanes are normally present in crude oils and its fractions in variable proportions. The presence of large amounts of these cyclic compounds in the naphtha range has its significance in the production of aromatic compounds. Naphtha cuts with a high percentage of naphthenes would make an excellent feedstock for aromatization.
- Alkenes or olefins—Alkenes are unsaturated hydrocarbon compounds having the general formula C_nH_n. They are practically not present in crude oils, but they are produced during processing of crude oils at high temperatures. Alkenes are very reactive compounds. Light olefinic hydrocarbons are considered the base stock for many petrochemicals. Ethylene, the simplest alkene, is an important monomer in this regard. For example, polyethylene is a well-known thermoplastic polymer and polybutadiene is the most widely used synthetic rubber.
- Aromatics—Aromatic compounds are normally present in crude oils. Only monomolecular compounds in the range of C6–C8 (known as B-T-X) have gained commercial importance. Aromatics in this range are not only important petrochemical feedstocks but are also valuable for motor fuels.

Dinuclear and polynuclear aromatic compounds are present in heavier petroleum fractions and residues. Asphaltenes, which are concentrated in heavy residues and in some asphaltic crude oils, are, in fact, polynuclear aromatics of complex structures. It has been confirmed by mass spectroscopic techniques that condensed-ring aromatic hydrocarbons and heterocyclic compounds are the major compounds of asphaltenes.

3.2.1.2 Second Nonhydrocarbon Compounds

So far, a brief review of the major classes of the hydrocarbon compounds that exist in crude oils and their products has been presented. For completeness, we should mention that other types of nonhydrocarbon compounds occur in crude oils and refinery streams. Most important are sulfur compounds, nitrogen compounds, oxygen compounds, and metallic compounds.

3.2.1.2.1 Sulfur Compounds

In addition to the gaseous sulfur compounds in crude oil, many sulfur compounds have been found in the liquid phase in the form of organosulfur. These compounds are generally not acidic. Sour crude oils are those containing a high percentage of hydrogen sulfide. However, many of the organic sulfur compounds are not thermally stable, thus producing hydrogen sulfide during crude processing.

High-sulfur crude oils are in less demand by refineries because of the extra cost incurred for treating refinery products. Naphtha feed to catalytic reformers is hydrotreated to reduce sulfur compounds to very low levels (1 ppm) to avoid catalyst poisoning.

The following sulfur compounds are typical:

- Mercaptans (H–S–R)—Hydrogen sulfide, H–S–H, may be considered as the simple form of mercaptan; however, the higher forms of the series are even more objectionable in smell. For example, butyl mercaptan ($H{-}S{-}C_4H_9$) is responsible for the unusual odor of the shank.
- Sulfides (R–S–R)—When an alkyl group replaces the hydrogen in the sulfur-containing molecule, the odor is generally less obnoxious. Sulfides could be removed by the hydrotreating technique, which involves the hydrogenation of the petroleum streams as follows:

$$R{-}S{-}R + 2H{-}H \rightarrow 2R{-}H + H{-}S{-}H$$

$$R{-}S{-}R + H{-}H \rightarrow R{-}R + H{-}S{-}H$$

  The hydrogen sulfide may be removed by heating and may be separated by using amine solutions.

- Polysulfides (R–S–S–R)—These are more complicated sulfur compounds and they may decompose, in some cases depositing elemental sulfur. They may be removed from petroleum fractions, similar to the sulfides, by hydrotreating.

3.2.1.2.2 *Nitrogen Compounds*

Nitrogen compounds in crude oils are usually low in content (about 0.1%–0.9%) and are usually more stable than sulfur compounds. Nitrogen in petroleum is in the form of heterocyclic compounds and may be classified as basic and nonbasic. Basic nitrogen compounds are mainly composed of pyridine homologs and have the tendency to exist in the high-boiling fractions and residues. The nonbasic nitrogen compounds, which are usually pyrrole and indole, also occur in high-boiling fractions and residues. Only a trace amount of nitrogen is found in light streams.

During hydrotreatment (hydrodesulfurization) of petroleum streams, hydrodenitrogeneation takes place as well, removing nitrogen as ammonia gas, thus reducing the nitrogen content to the acceptable limits for feedstocks to catalytic processes.

It has to be stated that the presence of nitrogen in petroleum is of much greater significance in refinery operations than might be expected from the very small amounts present. It is established that nitrogen compounds are responsible for the following:

- Catalyst poisoning in catalytic processes
- Gum formation in some products such as domestic fuel oils

3.2.1.2.3 *Oxygen Compounds*

Oxygen compounds in crude oils are more complex than sulfur compounds. However, oxygen compounds are not poisonous to processing catalysts. Most oxygen compounds are weakly acidic, such as phenol, cresylic acid, and naphthenic acid. The oxygen content of petroleum is usually less than 2%, although larger amounts have been reported.

3.2.1.2.4 *Metallic Compounds*

Many metals are found in crude oils; some of the more abundant are sodium, calcium, magnesium, iron, copper, vanadium, and nickel. These normally occur in the form of inorganic salts soluble in water (as in the case of sodium chloride) or in the form of organometallic compounds (as in the case of iron, vanadium, and nickel).

The occurrence of metallic constituents in crude oils is of considerably greater interest to the petroleum industry than might be expected from the very small amounts present. The organometallic compounds are usually concentrated in the heavier fractions and in crude oil residues. The presence of high concentration of vanadium compounds in naphtha streams for catalytic reforming feeds will cause permanent poisons. These feeds should be hydrotreated not only to reduce the metallic poisons but also to desulfurize and denitrogenate the sulfur and nitrogen compounds. Hydrotreatment may also be used to reduce the metal content in heavy feeds to catalytic cracking.

3.2.2 Physical Methods

Having discussed the various chemicals found in crude oils and realizing not only the complexity of the mixture but the difficulty of specifying a crude oil as a particular mixture of chemicals, we can understand why the early petroleum producers adopted the physical methods generally used for classification.

As may be seen, crude oils from different locations may vary in appearance and viscosity and also vary in their usefulness as producers for final products. It is possible by the use of certain basic tests to identify the quality of crude oil stocks. The tests included in the following list are primarily physical (except sulfur determination):

- Distillation
- Density, specific gravity, and American Petroleum Industry (API) gravity
- Viscosity
- Vapor pressure
- Flash and fire points
- Cloud and pour points
- Color
- Sulfur content
- Basic sediments and water (BS&W)
- Aniline point
- Carbon residue

The details of some of these tests are described next.

- *API gravity.* Earlier, density was the principal specification for petroleum products. However, the derived relationships between the density and its fractional composition were only valid if they were applied to a certain type of petroleum. Density is defined as the mass of a unit volume of material at a specified temperature. It has the dimensions of grams per cubic centimeter.

  Another general property, which is more widely used, is the specific gravity. It is the ratio of the density of oil to the density of water and is dependent on two temperatures, those at which the densities of the oil sample and the water are measured. When the water temperature is 4°C (39°F), the specific gravity is equal to the density in the cgs system, because the volume of 1 g of water at that temperature is, by definition, 1 mL. Thus, the density of water, for example, varies with temperature, whereas its specific gravity is always unity at equal temperatures. The standard temperatures for specific gravity in the petroleum industry in North America are 60/60°F and 15.6/15.6°C.

Although density and specific gravity are used extensively in the oil industry, the API gravity is considered the preferred property. It is expressed by the following relationship:

$$°API = \frac{141.5}{\gamma} - 131.5$$

where γ is the oil specific gravity at 60°F. Thus, in this system, a liquid with a specific gravity of 1.00 will have an API of 10 deg. A higher API gravity indicates a lighter crude or oil product, whereas a low API gravity implies a heavy crude or product.

- *Carbon residue.* Carbon residue is the percentage of carbon by weight for coke, asphalt, and heavy fuels found by evaporating oil to dryness under standard laboratory conditions. Carbon residue is generally referred to as Conradson carbon residue (CCR). It is a rough indication of the asphaltic compounds and the materials that do not evaporate under conditions of the test, such as metals and silicon oxides.
- *Viscosity.* The viscosity is the measure of the resistance of a liquid to flow, hence indicating the *pumpability* of oil.
- *Pour point.* Pour point is defined as the lowest temperature (5°F) at which the oil will flow. The lower the pour point, the lower the paraffin content of the oil.
- *Ash content.* Ash content is an indication of the contents of metal and salts present in a sample. The ash is usually in the form of metal oxides, stable salts, and silicon oxides. The crude sample is usually burned in an atmosphere of air and the ash is the material left unburned.
- *Reid vapor pressure.* The Reid vapor pressure (RVP) is a measure of the vapor pressure exerted by oil or by light products at 100°F.
- *Metals.* In particular, arsenic, nickel, lead, and vanadium are potential poisons for process catalysts. Metal contents are reported in parts per million (ppm).
- *Nitrogen.* It is the weight of total nitrogen determined in a liquid hydrocarbon sample (in parts per million). Nitrogen compounds contribute negatively to process catalysts.
- *Salt content.* Salt content is typically expressed as pounds of salt (sodium chloride, NaCl) per 1000 barrels of oil (PTB). Salts in crude oil and in heavier products may create serious corrosion problems, especially in the top-tower zone and the overhead condensers in distillation columns.
- *Sulfur.* This is the percentage by weight (or parts per million) of total sulfur content determined experimentally in a sample of oil or its product. The sulfur content of crude oils is taken into consideration in addition to the API gravity in determining their commercial values.

It has been reported that heavier crude oils may have a high sulfur content.

- *Hydrogen sulfide.* Hydrogen sulfide dissolved in a crude oil or its products is determined and measured in parts per million. It is a toxic gas that can evolve during storage or in the processing of hydrocarbons.

  The aforementioned tests represent many properties for the crude oils that are routinely measured because they affect the transportation and storage facilities. In addition, these properties define what products can be obtained from a crude oil and contribute effectively to safety and environmental aspects. The price of a crude oil is influenced by most of these properties.

  To conclude, it can be stated that light and low sulfur crude oils are worth more than heavy and high sulfur ones as illustrated by the next diagram.

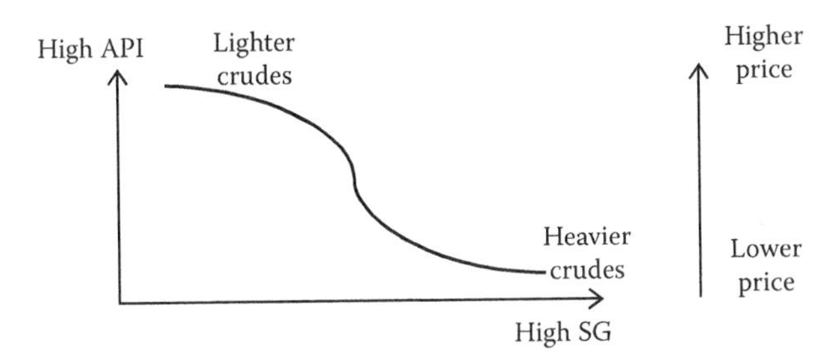

One can summarize the two approaches of examining crude oils as follows:

1. Chemical composition
2. Physical properties
   a. API, S, salt, metals, nitrogen, and so forth
   b. Distillation: ASTM, TBP, EFV
   c. Correlations: Kw, Ind

TBP is true boiling point, EFV is equilibrium flash vaporization, Kw is the Watson characterization factor, and Ind is U.S. Bureau of Mines correlation index.

3.3 Classification of Crude Oils

3.3.1 Broad Classification (Based on Chemical Structures)

Although there is no specific method for classifying crude oils, it would be useful for a refiner to establish some simple criteria by which the crude in hand would be classified. A broad classification of crudes has been developed based on some simple physical and chemical properties. Crude oils are generally classified into three types depending on the relative amount of the hydrocarbon class that predominates in the mixture.

1. *Paraffinic* constituents are predominantly paraffinic hydrocarbons with a relatively lower percentage of aromatics and naphthenes.
2. *Naphthenics* contain relatively a higher ratio of cycloparaffins and a higher amount of asphalt than in paraffinic crudes.
3. *Asphaltics* contain relatively a large amount of fused aromatic rings and a high percentage of asphalt.

3.3.2 Classification by Chemical Composition

Petroleum contains a large number of chemicals with different compositions depending on the location and natural processes involved. Petroleum composition varies (molecular type and weight) from one oil field to another, from one well to another in the same field and even from one level to another in the same well.

A correlation index was introduced to indicate the crude type or class. The following relationship between the mid-boiling point of the fraction and its specific gravity gives the correlation index, known as the Bureau of Mines Correlation Index

$$BMCI = \frac{48,640}{K} + (473.7d - 456.8)$$

where K is the mid-boiling point of a fraction in Kelvin degrees and d is the specific gravity of the fraction at 60/60°F. It is possible to classify crudes as paraffinic, naphthenic (mixed) or asphaltic according to the calculated values using the preceding relationship. A zero value has been assumed for paraffins and 100 for aromatics.

3.3.3 Classification by Density

Density gravity (specific gravity) has been extensively applied to specify crude oils. It is a rough estimation of the quality of a crude oil. Density is expressed in terms of API gravity by the following relationship:

$$°API = [141.5/\text{Specific gravity}] - 131.5$$

Another index used to indicate the crude type is the Watson characterization (UOP) factor. This also relates the mid-boiling point of the fraction in Kelvin degrees to the density.

TABLE 3.2

General Properties of Crude Oils

Property	Paraffin Base	Asphalt Base
API gravity	High	Low
Naphtha content	High	Low
Naphtha octane number	Low	High
Naphtha odor	Sweet	Sour
Kerosene smoking tendency	Low	High
Diesel-fuel knocking tendency	Low	High
Lube-oil pour point	High	Low
Lube-oil content	High	Low
Lube-oil viscosity index	High	Low

$$\text{Watson correlation factor} = \frac{(K^{1/3})}{d}$$

A value higher than 10 indicates predominance of paraffins, and a value lower than 10 indicates predominance of aromatics. Properties of crude oils will thus vary according to their base type, as shown in Table 3.2.

3.4 Crude Oil Comparisons and Crude Oil Assay

In order to establish a basis for the comparison between different types of crude oil, it is necessary to produce experimental data in the form of what is known as an *assay*. The assay can be an inspection assay or comprehensive assay. Crude assays are described as the systematic compilation of data for the physical properties of the crude and its fractions, as well as the yield. In other words, a crude assay involves the determination of the following: the properties of crude oil and the fractions obtained (i.e., their percentage yield and properties). Analytical testing only without carrying out distillation may be considered an assay. However, the most common assay is a comprehensive one that involves all of the aforementioned parameters.

The basis of the assay is the distillation of a crude oil under specified conditions in a batch laboratory distillation column, operated at high efficiency (column with 14 plates and reflux ratio [RR]). Pressure in the column is reduced in stages to avoid thermal degradation of high boiling components.

A comparison of the characteristics of different types of crude oil over the distillation range could be made via a graph that relates the density of distillate fractions and their mid-boiling points. Such a comparison is illustrated

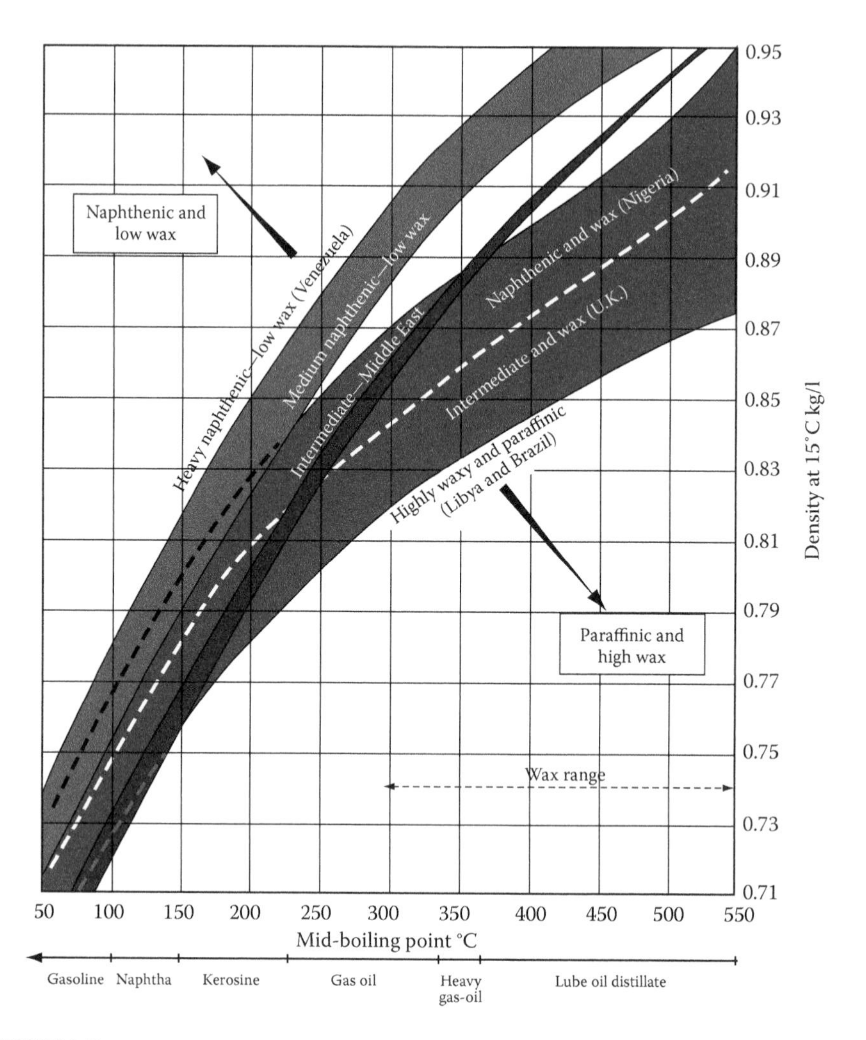

FIGURE 3.2
Comparisons of types of crude oils based on density/mid-boiling point.

in Figure 3.2. The density level of crude at a given boiling point on the curve is a function of the relative proportions of the main three hydrocarbon series: aromatics, cycloparaffins, paraffins; their densities decrease in that order.

In addition, a comparison with some standard crude is usually recommended. A review of some of the crudes that are adopted in the oil industry is given next.

3.4.1 Benchmark

The three most quoted oil products are North America's West Texas Intermediate Crude (WTI), North Sea Brent Crude, and the UAE Dubai Crude, and their pricing is used as a barometer for the entire petroleum industry, although, in total, there are 46 key oil-exporting countries. Brent Crude is typically priced at about $2 over the WTI Spot price, which is typically priced $5 to $6 above the EIA's Imported Refiner Acquisition Cost (IRAC) and OPEC Basket prices.

3.4.1.1 Benchmark Crude

A benchmark crude or marker crude is a crude oil that serves as a reference price for buyers and sellers of crude oil. There are three primary benchmark crude oils: WTI, Brent Blend, and Dubai. Other well-known blends include:

- Opec basket used by OPEC
- Tapis Crude, which is traded in Singapore
- Bonny Light used in Nigeria and Mexico's Isthmus

The Energy Intelligence Group publishes a handbook that identified 195 major crude streams or blends in its 2011 edition (Energy Intelligence Group 2011). Benchmark crude oil that is traded so regularly in the spot market that its price quotes are relied upon by sellers of other crude oils as a reference point for setting term or spot prices.

3.4.1.2 Heavy Crude

Heavy crude has API gravity lower than 28°C. The lower the API gravity, the heavier the oil should be. It usually contains high concentrations of sulfur and several metals, particularly nickel and vanadium (high amount of wax). These are the properties that make them difficult to pump out of the ground or through a pipeline and interfere with refining. These properties also present serious environmental challenges to the growth of heavy oil production and use.

3.4.1.3 Light Crude

Light crude has API gravity higher than 33°C. The higher the API gravity, the lighter the crude oil shall be. Light crude is defined as having a low concentration of wax. This classification of oil is easier to pump and transport.

3.4.1.4 Sweet Crude

Sweet crude has small amounts of hydrogen sulfide and carbon dioxide, and is used primarily in gasoline. Sweet crude is usually a crude oil that has a sulfur content that is 0.5% or less by weight. Lower sulfur content improves the quality of the resulting refined products, and sweet crudes do not require as much processing as sour crudes. They are referred to as sweet because of the absence of an unpleasant sulfur smell.

3.4.1.5 Sour Crude

Usually crude oil that has a sulfur content that is greater than 0.5% is considered sour. This higher sulfur content affects the quality of the resulting refined products and sometimes means extra processing is required. It is referred to as sour because of the unpleasant smell of the sulfur.

3.5 Crude Oil Products

Petroleum is of little use as it comes from the ground. Thus crude oil must be put through a series of processes in order to be converted into the many hundreds of finished oil products that are derived from the crude. Crude oil separation is accomplished in two consecutive steps: first by fractionating the total crude oil at essentially atmospheric pressure; then feeding the bottom residue from the atmospheric tower to a second fractionator, operating at high vacuum.

As applied to crude oil, the distillation process progressively *sifts out* such components as gas, gasoline, kerosene, home heating oil, lubricating oils, heavy fuel oils, and asphalt. Distillation is a physical process. It can separate crude into various cuts, but it cannot produce more of a particular cut than existed in the original crude. This leads us to the next step in refining, which involves chemical conversion to produce more high-octane gasoline and many other products.

It should be mentioned that refineries rely on four major chemical processing operations in addition to the backbone physical operation of fractional distillation in order to alter the ratios of the different fractions. This makes a total of what is called the *Five Pillars* of petroleum refining.

- Pillar No. 1: Fractional Distillation
- Pillar No. 2: Cracking
- Pillar No. 3: Unification (Alkylation)
- Pillar No. 4: Alteration (Catalytic Reforming)
- Pillar No. 5: Hydro-processing

Figure 3.3 is an illustration of an atmospheric distillation column along with petroleum products.

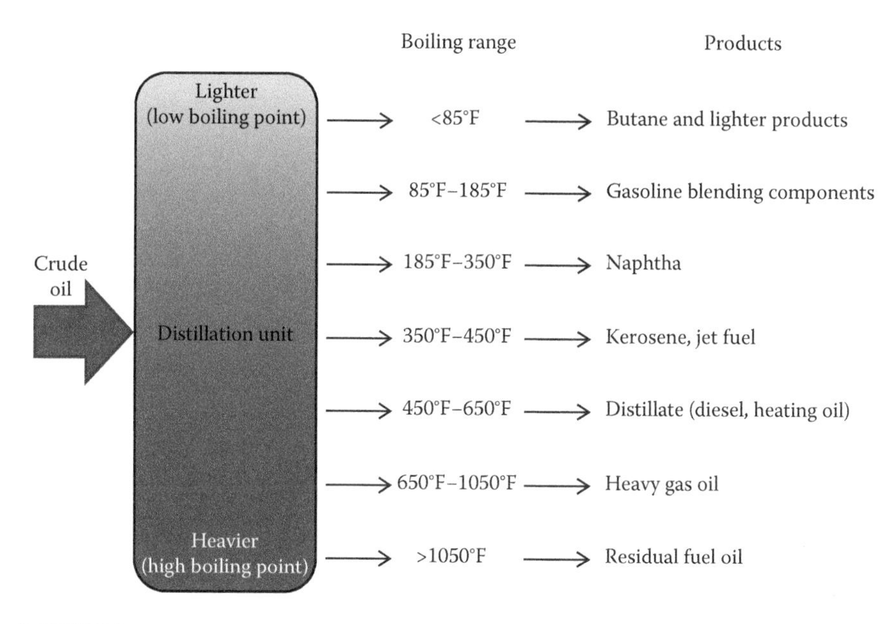

FIGURE 3.3
Crude distillation and petroleum products. (From U.S. Energy Information Administration, Today in Energy, July 5, 2012. With permission.)

4

Composition and Characteristics of Natural Gas

Natural gas, as it is used by consumers, is much different from the natural gas that is brought from underground up to the wellhead. Although the processing of natural gas is in many respects less complicated than the processing and refining of crude oil, it is equally as necessary before its use by end users.

Natural gas is stripped down to methane before being used by consumers. However, natural gas found at the wellhead, although still composed primarily of methane, is by no means as pure.

This chapter provides an overview of some of the characteristics of natural gas as a potential source of fossil energy. In particular, the following are covered:

- Background
- Sources and origin of natural gas
- Composition
- Properties and gas specs
- Applications and uses

4.1 Background

Gas is often referred to as *natural gas* because it is a naturally occurring hydrocarbon. It is colorless and consists mainly of methane, which is the simplest hydrocarbon. Natural gas is extracted by drilling wells into the ground, through the geographic layers, to reach the gas deposits.

Natural gas travels from the wellhead to end consumers through a series of pipelines. These pipelines, including flow lines, gathering lines, transmission lines, distribution lines, and service lines, carry gas at varying rates of pressure.

Raw natural gas comes from three types of wells:

1. Oil wells, where natural gas is typically termed *associated gas*. This gas can exist separate from oil in the formation (free gas) or dissolved in the crude oil (dissolved gas). Dissolved gas is that portion of the gas dissolved in the crude, and associated gas (called gas cap) is free gas in contact with the crude oil. Natural gas extracted from oil wells is called casinghead gas.
2. Gas wells, in which there is little or no crude oil, is termed *nonassociated gas*. Gas wells typically produce raw natural gas by itself (dry gas).
3. Condensate wells, producing free natural gas along with a semiliquid hydrocarbon condensate (wet gas).

All crude oil reservoirs contain dissolved gas, and may or may not contain associated gas.

Whatever the source of the natural gas, once separated from crude oil (if present) it commonly exists in mixtures of methane with other hydrocarbons, principally ethane, propane, butane, and pentanes. In addition, raw natural gas contains water vapor along with some nonhydrocarbons such as hydrogen sulfide (H_2S), carbon dioxide (CO_2), helium, nitrogen, and other compounds. These characteristics of natural gas are illustrated as shown in Figure 4.1.

As the reservoir pressure drops when nonassociated gas is extracted from a field under supercritical (pressure/temperature) conditions; higher molecular weight components (C_3^+) may partially condense upon reducing the pressure, under isothermal conditions. This effect is called retrograde condensation. The liquid thus formed may get trapped as the pores of the gas reservoir get depleted. More frequently, the liquid condenses at the surface producing what is called natural gas liquid (NGL). Once separated from the gas stream, the NGLs can be further separated into fractions, ranging from the heaviest condensates (hexanes, pentanes, and butanes) through liquefied petroleum gas (LPG; essentially butane and propane) to ethane. This source of light hydrocarbons is especially prominent in the United States, where natural gas processing provides a major portion of the ethane feedstock for olefin manufacture and the LPG for heating and commercial purposes.

The world's largest gas field is the offshore South Pars/North Dome Gas-Condensate field, shared between Iran and Qatar. It is estimated to have 51 trillion cubic meters of natural gas and 50 billion barrels of natural gas condensates. Trends in the top five natural gas-producing countries (U.S. Energy Information Administration data) is shown in Figure 4.2.

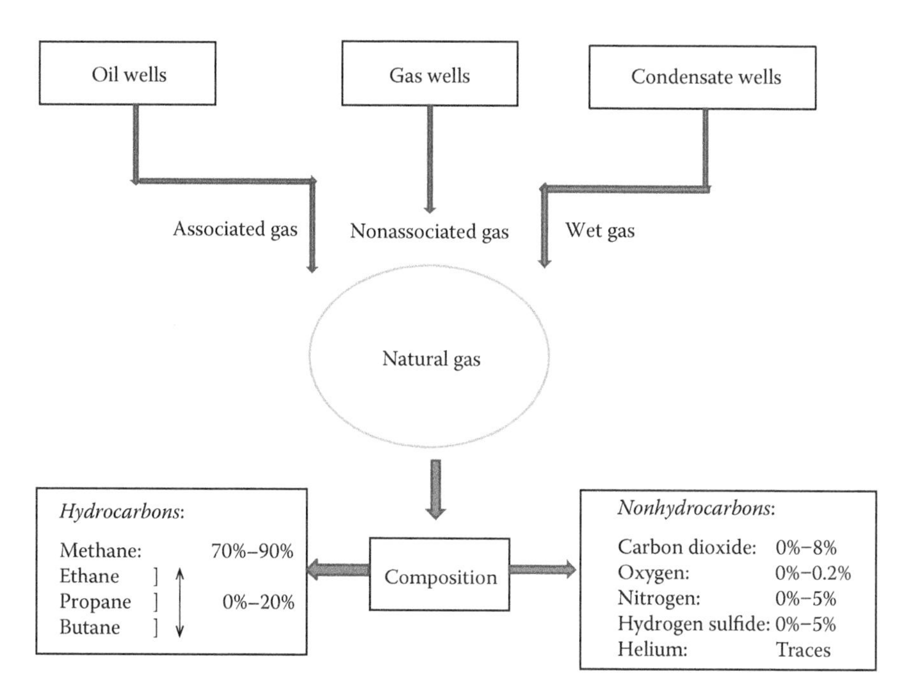

FIGURE 4.1
Some characteristics of natural gas.

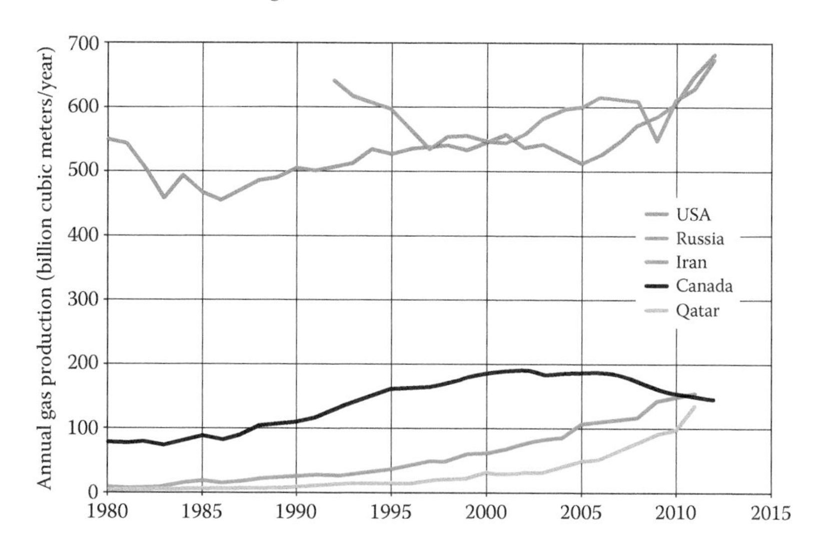

FIGURE 4.2
Top five natural gas producing countries: United States, Russia, Iran, Canada, and Qatar. (Data from U.S. Energy Information Administration, https://en.wikipedia.org/wiki/List_of_countries_by_natural_gas_production.)

4.2 Sources and Origin of Natural Gas

Natural gas is a fossil fuel, like oil. It was formed over hundreds of millions of years from organic matter, such as plankton, plants, animals, and other life forms. Over time, sand, sediment, and rock buried the organic matter. This sediment and debris puts a great deal of pressure on the organic matter, which is compressed under the earth, at very high pressure and for a very long time. This leads to the formation of what is called *thermogenic* methane, reference to microorganisms that create heat within organic wastes.

Now, this compression, combined with high temperatures found deep underneath the earth, breaks down the carbon bonds in the organic matter, because as one gets deeper under the earth's crust, the temperature shoots higher. On the other hand, at shallower deposits (where temperature is lower), more oil is produced relative to natural gas.

In general at higher temperatures, however, more natural gas is created, as opposed to oil. That is why natural gas is usually associated with oil in deposits that are 1 to 2 miles below the earth's crust. Deeper deposits, very far underground, usually contain primarily natural gas and, in many cases, pure methane.

Natural gas can also be formed through the transformation of organic matter by tiny microorganisms. This type of methane is referred to as *biogenic methane*, which is produced by tiny methane-producing microorganisms (methanogens). They are capable of chemically breaking down organic matter to produce methane. These microorganisms are commonly found in areas near the surface of the earth that are void of oxygen. These microorganisms also live in the intestines of most animals, including humans. Formation of methane in this manner usually takes place close to the surface of the earth, and the methane produced is usually lost into the atmosphere.

In some cases, this methane can be trapped underground and become recoverable as natural gas. An example of biogenic methane is landfill gas. Waste-containing landfills produce a relatively large amount of natural gas from the decomposition of the waste materials that they contain. New technologies are allowing this gas to be harvested and used to add to the supply of natural gas.

4.3 Composition

Natural gas is a vital component of the world's supply of energy. It is one of the cleanest, safest, and most useful of all energy sources. Natural gas is a combustible mixture of hydrocarbon consisting primarily of saturated light paraffin such as methane and ethane, both of which are gaseous under atmospheric conditions. The mixture also may contain other hydrocarbons such as propane, butane, pentane, and hexane. In natural gas reservoirs even the heavier hydrocarbons occur for the most part in gaseous form because of the higher pressures. The composition of natural gas can vary widely. A typical composition of natural gas, exit a gas well is shown next:

Typical Composition of Natural Gas

Methane	CH_4	70%–90%
Ethane	C_2H_6	0%–20%
Propane	C_3H_8	0%–20%
Butane	C_4H_{10}	0%–20%
Carbon dioxide	CO_2	0%–8%
Oxygen	O_2	0%–0.2%
Nitrogen	N_2	0%–5%
Hydrogen sulfide	H_2S	0%–5%
Rare gases	A, He, Ne, Xe	Trace

In its purest form, natural gas, such as that used domestically, is almost pure methane. The distinctive *rotten egg* smell that we often associate with natural gas is actually an odorant called *mercaptan*, added to the gas before it is delivered to the end user. Mercaptans make it safe in handling natural gas by detecting any leaks. Natural gas is considered *dry* when it is almost pure methane, having had most of the other commonly associated hydrocarbons removed. When other hydrocarbons are present, the natural gas is called *wet*.

4.4 Properties and Gas Specs

Natural gas is a combustible gas, about 40% lighter than air, so should it ever leak, it can dissipate into the air. Other positive attributes of natural gas are a high ignition temperature and a narrow flammability range, meaning natural gas will ignite at temperatures above 1100°F and burn at a mix of 4%–15% volume in air.

Other properties are summarized as follows:

- Clean burning (The simpler molecular structure of natural gas enables a clean burn, so its combustion does not produce solid particles or sulfur.)
- Efficient
- Abundant

TABLE 4.1

Typical Raw Gas Compositions

	Casinghead (Wet) Gas (Mol %)	Gas Well (Dry) Gas (Mol %)	Condensate Well Gas (Mol %)
Carbon dioxide	0.63	–	–
Nitrogen	3.73	1.25	0.53
Hydrogen sulfide	0.57	–	–
Methane	68.48	91.01	94.87
Ethane	11.98	4.88	2.89
Propane	8.75	1.69	0.92
Isobutane	0.93	0.14	0.31
n-Butane	2.91	0.52	0.22
Isopentane	0.54	0.09	0.09
n-Pentane	0.80	0.18	0.06
Hexanes	0.37	0.13	0.05
Heptanes plus	0.31	0.11	0.06
Total	100	100	100

TABLE 4.2

Representative Pipeline Specs for Natural Gas

	Minimum	Maximum
Major and Minor Components (Mol%)		
Methane	75	–
Ethane	–	10
Propane	–	5
Butanes	–	2
Pentanes plus	–	0.5
Nitrogen and other inerts	–	3–4
Carbon dioxide		3–4
Trace Components		
Hydrogen sulfide	–	0.25–1.0 gr/100 SCF
Mercaptan sulfur	–	0.25–1.0 gr/100 SCF
Total sulfur	–	5–20 gr/100 SCF
Water vapor	–	7.0 lbs/mmcf
Oxygen	–	0.2–1.0 ppmv
Heating Value		
Heating value, Btu/SCF gross saturated	950	1150

Note: Liquids: free of liquid water and hydrocarbons at delivery temperature and pressure; solids: free of particulates in amounts deleterious to transmission and utilization equipment.

- Odorless (Because it is odorless, mercaptan is added to the gas as explained in Section 4.2.)
- Tasteless and nontoxic
- Nonabsorbing
- Noncorrosive
- Explosive under pressure

As far as the specifications of natural gas, the composition of *raw natural gas* for the three cases of casinghead (wet gas), gas well (dry gas), and condensate well gas are presented first as shown in Table 4.1, followed by the standard pipeline quality specs for natural gas shown in Table 4.2.

4.5 Natural Gas Processing

The properties of the produced fluids from crude oil and natural gas wells vary. Nonassociated gas (free gas) may be produced at a high pressure. It may require minimal treatment before it is transferred to a pipeline. Associated gas however, undergoes systematic processing and treatment, as detailed in Section IV.

4.6 Applications and Uses

There are so many different applications for this fossil fuel that it is hard to provide an exhaustive list of everything it is used for. Natural gas is used across all sectors, in varying amounts. Most important are the following:

- Residential use
- Commercial use
- Use in industry
- Transportation sector
- Electric generation using natural gas

In Figure 4.3, utilization of natural gas used per sector for the United States is shown.

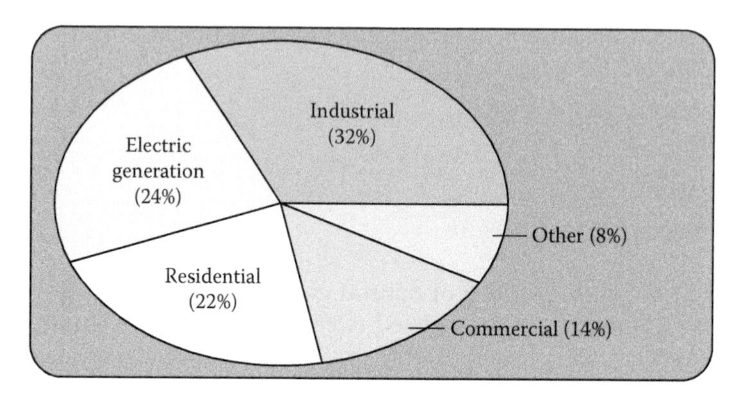

FIGURE 4.3
Natural gas use by sector. (From U.S. Energy Information Administration, Annual Energy Outlook 2002. With permission.)

5

Role of Economics in Oil and Gas Field Operations

The oil and gas industry has invested billions of dollars in finding, discovering, developing, producing, transporting, and refining hydrocarbons for more than a century and has long been an enormous source of wealth creation.

Capital expenditure proposals must be sufficiently specific to permit their justification for exploration and production (E&P) operations, surface petroleum operations (SPO), petroleum refining (PR), and expansion purposes, or for cost reduction improvements and necessary replacements. In reality, an evaluation of capital expenditure proposals is both technical and economic in nature. First, there are the technical feasibilities and validities associated with a project; next comes the economic evaluation and justification.

In the economic phase of evaluation, oil management may find that it has more investment opportunities than capital to invest, or more capital to invest than investment opportunities. Whichever situation exists, oil management needs to resort to some economic criteria for selecting or rejecting investment proposals. Management's decision is, in either case, likely to be largely based on the measures of financial return on the investment or its economic attractiveness.

In this chapter, we will present some economic fundamentals that are applied to gas and oil field operations. In oil and gas production operations, one is interested in finding answers to the following questions:

1. Where and how is a capital investment spent?
2. How much does it cost to produce crude oil and natural gas? How do you calculate the total cost of producing one barrel of oil?
3. How do you apply some economic indicators to judge the profitability of an investment in oil and gas projects?

5.1 Depreciation and Depletion in Oil Production

The answer to where and how a capital investment is spent is found through the following discussion. Economic analysis of the expenditures and revenues for oil operations requires recognition of two important facts:

1. Physical assets decrease in value with time, i.e., they *depreciate.*
2. Oil resources, like other natural resources, cannot be renewed over the years and they are continuously *depleted.*

Depreciation or amortization is described as the systematic allocation of the cost of an asset from the balance sheet to a depreciation expense on

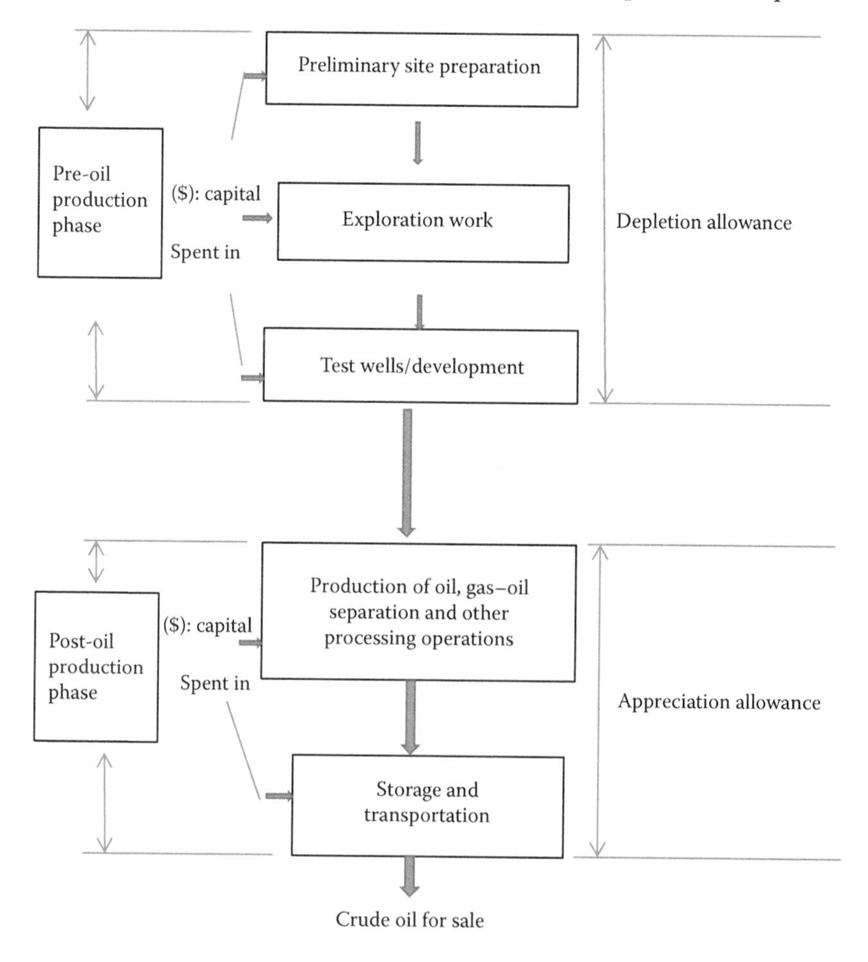

FIGURE 5.1
Contribution of depletion and depreciation allowances in oil production cost.

the income statement over the useful life of an asset. On the other hand, depletion allowance is a depreciation-like charge applied to account for the exhaustion of natural resources.

As shown in Figure 5.1, for oil production operations, we have two phases where capital investment has to be spent. The first phase, called the *pre–oil-production phase,* involves preliminary preparation, exploration, dry-well drilling, and development. The property is now ready for the second phase, where money is spent in providing necessary assets and equipment for the production stage and post–oil production. The question is: How can we recover the capital spent in the pre–oil-production phase and the *production/postproduction phase* as well?

For oil production operation, as seen in Section 1.9, capital investment is spent in two consecutive phases: the pre-production oil phase and the production/post-production oil phase. As far as the second phase, physical assets can be tangibly verified in a property; hence *depreciation accounting* can be applied to recover this capital investment. The *first* phase, on the other hand, exhibits the contrary: intangible costs were invested, because no physical assets can count for them. In this case, *depletion accounting* is introduced to recover the development costs that were spent for exploration and other preliminary operations prior to the actual production of oil and gas. In other words, depletion allowance is a depreciation-like charge applied to an account for the exhaustion of natural resources.

5.1.1 Methods of Determining Depreciation

There are several methods for determining depreciation for a given period of time. The *straight-line method* is widely used by engineers and economists working in the oil industry because of its simplicity. The method is explained as follows. Mathematically speaking, it is assumed that the value of the asset decreases linearly with time. Now, if the following variables are defined:

d = annual depreciation rate ($/year)

V_o and V_s = original value and salvage values of asset respectively ($)

n = service life (years)

$$\text{then the annual depreciation cost} = \frac{\text{depreciable capital}}{n}, \text{ or}$$

$$d = \frac{V_o - V_s}{n}$$

The asset value V_a, at year a, is given by

$$V_a = V_o - (a)(d)$$

Example 5.1

Assume that a heat exchanger needed in the field has a depreciable cost of $100,000. It will last in service after processing, say, 20 million bbl. Calculate the annual depreciation cost of the heat exchanger if it is processing 600,000 bbl yearly.

Solution

Depreciation factor = 100,000/20,000,000 = $0.005 per bbl. The annual depreciation is d = (600,000)(0.005) = $30,000.

5.1.2 Methods of Determining Depletion

If a depletion allowance is to be used in an oil field, there are two possible methods of calculating its value: (1) fixed percentage method and (2) cost-per-unit basis. For the fixed percentage method, the percentage depletion is usually set by the ruling government (in the United States it has been 22% of net sales), but in no case can the fixed percentage exceed 50% of net income before deduction of depletion. For the cost-per-unit basis, the reader may refer to Chapter 5, "Petroleum Economics and Engineering, 3rd edition, CRC press 2014," for information on using this method.

5.2 Total Production Costs of Crude Oil and Natural Gas

This section represents the answer to the second question: How much does it cost to produce crude oil and natural gas? A measure of the total production costs to produce crude oil and natural gas is the upstream costs, which are made up of two components: *finding* costs and lifting costs. Finding costs are the costs of exploring for and developing reserves of oil and gas, and the costs to purchase properties or acquire leases that might contain oil and gas reserves. Lifting costs, on the other hand, are the costs to operate and maintain oil and gas wells, and related equipment and facilities to bring oil and gas to the surface.

Now, having calculated the depreciation costs and the depletion allowance for an oil field as shown in Section 5.1.2, the total production cost is readily calculated:

Total production costs = Sum of all costs incurred in the oil field operation
= Depletion costs + Depreciation costs + Operating expenses

Table 5.1 reports data on the total production costs of oil and gas for the United States and other countries.

TABLE 5.1
Production Costs of Crude Oil and Gas 2007–2009 (2009 $/Barrel of Oil Equivalent[a])

	Lifting Costs	Finding Costs	Total Upstream Costs
United States (Average)	$12.18	$21.58	$33.76
On-shore	$12.73	$18.65	$31.38
Off-shore	$10.09	$41.51	$51.60
All Other Countries (Average)	$9.95	$15.13	$25.08
Canada	$12.69	$12.07	$24.76
Africa	$10.31	$35.01	$45.32
Middle East	$9.89	$6.99	$16.88
Central and South America	$6.21	$20.43	$26.64

Source: U.S. Energy Information Administration (last reviewed: January 15, 2014).
[a] 5.618 cubic feet of natural gas are equivalent to one barrel.

5.3 Financial Measures and Profitability Analysis

This section provides the answer to our third question: How do you apply some economic indicators to judge the profitability of a project? The basic aim of financial measures and profitability analysis is to provide some yardsticks for the attractiveness of a venture or a project, where the expected benefits (revenues) must exceed the total production costs. There are many different ways to measure financial performance, but all measures should be taken in aggregation.

Profitability measures the extent to which a business generates a profit from the use of resources, land, labor, or capital. Behind the need for profitability is the fact that any business enterprise makes use of invested money to earn profits. In very simple language, profitability is measured by dividing the profits earned by the company by the investment (or money) used.

The most common measures, methods, and economic indicators of economically evaluating the return on capital investment to be discussed are:

- Rate of return, or return on investment (ROI)
- Payment period (PP)
- Discounted cash-flow rate of return (DCFR) and present value index (PVI)
- Net present value (NPV)

5.3.1 Annual Rate of Return (ROI)

The annual rate of return is defined by the equation

$$\text{ROI} = (\text{Annual profit}/\text{Capital investment})(100)$$

For oil ventures, where the cash flow extends over a number of years, the average rate of return is calculated using an average value for the profit, then dividing the sum of the annual profits by the useful lifetime:

$$ROI = \left[\frac{\sum_{1=y}^{n} \text{annual profits}}{n}\right] \Big/ (\text{capital investment})(100) \qquad (5.1)$$

Example 5.2

It is necessary to calculate the ROI for two projects involving the desalting of crude oil; each has an initial investment of $1 million. The useful life of project 1 is 4 years, and for project 2 it is 5 years. The earnings pattern is given in Table 5.2.

Solution

The average rate of return is calculated for both projects as shown in Table 5.2. The final answers are the ROI for project 1 is 16.25% and the ROI for project 2 is 22.2%.

5.3.2 Payout Period (PP), Payback Time, or Cash Recovery Period

Payout period is defined as the time required for the recovery of the depreciable capital investment in the form of cash flow to the project. Cash flow would imply the total income minus expenses.

Payout period (years) = Depreciable capital investment/Average annual cash

Example 5.3

Calculate the payout period for the two alternatives of capital expenditures involving an investment of $2 million each for a sulfur removal plant, as given in Table 5.3. The life of project 1 and project 2 is 6 and 10 years, respectively.

Solution

From the cash flow given in Table 5.2, the payout period (PP) is calculated as follows:

$(PP)1 = 2 \times 10^6/471{,}429 = 4.24$ years, where $471,429 is the average annual cash flow

$(PP)2 = 2 \times 10^6/370{,}000 = 5.41$ years, where 370,000 is the average annual cash

TABLE 5.2

Average Return on Investment for Crude Oil Desalting (Solution of Example 5.2)

	Project 1			Project 2		
Year	Income before Depreciation	Depreciation Allowance	Net Earning after Depreciation	Income before Depreciation	Depreciation Allowance	Net Earning after Depreciation
1	$400,000	$250,000	$150,000	$75,000	$200,000	–$125,000
2	$350,000	$250,000	$100,000	$180,000	$200,000	–$20,000
3	$300,000	$250,000	$50,000	$300,000	$200,000	$100,000
4	$275,000	$250,000	$25,000	$400,000	$200,000	$200,000
5				$600,000	$200,000	$400,000
Sum			$325,000			$555,000
Average Investment			$500,000			$500,000
Average Earning			$81,250			$111,000
Average Rate of Return		16.25%			22.20%	

TABLE 5.3
Cash Flow for the Sulfur Removal Plant (Example 5.3)

	Cash Flow ($)	
Year	**Project 1**	**Project 2**
0	2,000,000	2,000,000
1	1,500,000	200,000
2	500,000	300,000
3	400,000	400,000
4	350,000	400,000
5	250,000	400,000
6	200,000	400,000
7	100,000	400,000
8	–	400,000
9	–	400,000
10	–	400,000
Cash Flow	$3,300,000	$3,700,000
Annual Cash Flow ($/yr)	471,429	370,000
PP (yr)	4.24	5.41

5.3.3 Discounted Cash Flow Rate of Return (DCFR) and Present Value Index (PVI)

If we have an oil asset (oil well, surface treatment facilities, a refining unit, etc.) with an initial capital investment, P, generating annual cash flow over a lifetime, n, then the DCFR is defined as the rate of return, or interest rate that can be applied to yearly cash flow, so that the sum of their present value equals P.

From the computational point of view, DCFR cannot be expressed by an equation or formula, similar to the previous methods. The following example illustrates the basic concepts involved in calculating the DCFR.

Example 5.4

Assume an oil company is offered a lease of oil wells that would require a total capital investment of $110,000 for equipment used for production. This capital includes $10,000 working money, $90,000 depreciable investment, and $10,000 salvage value for a lifetime of 5 years. Cash flow to the project (after taxes) gained by selling the oil is as given in Table 5.4. Based on calculating the DCFR, a decision has to be made: Should this project be accepted?

Solution

Two approaches are presented to handle the DCFR.

TABLE 5.4

Cash Flow Pattern (Example 5.4)

Year	Cash Flow ($$10^3$)
0	–110
1	30
2	31
3	36
4	40
5	43

APPROACH 1: USING THE FUTURE WORTH

Our target is to set the following equity: By the end of 5 years, the future worth of the cash flow recovered from oil sales, F_o, should break even with the future worth of the capital investment, had it been deposited for compound interest in a bank, F_B, at an interest rate, i.

That is to say

$$F_o = F_B \tag{5.2}$$

where $F_B = 110{,}000(1 + i)^5$, for banking, and $F_o = \sum_{i=1}^{5} F_i$, for oil investment, which represents the cash flow to the project, compounded on the basis of end-of-year income.

Hence,

$$F_o = 30{,}000(1+i)^4 + 31{,}000(1+i)^3 + 36{,}000(1+i)^2 + 40{,}000(1+i) + 43{,}000 + 20{,}000$$

Notice that the $20,000 represents the sum of working capital and salvage value; both are released by the end of the fifth year.

Setting up $F_B = F_o$, we have one equation involving i as the only unknown, which could be calculated by trial and error. The value of i is found to be 0.207, that is, the DCFR is 20.7%.

APPROACH 2: USING THE DISCOUNTING TECHNIQUE

Our objective here is to discount the annual cash flow to present values using an assumed value of i. The correct i is the one that makes the sum of the discounted cash flow equal to the present value of capital investment, P. The solution involves using the following equation:

$$P = \sum_{y=1}^{5} p_y \tag{5.3}$$

where

$$p_y = (\text{annual cash flow})_y \; d_y = (\text{ACF})_y \left(\frac{1}{1+i}\right)^y$$

for the year y, between 1 and 5.

Another important criterion that can be used to arrive at the correct value of i in the discounting of the cash flow is given by the following relationship. DCFR is the value that makes PVI = 1.0, where PVI stands for the present value index and is defined by

$$\text{PVI} = \frac{\text{sum of discounted cash flow (present value)}}{\text{initial capital investment}} \tag{5.4}$$

The solution of this example applying the discount factor is illustrated in Table 5.5.

If the annual cash flow has been constant from year to year, say, A $/yr, then the following can be applied:

$$A\left[\frac{1}{(1+i)} + \frac{1}{(1+i)^2} + \ldots + \frac{1}{(1+i)^n}\right] = P \tag{5.5}$$

Multiplying both sides of Equation 5.5 by $(1 + i)^n$, we get

$$A[(1 + i)^{n-1} + (1 + i)^{n-2} + \ldots + 1] = P(1 + i)^n \tag{5.6}$$

The sum of the geometric series in the left-hand side is given by

$$\frac{(1+i)^n - 1}{i}$$

Hence, Equation 5.6 can be rewritten in the form

$$P(1+i)^n = A\frac{(1+i)^n - 1}{i} \tag{5.7}$$

It is interesting to point out that this equation satisfies both sides of Equation 5.2; that is, the future worth of P, if invested in the bank, is given by:

$$F_b = P(1 + i)^n$$

The future worth of the annual cash flow received from oil investment (A), if compounded in a sinking-fund deposit, is given by

$$F_o = A\frac{(1+i)^n - 1}{i}$$

TABLE 5.5

DCFR for Investment in a Lease of Oil Wells

Year (y)	Cash Flow	i = 15%		i = 20%		i = 25%		i = 20.7%	
		d_y	Present Value ($)	d_y	Present Value ($)	d_y	Present Value ($)	d_y	Present Value ($)
0	110,000								
1	30,000	0.8696	26,088	0.8333	24,999	0.8000	24,000	0.8290	24,870
2	31,000	0.7561	23,439	0.6944	21,526	0.6400	19,840	0.6870	21,297
3	36,000	0.6575	23,670	0.5787	20,833	0.5120	18,432	0.5700	20,520
4	40,000	0.5718	22,872	0.4823	19,292	0.4096	16,384	0.4720	18,880
5	43,000	0.4971	21,375	0.4019	17,282	0.3277	14,091	0.3910	16,813
	20,000								
Total			117,444		103,932		92,747		102,380
PVI			1.07		0.94		0.84		0.93

Now, Equation 5.7 can be used to directly calculate the DCFR by trial and error knowing the values of A, P, and n. The DCFR thus represents the maximum interest rate at which money could be borrowed to finance an oil project.

5.3.4 Net Present Value (NPV)

The DCFR method is based on finding the interest rate that satisfies the conditions implied by the method. Here we provide a value for i that is an acceptable rate of return on the investment and then calculate the discounted value (present value) of the cash flow using this i. The net present value is then given by

$$\text{NPV} = \text{Present value of cash flow discounted at a given i} - \text{Capital investment} \quad (5.8)$$

Example 5.5

Calculate the NPV of the cash flow for the oil lease described in Example 5.4, if money is worth 15%.

Solution

At i = 0.15, the annual cash flow is discounted. The present value of the sum of the cash flows = \$127,000. The NPV is directly calculated using Equation 5.8:

$$\text{NPV} = 127{,}000 - 110{,}000 = \$17{,}000$$

That is, the oil lease can generate \$17,000 (evaluated at today's dollar value) over and above the totally recovered capital investment. The solution is presented as given in Table 5.6 and is illustrated in this graph.

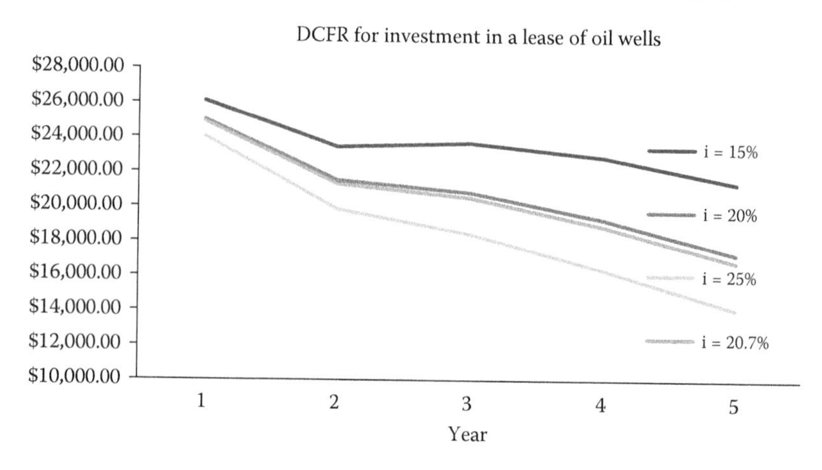

TABLE 5.6

DCFR for a Lease of Oil Wells

		i = 5%		i = 20%		i = 25%		i = 20.7%	
Year (y)	Cash Flow	d_y	Present Value ($)	d_y	Present Value ($)	d_y	Present Value ($)	d_y	Present Value ($)
0	110,000								
1	30,000	0.8696	26,088	0.8333	24,999	0.8000	24,000	0.8290	24,870
2	31,000	0.7561	23,439	0.6944	21,526	0.6400	19,840	0.6870	21,297
3	36,000	0.6575	23,670	0.5787	20,833	0.5120	18,432	0.5700	20,520
4	40,000	0.5718	22,872	0.4823	19,292	0.4096	16,384	0.4720	18,880
5	43,000	0.4971	21,375	0.4019	17,282	0.3277	14,091	0.3910	16,813
	20,000								
Total			117,444		103,932		92,747		102,380
PVI			1.07		0.94		0.84		0.93
NPV			$7444.40		–$6067.70		–$17,252.90		–$7620.00

Example 5.6

An oil company expects a cash flow of $800,000 by the end of 10 years, and 10% is the current interest rate on money. Calculate the NPV of this venture.

Solution

No capital investment is involved here, so the problem is simply a discounting procedure.

$$\begin{aligned}\text{Present value of the cash flow} &= 800{,}000(1 + 0.1) - 10 \\ &= \$308{,}000\end{aligned}$$

Section II

Separation of Produced Fluids

Once the oil and gas mixture is brought to the surface, our main goal becomes that of directing the wellhead stream to the gas-oil separation plant (GOSP) to separate the crude oil, natural gas, and water phases into separate streams. Chapters 6 and 7 constitute this section. This signals the start of the surface production operations, which is illustrated by the next sketch.

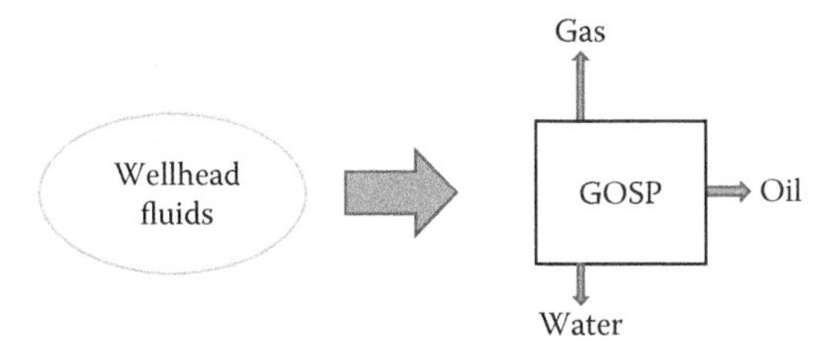

To meet process requirements, the oil/gas separators are normally designed in stages, in which the first stage separator is used for preliminary phase separation, and the second and third stage separators are applied for further treatment of each individual phase (gas, oil, and water).

6

Two-Phase Gas–Oil Separation

Separation of produced fluids (gas, oil, water) represent the very first task in crude oil field processing. This puts the gas–oil separation step as the initial one in the series of field treatment operations of crude oil. Here, the primary objective is to allow most of the gas to free itself from these valuable hydrocarbons, hence increasing the recovery of crude oil. Oil field separators can be classified into two types based on the number of phases to separate:

1. Two-phase separators, which are used to separate gas from oil in oil fields, or gas from water for gas fields
2. Three-phase separators, which are used to separate the gas from the liquid phase, and water from oil

A gas–oil separator is a vessel that does this job.

This chapter handles two-phase separation, where the gas is separated from the liquid (oil), with the gas and liquid being discharged separately. The theory of gas–oil separation, methods, and equipment used are fully explained. Sizing of gas–oil separators is handled using fundamental concepts of unit operations. Determination of optimum pressure for a gas–oil separator plant and flash equilibrium calculations are presented.

6.1 Introduction

At the high pressure existing at the bottom of the producing well, crude oil contains great quantities of dissolved gases. When crude oil is brought to the surface, it is at a much lower pressure. Consequently, the gases that were dissolved in it at the higher pressure tend to come out from the liquid. Some means must be provided to separate the gas from oil without losing too much oil.

In general, well effluents flowing from producing wells come out in two phases: vapor and liquid under a relatively high pressure. The fluid emerges as a mixture of crude oil and gas that is partly free and partly in solution. Fluid pressure should be lowered and its velocity should be reduced in order to separate the oil and obtain it in a stable form. This is usually done by admitting the well fluid into a gas–oil separator plant (GOSP) through which the pressure of the gas–oil mixture is successively reduced to atmospheric pressure in a few stages. Upon decreasing the pressure in the GOSP, some of the lighter and more valuable hydrocarbon components that belong to oil will be unavoidably lost along with the gas into the vapor phase.

Crude oil as produced at the wellhead varies considerably from field to field due not only to its physical characteristics (as explained in Chapter 3) but also to the amount of gas and salt water it contains. In some fields, no salt water will flow into the well from the reservoir along with the produced oil. This is the case we are considering in this chapter, where it is only necessary to separate the gas from the oil (i.e., two-phase separation). When, on the other hand, salt water is produced with the oil, it is then essential to use a three-phase separator, the subject matter of Chapter 7.

Oil from each producing well is conveyed from the wellhead to a gathering center through a flow line. The gathering center, usually located in some central location within the field, will handle the production from several wells in order to process the produced oil–gas mixture.

Separation of the oil phase and the gas phase enables the handling, metering, and processing of each phase independently, hence producing marketable products.

6.1.1 Some Basic Fundamentals

Gas–oil separators are commonly classified by *operating configuration* into:

- Horizontal separators—They vary in size from 10 to 12 in up to 15 to 16 ft diameter, and 4 to 5 ft seam-to-seam (S to S) up to 60 to 70 ft. Separators are manufactured with monotube and dual tube shell.
- Vertical separators—They vary in size from 10 to 12 in up to 10 to 12 ft, diameter and 4 to 5 ft S to S up 15 to 25 ft.
- Spherical separators—They are usually available in 24 or 30 in up to 66 to 72 in diameter.

Based on *separation functions*, oil–gas separators can be grouped into gas–liquid two-phase separators or oil–gas–water three-phase separators. They can also be classified into primary phase separator, test separator, high-pressure separator, low-pressure separator, degasser, and others.

- Separation performance can be evaluated by liquid carrying over and gas carrying down rates, which are affected by many factors such as flow rates, fluid properties, vessel configuration, internals. and control system.

- The gas capacity of most gas–liquid separation vessels is sized on the basis of removing a certain size of liquid droplets.
- The liquid capacity of most separators is sized to provide enough retention time to allow gas bubbles to form and separate out.
- Vessel internals could significantly affect the operating performance of an oil–gas separator through the following factors: flow distribution, drop/bubble shearing and coalescence, foam creation, mixing, and level control.
- Unit operation(s) underlying gas–oil separation that include equilibrium (flash) distillation are indicated as shown in the following diagram.

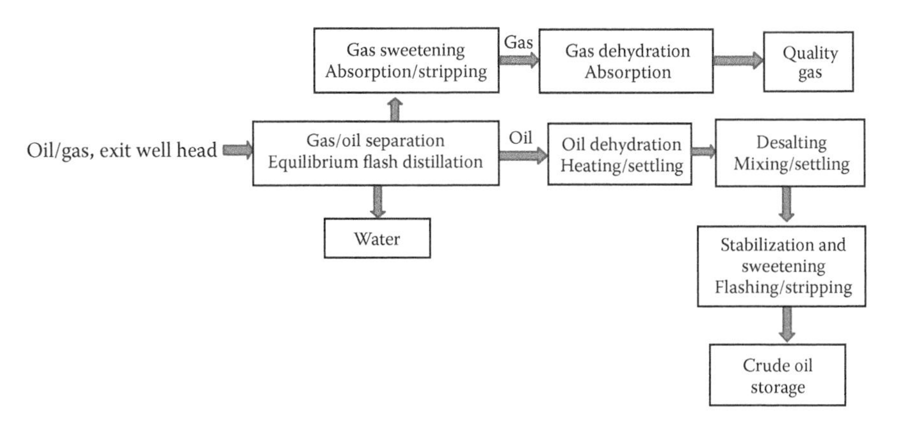

6.2 How to Handle the Separation Problem

High-pressure crude oils containing large amounts of free and dissolved gas flow from the wellhead into the flow line, which routes the mixture to the GOSP. In the separator, crude oil separates out, and settles in the lower part of the vessel. The gas, lighter than oil, fills the upper part of the vessel. Crude oils with a high gas–oil ratio (GOR) must go through two or more stages of separation.

Gas goes out the top of the separators to a gas collection system, a vapor recovery unit (VRU), or a gas flow line. Crude oil, on the other hand, goes out the bottom and is routed to other stages of separation, if necessary, and then to the stock tank (Figure 6.1).

Movement of the crude oil within the GOSP takes place under the influence of its own pressure. Pumps, however, are used to transfer the oil in its final trip to the tank farm or pipeline (Figure 6.2). Pressure reduction in moving the oil from stage to stage is illustrated in Figure 6.3.

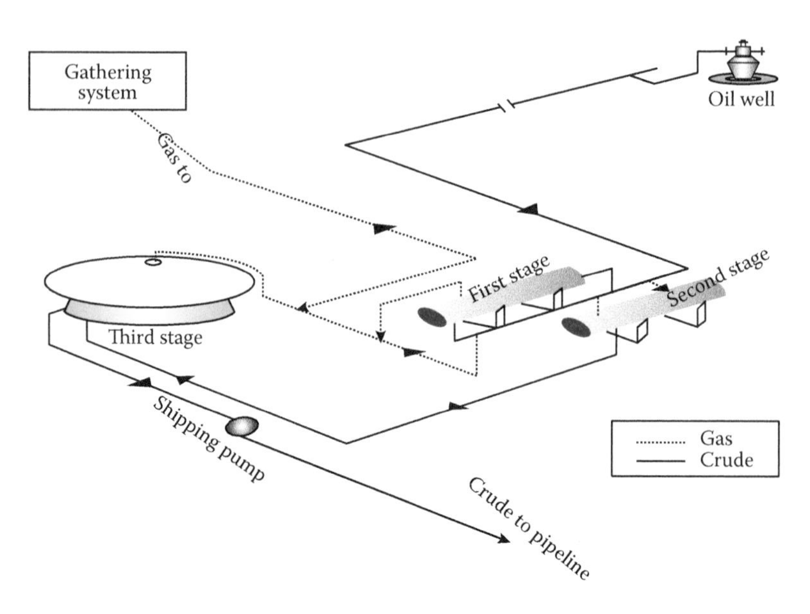

FIGURE 6.1
Flow of crude oil from oil well through GOSP.

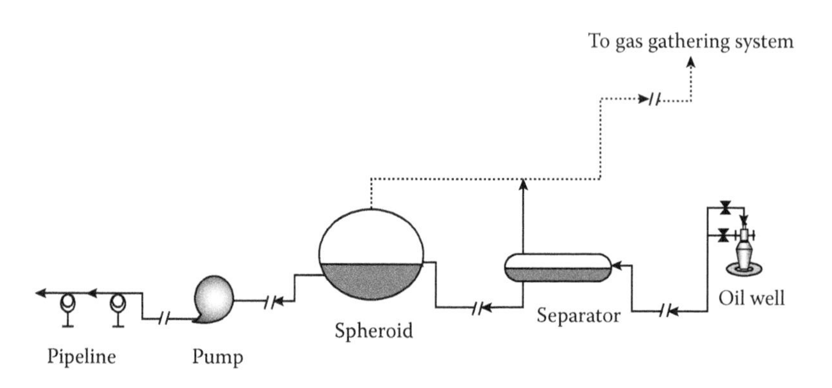

FIGURE 6.2
Transfer of oil from GOSP.

In order to visualize the changes that occur during the gas–oil separation process, Figure 6.4 is presented, which summarizes the results of a three-stage gas–oil separation pilot plant. The results are reported for an initial reservoir oil mixture of 1.32 bbl in place. Data reported for each stage show the flow rates as well as gas composition. One barrel of oil is collected in the storage tank, and the loss of 0.32 bbl represents the amount of gas separated.

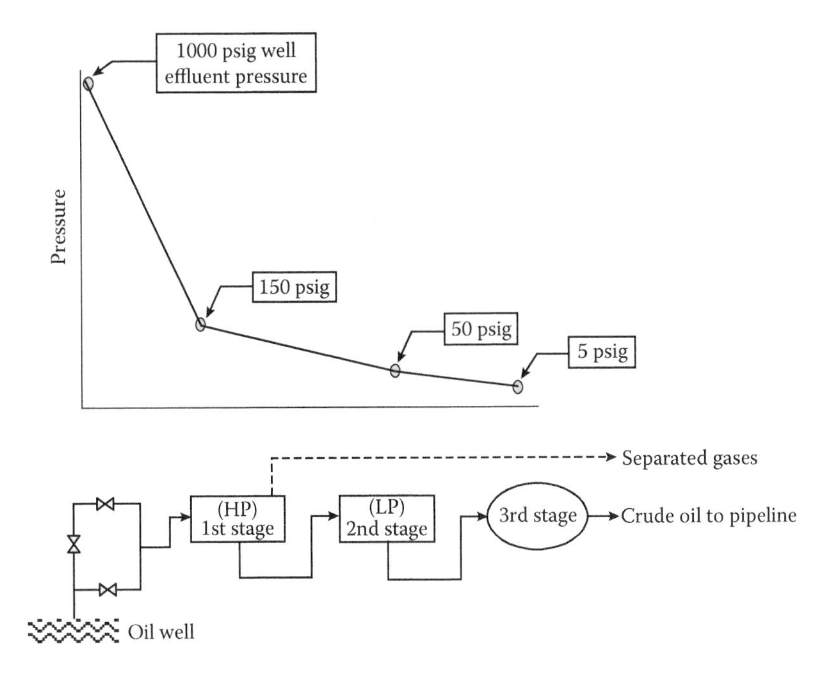

FIGURE 6.3
Pressure drop profile for a typical GOSP in a field in the Middle East.

6.3 Theory of Gas–Oil Separation

In order to understand the theory underlying the separation of well-effluent hydrocarbon mixtures into a gas stream and oil product, it is assumed, for simplicity, that such mixtures contain essentially three main groups of hydrocarbon, as illustrated in Figure 6.5:

1. Light group, which consists of CH_4 (methane) and C_2H_6 (ethane)
2. Intermediate group, which consists of two subgroups—the propane/butane (C_3H_8/C_4H_{10}) group and the pentane/hexane (C_5H_{12}/C_6H_{14}) group
3. Heavy group, which is the bulk of crude oil and is identified as C_7H_{16}

In carrying out the gas–oil separation process, the main objective is to try to achieve the following:

- Separate the C_1 and C_2 light gases from oil
- Maximize the recovery of heavy components of the intermediate group in crude oil
- Save the heavy group components in liquid product

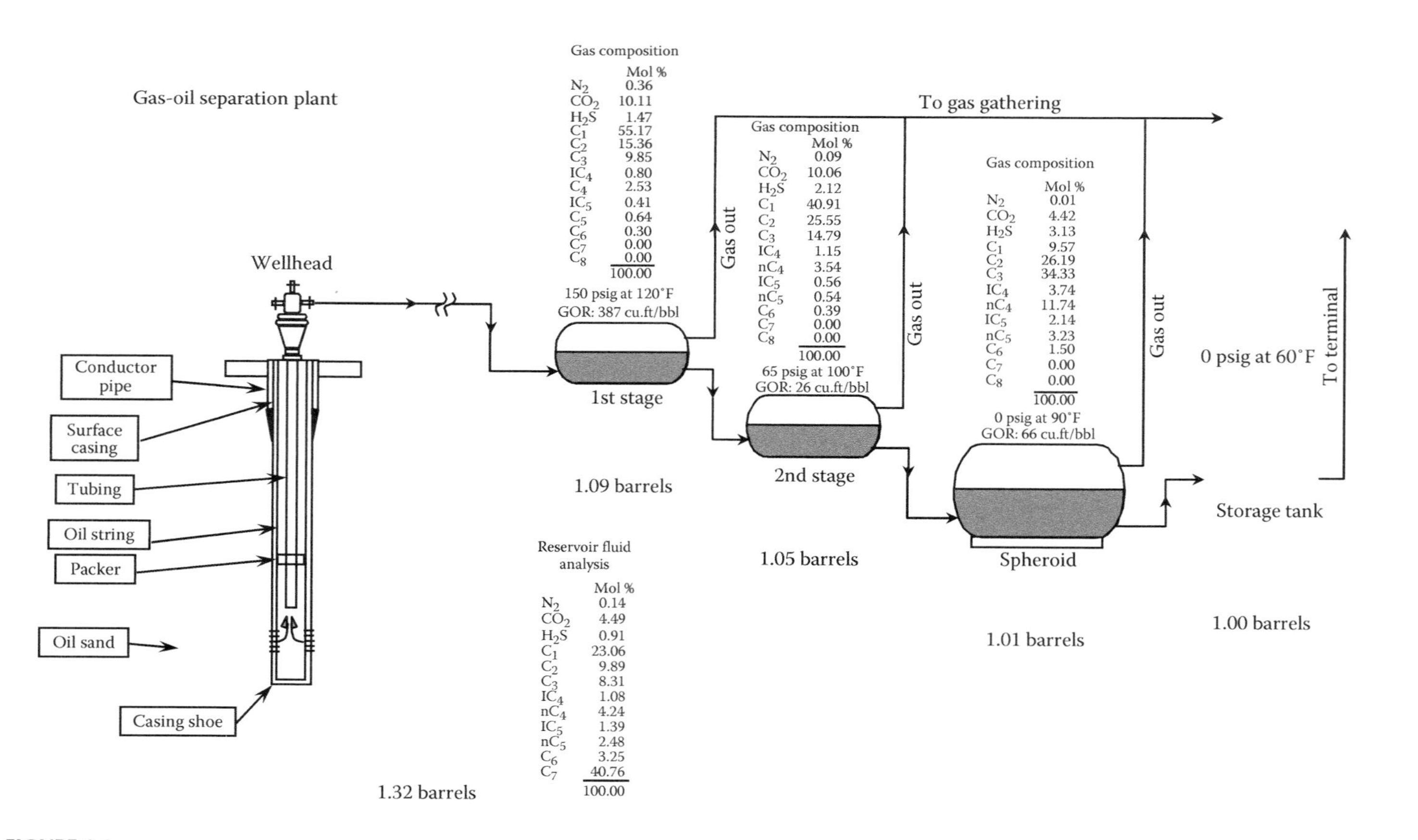

FIGURE 6.4
Performance data for a GOSP.

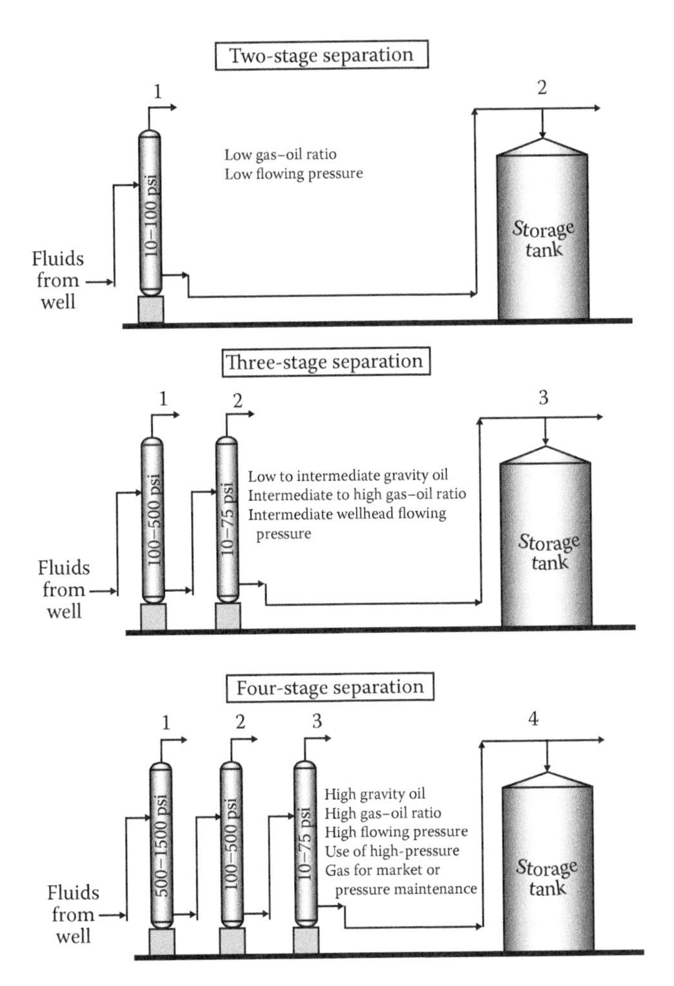

FIGURE 6.5
Stage separation flow diagram.

To accomplish these objectives, some hydrocarbons of the intermediate group are unavoidably lost in the gas stream. In order to minimize this loss and maximize liquid recovery, two methods representing the mechanics of separation are compared: (1) differential or enhanced separation and (2) flash or equilibrium separation. In differential separation, light gases (light group) are gradually and almost completely separated from oil in a series of stages, as the total pressure on the well-effluent mixture is reduced. Differential separation is characterized by the fact that light gases are separated as soon as they are liberated (due to reduction in pressure). In other words, light components do not come into contact with heavier hydrocarbons; instead, they find their way out. For flash separation, on the other hand, gases liberated from the oil are kept in intimate contact with the liquid phase. As a result, thermodynamic equilibrium is established between the two phases and separation takes place at the required pressure.

Comparing the two methods, one finds that in differential separation, the yield of heavy hydrocarbons (intermediate and heavy groups) is maximized and oil-volume shrinkage experienced by crude oil in the storage tank is minimized. This could be explained by the fact that separation of most of the light gases takes place at the earlier high-pressure stages; hence, the opportunity of loosing heavy components with the light gases in low-pressure stages is greatly minimized.

As a result, it may be concluded that flash separation is inferior to differential separation because the former experiences greater losses of heavy hydrocarbons that are carried away with the light gases due to equilibrium conditions. Nevertheless, commercial separation based on the differential concept is very costly and is not a practical approach because of the many stages required. This would rule out differential separation, leaving the flash process as the only viable scheme to affect gas–oil separation using a small number of stages. As illustrated in Table 6.1, a close approach to differential separation is reached by using four to five flash separation stages. A comparison between the mechanisms of separation by the two methods is schematically simplified and presented in Table 6.2.

TABLE 6.1

Flash and Differential Separation

Number of Flash Stages	Percent Approach to Differential
2	0
3	75
4	90
5	96
6	98.5

TABLE 6.2

Comparison between Separation Mechanisms

Parameter	Differential Separation	Flash Separation
Process arrangement	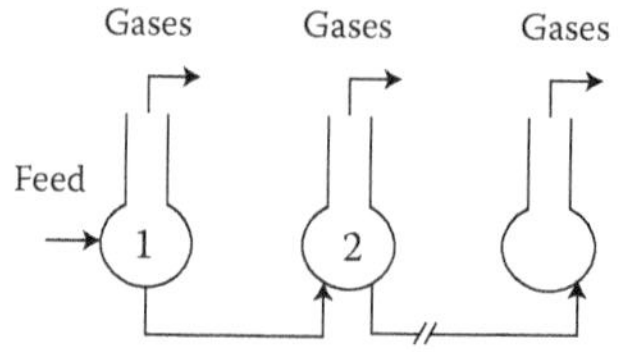 Gases Gases Gases Feed 1 2	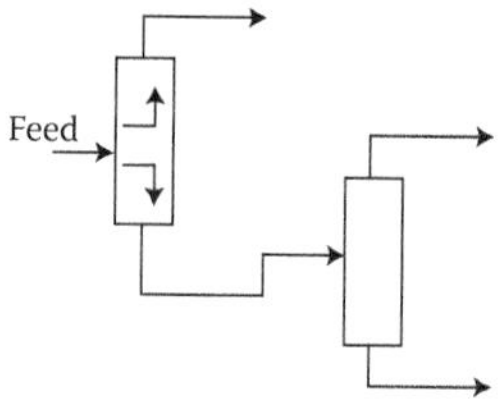 Feed
Losses of hydrocarbon	Low	High
No. of stages	Too many (can reach 100 stages)	Few (2–4 stages)
Commercial application	Not applied	Applied

6.4 Methods Used in Separation

The traditional process for recovering crude oil from high-pressure well streams is based on the *flash separation concept* explained in Section 6.3, which consists of a series of flash separators operating over a pressure range from wellhead pressure to atmospheric pressure. However, with the increased desirability of recovering natural gas and natural gas liquids (NGLs), other methods have been proposed as modification to the basic flash separation technique.

Separation methods could be broadly classified into two main categories:

1. Conventional methods
2. Modified methods, which is accomplished by
   a. Adding a vapor recompression unit to the conventional methods
   b. Replacing the conventional methods by a stabilizer and a recompression unit

The conventional method is a multistage flash separation system and is recommended for comparatively high-pressure fluids. Several stages operated at successively lower pressures affect the separation of oil from gas, thus increasing the oil recovery. In general, the number of stages in a multistage conventional separation process is a function of the following the API gravity of the oil, the gas–oil ratio (GOR), and the flowing pressure. Consequently, high API gravity oils with high GOR flowing under high pressure would require the greatest number of stages (from three to four), as depicted in Figure 6.5.

The first modified method of separation implies adding several stages of gas compression to recompress the separated gas from each flash stage. Liquids from interstage vessels between the compressors can be collected and processed as liquid natural gas (LNG) stock. Natural gas will be delivered at the desired pressure depending on its usage. The second modified method of separation is different in concept from the conventional (flash) separation. It makes use of crude stabilizer columns. Normally, these columns have top feed trays with no rectifying section and no condenser, but are provided with interstage reboilers and feed preheaters. Crude stabilization systems are advantageous as GOSPs if space is critical, as may be encountered on an offshore platform, because they occupy less space than conventional GOSPs.

6.5 Gas–Oil Separation Equipment

The conventional separator is the very first vessel through which the well-effluent mixture flows. In some special cases, other equipment (heaters, water knockout drums) may be installed upstream of the separator.

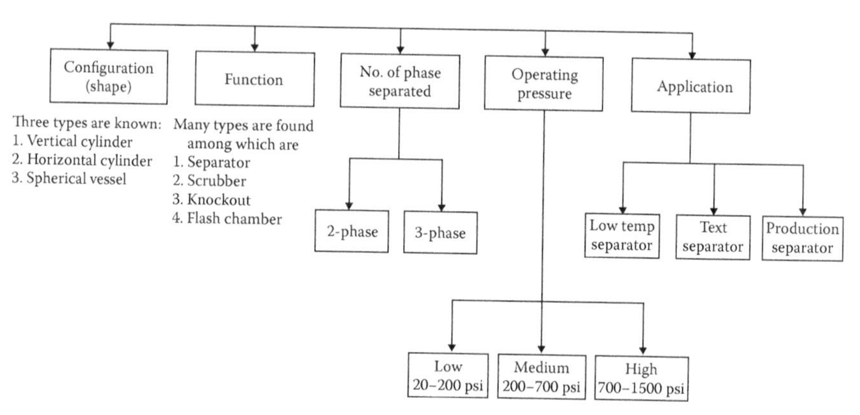

FIGURE 6.6
Classifications of gas–oil separators.

The essential characteristics of the conventional separator are it causes a decrease in the flow velocity, permitting separation of gas and liquid by gravity; and it always operates at a temperature above the hydrate point of the flowing gas. The choice of a separator for the processing of gas–oil mixtures containing water or without water under given operating conditions and for a specific application normally takes place guided by the general classification illustrated in Figure 6.6.

6.5.1 Functional Components of a Gas–Oil Separator

Regardless of their configuration, gas–oil separators usually consist of four functional sections:

- Section A—Initial bulk separation of oil and gas takes place in this section. The entering fluid mixture hits the inlet diverter. This causes a sudden change in momentum and, due to the gravity difference, results in bulk separation of the gas from the oil. The gas then flows through the top part of the separator and the oil through the lower part.
- Section B—Gravity settling and separation is accomplished in this section of the separator. Oil droplets settle and separate from the gas because of the substantial reduction in gas velocity and the density difference.
- Section C—Known as the mist extraction section, it is capable of removing the very fine oil droplets that did not settle in the gravity settling section from the gas stream.
- Section D—This is known as the liquid sump or liquid collection section. Its main function is collecting the oil and retaining it for a sufficient time to reach equilibrium with the gas before it is discharged from the separator.

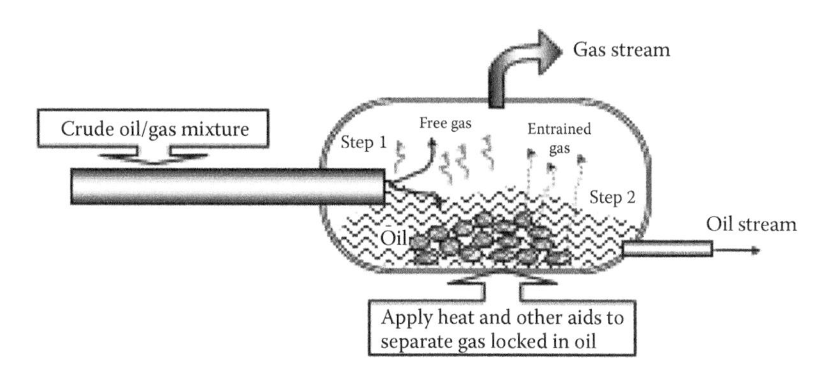

FIGURE 6.7
Proposed two-step mechanism for separating gas from oil.

In separating the gas from oil, the following two-step mechanism could be suggested as shown in Figure 6.7:

1. To separate oil from gas—Here, we are concerned primarily with recovering as much oil as we can from the gas stream. Density difference or gravity differential is responsible for this separation. At the separator's operating condition of high pressure, this difference in density between oil and gas becomes small (gas law). Oil is about eight times as dense as the gas. This could be a sufficient driving force for the liquid particles to separate and settle down. This is especially true for large-sized particles, having diameters of 100 μm or more. For smaller ones, mist extractors are needed.
2. To remove gas from oil—The objective here is to recover and collect any nonsolution gas that may be entrained or *locked* in the oil. Recommended methods to achieve this are settling, agitation, and applying heat and chemicals.

6.5.2 Commercial Gas–Oil Separators

Based on the configuration, the most common types of separators are horizontal, vertical, and spherical, as illustrated in Figures 6.8, 6.9, and 6.10, respectively. A concise comparison among these three types is presented in Table 6.3. Large horizontal gas–oil separators are used almost exclusively in processing well fluids in the Middle East, where the gas–oil ratio of the producing fields is high. Multistage GOSPs normally consists of three or more separators. Following is a brief description of separators for specific applications. In addition, the features of what is known as a *modern* GOSP are highlighted.

6.5.2.1 Test Separators

Test separators are used to separate and measure at the same time the well fluids. Potential test is one of the recognized tests for measuring the

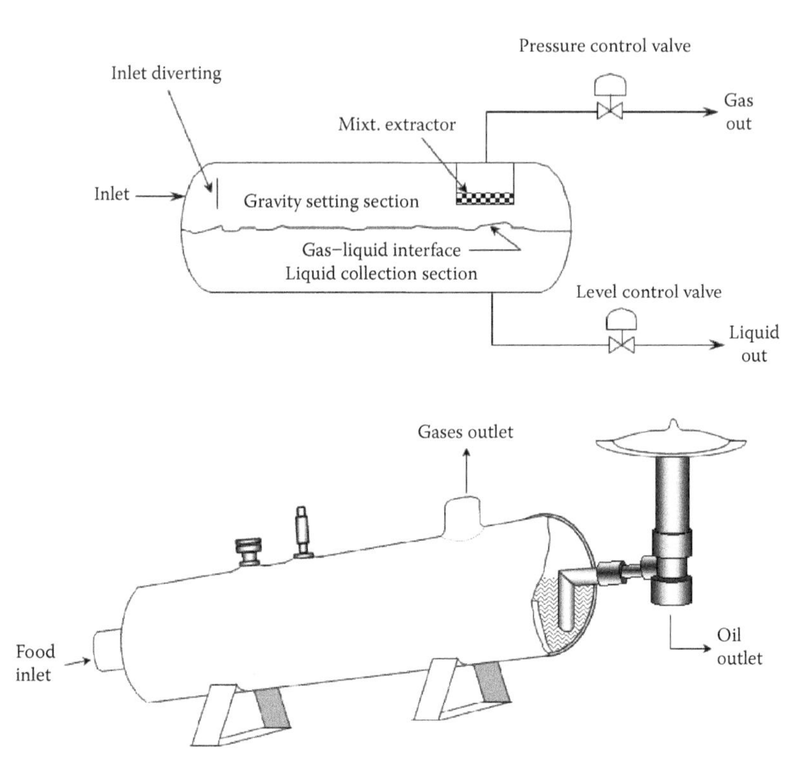

FIGURE 6.8
Single-barrel horizontal separator.

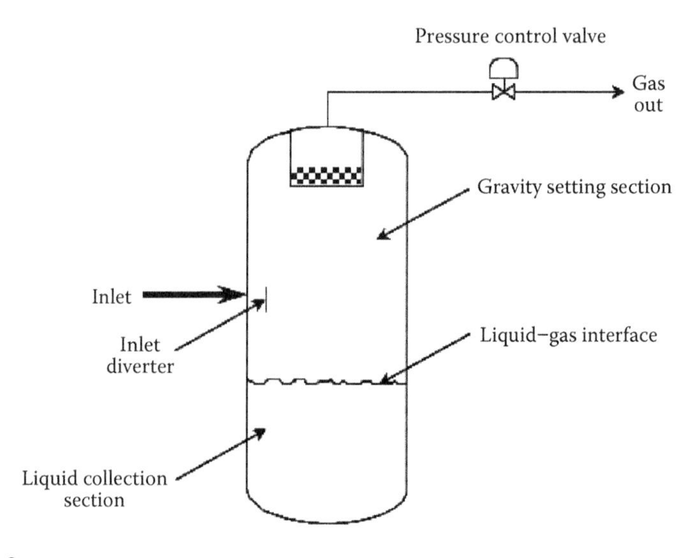

FIGURE 6.9
Vertical separator.

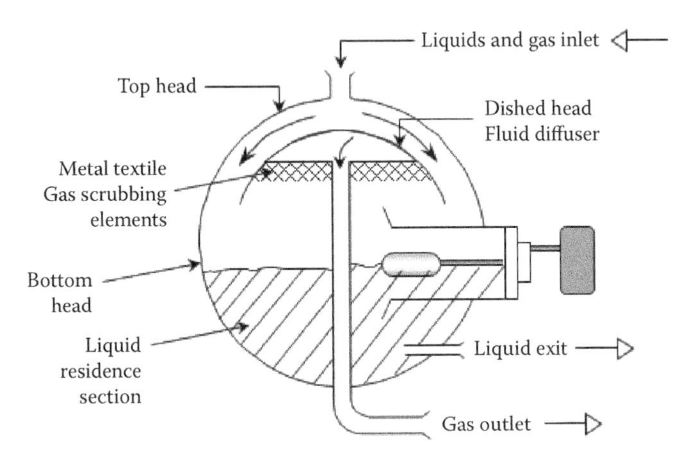

FIGURE 6.10
Spherical separator.

quantity of both oil and gas produced by the well in a 24-hour period under steady state of operating conditions. The oil produced is measured by a flow meter (normally a turbine meter) at the separator's liquid outlet and the cumulative oil production is measured in the receiving tanks. An orifice meter at the separator's gas outlet measures the produced gas. Physical properties of the oil and GOR are also determined. Equipment for test units is shown in Figure 6.11.

6.5.2.2 Low-Temperature Separators

Low-temperature separators (LTSs) are used to effectively remove light condensable hydrocarbons from a high-pressure gas stream (gas condensate feed). Liquid (condensate) separation is made possible by cooling the gas stream before separation. Temperature reduction is obtained by what is known as the Joule–Thomson effect of expanding the well fluid as it flows through the pressure-reducing choke or valve into the separator. Condensation of the vapors takes place accordingly, where the temperature is in the range 0°F–10°F.

6.5.2.3 Modern Gas–Oil Separators

Safe and environmentally acceptable handling of crude oils is ensured by treating the produced crude in the GOSP and related crude-processing facilities. The number one function of the GOSP is to separate the associated gas from oil. As the water content of the produced crude increases, field facilities for control or elimination of water are to be added. This identifies the second function of a GOSP. If the effect of corrosion due to high salt content in the

TABLE 6.3

Comparison among Different Configurations of Gas–Oil Separators

Function	Vertical	Horizontal	Spherical
Usage	For low G/O	For high G/O	For small leases operating at moderate pressure
Location of inlet and outlet stream	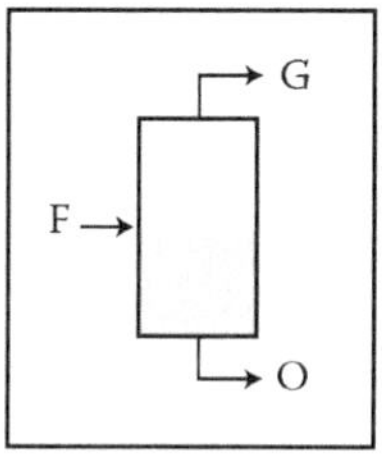 F G O	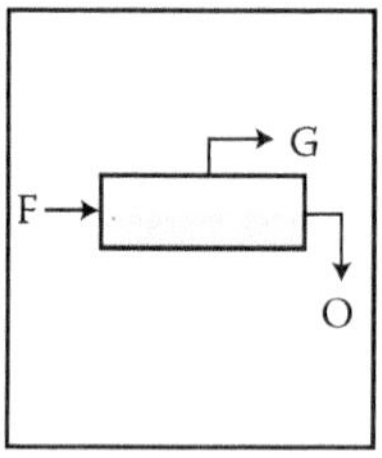 F G O	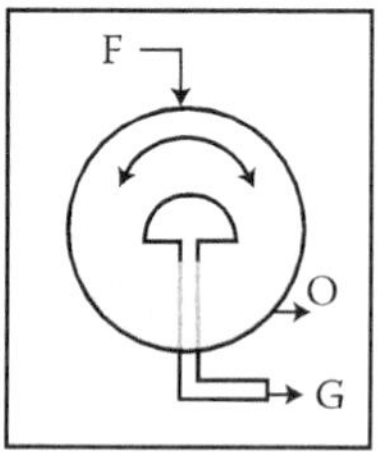 F O G
Capacity or efficiency	Large fluid capacity	Large gas capacity (handles high GOP)	Capacity rated less (low efficiency)
Handling foreign material	Rated no. 1	Rated 3rd	Rated 2nd
Separation efficiency	Rated 2nd	Rated no. 1	Rated 3rd
Ranking in use in Middle East	Rated 2nd	Rated no. 1	Rated 3rd
Handling foaming oil	Rated 2nd	Rated no. 1	Rated 3rd
Maintenance and inspection	Very difficult	Accessible	Average
Cost per unit capacity	Average	The cheapest	Most expensive
Installation	Most difficult	Average	Easy to install

crude is recognized, then modern desalting equipment could be included as a third function in the GOSP design.

One has to differentiate between *dry* crude and *wet* crude. The former is produced with no water, whereas the latter comes along with water. The water produced with the crude is a brine solution containing salts (mainly sodium chloride) in varying concentrations.

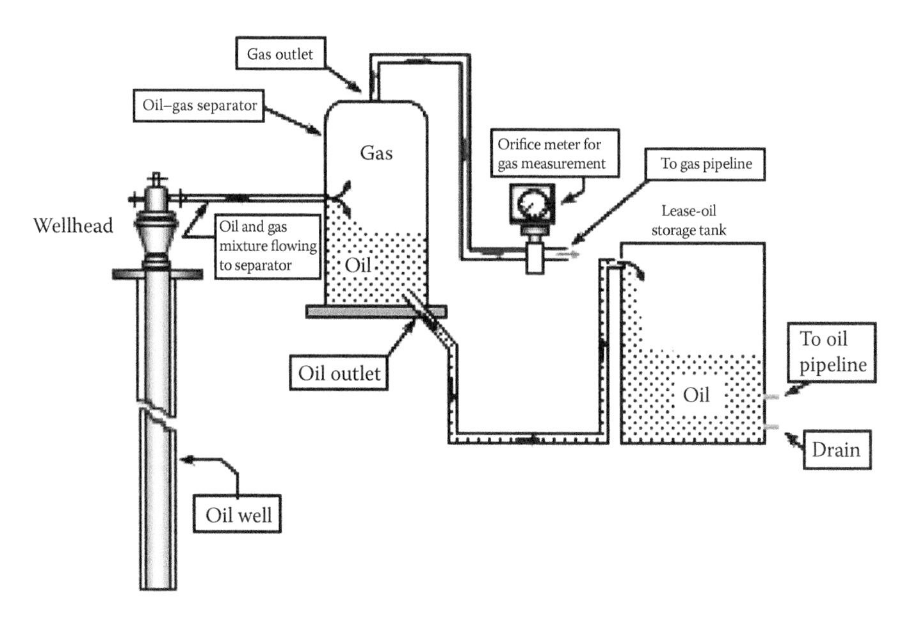

FIGURE 6.11
Test gas–oil separator.

The input of wet crude oil into a modern GOSP consists of the following:

- Crude oil.
- Hydrocarbon gases.
- Free water dispersed in oil as relatively large droplets, which will separate and settle out rapidly when wet crude is retained in the vessel.
- Emulsified water, dispersed in oil as very small droplets that do not settle out with time. Each of these droplets is surrounded by a thin film and held in suspension.
- Salts dissolved in both free water and in emulsified water.

The functions of a modern GOSP could be summarized as follows:

- Separate the hydrocarbon gases from crude oil
- Remove water from crude oil
- Reduce the salt content to the acceptable level (basic sediments and water)

It should be pointed out that some GOSPs do have gas compression and refrigeration facilities to treat the gas before sending it to gas processing plants. In general, a GOSP can function according to one of the following process operations:

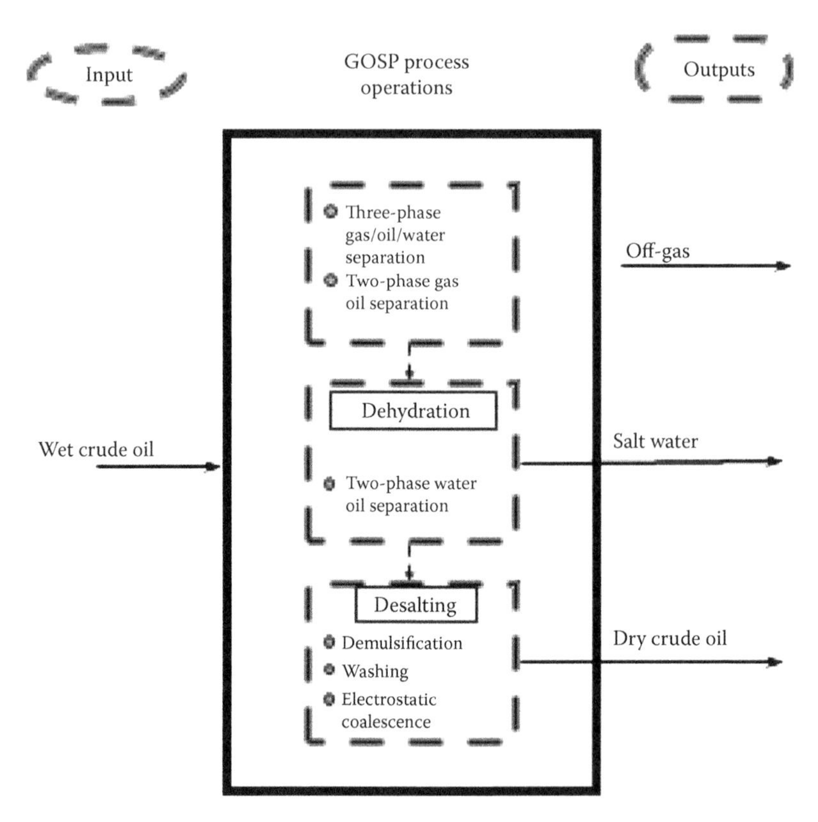

FIGURE 6.12
Functions of a modern GOSP.

- Three-phase, gas–oil–water separation (see Chapter 7)
- Two-phase, gas–oil separation
- Two-phase, oil–water separation
- De-emulsification
- Washing
- Electrostatic coalescence

To conclude, the ultimate result in operating a modern three-phase separation plant is to change wet crude input into the desired outputs, as given in Figure 6.12.

6.5.3 Controllers and Internal Components of Gas–Oil Separators

Gas–oil separators are generally equipped with the following control devices and internal components.

6.5.3.1 Liquid Level Controller

The liquid level controller (LLC) is used to maintain the liquid level inside the separator at a fixed height. In simple terms, it consists of a float that exists at the liquid–gas interface and sends a signal to an automatic diaphragm motor valve on the oil outlet. The signal causes the valve to open or close, thus allowing more or less liquid out of the separator to maintain its level inside the separator.

6.5.3.2 Pressure Control Valve

The pressure control valve (PCV) is an automatic backpressure valve that exists on the gas stream outlet. The valve is set at a prescribed pressure. It will automatically open or close, allowing more or less gas to flow out of the separator to maintain a fixed pressure inside the separator.

6.5.3.3 Pressure Relief Valve

The pressure relief valve (PRV) is a safety device that will automatically open to vent the separator if the pressure inside the separator exceeded the design safe limit.

6.5.3.4 Mist Extractor

The function of the mist extractor is to remove the very fine liquid droplets from the gas before it exits the separator. Several types of mist extractors are available, including wire-mesh, vane and centrifugal. Wire-mesh mist extractors are made of finely woven stainless-steel wire wrapped into a tightly packed cylinder of about 6 in thickness. The liquid droplets that did not separate in the gravity settling section of the separator coalesce on the surface of the matted wire, allowing liquid-free gas to exit the separator. As the droplets size grows, they fall down into the liquid phase. Provided that the gas velocity is reasonably low, wire-mesh extractors are capable of removing about 99% of the 10 μm and larger liquid droplets. It should be noted that this type of mist extractor is prone to plugging. Plugging could be due to the deposition of paraffin or the entrainment of large liquid droplets in the gas passing through the mist extractor (this will occur if the separator was not properly designed). In such cases, the vane-type mist extractor, described next, should be used.

Vane mist extractors consist of a series of closely spaced, parallel, corrugated plates. As the gas and entrained liquid droplets flowing between the plates change flow direction, due to corrugations, the liquid droplets impinge on the surface of the plates, where they coalesce and fall down into the liquid collection section.

Centrifugal mist extractors extractor use centrifugal force to separate the liquid droplets from the gas. Although it is more efficient and less susceptible to plugging than other extractors, it is not commonly used because of its performance sensitivity to small changes in flow rate.

6.5.3.5 Inlet Diverters

Inlet diverters are used to cause the initial bulk separation of liquid and gas. The most common type is the baffle plate diverter, which could be in the shape of a flat plate, a spherical dish, or a cone. Another type is the centrifugal diverter; it is more efficient but more expensive. The diverter provides a means to cause a sudden and rapid change of momentum (velocity and direction) of the entering fluid stream. This, along with the difference in densities of the liquid and gas, causes fluid separation.

6.5.3.6 Wave Breakers

In long horizontal separators, waves may develop at the gas–liquid interface. This creates unsteady fluctuations in the liquid level and would negatively affect the performance of the liquid level controller. To avoid this, wave breakers that consist of vertical baffles installed perpendicular to the flow direction are used.

6.5.3.7 Defoaming Plates

Depending on the type of oil and presence of impurities, foam may form at the gas–liquid interface. This results in the following serious operational problems:

- Foam will occupy a large space in the separator that otherwise would be available for the separation process; therefore, the separator efficiency will be reduced unless the separator is oversized to allow for the presence of foam.
- The foam, having a density between that of the liquid and gas, will disrupt the operation of the level controller.
- If the volume of the foam grows, it will be entrained in the gas and liquid streams exiting the separator; thus, the separation process will be ineffective. The entrainment of liquid with the exiting gas is known as liquid carryover. Liquid carryover could also occur as a result of a normally high liquid level, a plugged liquid outlet, or an undersized separator with regard to liquid capacity. The entrainment of gas in the exiting liquid is known as gas blowby. This could

also occur as a result of a normally low liquid level, an undersized separator with regard to gas capacity, or formation of a vortex at the liquid outlet.

Foaming problems may be effectively alleviated by the installation of defoaming plates within the separator. Defoaming plates are basically a series of inclined, closely spaced parallel plates. The flow of the foam through such plates results in the coalescence of bubbles and separation of the liquid from the gas.

In some situations, special chemicals known as foam depressants may be added to the fluid mixture to solve foaming problems. The cost of such chemicals could, however, become prohibitive when handling high production rates.

6.5.3.8 *Vortex Breaker*

A vortex breaker, similar in shape to those used in bathroom sink drains, is normally installed on the liquid outlet to prevent formation of a vortex when the liquid outlet valve is open. The formation of a vortex at the liquid outlet may result in withdrawal and entrainment of gas with the exiting liquid (gas blowby).

6.5.3.9 *Sand Jets and Drains*

As explained in Chapter 1, formation sand may be produced with the fluids. Some of this sand will settle and accumulate at the bottom of the separator. This takes up separator volume and disrupts the efficiency of separation. In such cases, vertical separators will be preferred over horizontal separators. However, when horizontal separators are needed, the separator should be equipped with sand jets and drains along the bottom of the separator. Normally, produced water is injected though the jets to fluidize the accumulated sand, which is then removed through the drains.

6.6 Design Principles and Sizing of Gas–Oil Separators

In this section, some basic assumptions and fundamentals used in sizing gas–oil separators are presented first. Next, the equations used for designing vertical and horizontal separators are derived. This will imply finding the diameter and length of a separator for given conditions of oil and gas flow rates, or vice versa. Solved examples are also given to illustrate the use of these equations.

6.6.1 Basic Assumptions

- No oil foaming takes place during the gas–oil separation (otherwise retention time has to be drastically increased as explained in Section 6.5.2).
- The cloud point of the oil and the hydrate point of the gas are below the operating temperature.
- The smallest separable liquid drops are spherical ones having a diameter of 100 μm.
- Liquid carryover with the separated gas does not exceed 0.10 gallon/MMSCF (M = 1000).

6.6.2 Fundamentals

- The difference in densities between liquid and gas is taken as a basis for sizing the gas capacity of the separator ($\rho_o - \rho_g$).
- A normal liquid (oil) retention time for gas to separate from oil is between 30 s and 3 min. Under foaming conditions, more time is considered (5–20 min). Retention time is known also as the residence time, V/Q, where V is the volume of vessel occupied by oil and Q is the liquid flow rate.
- In the gravity settling section, liquid drops will settle at a terminal velocity that is reached when the gravity force, F_g, acting on the oil drop balances the drag force, F_d, exerted by the surrounding fluid or gas.
- For vertical separators, liquid droplets (oil) separate by settling downward against an up-flowing gas stream; for horizontal ones, liquid droplets assume a trajectory-like path while it flows through the vessel (the trajectory of a bullet fired from a gun).
- For vertical separators, the gas capacity is proportional to the cross-sectional area of the separator, whereas for horizontal separators, gas capacity is proportional to area of disengagement (LD; i.e., length × diameter).

6.6.3 Settling of Oil Droplets

In separating oil droplets from the gas in the gravity settling section of a separator, a relative motion exists between the particle, which is the oil droplet, and the surrounding fluid, which is the gas.

An oil droplet, being much greater in density than the gas, tends to move vertically downward under the gravitational or buoyant force, F_g. The fluid (gas), on the other hand, exerts a drag force, F_d, on the oil droplet in the opposite direction. The oil droplet will accelerate until the frictional resistance of the fluid drag force approaches and balances the gravitational force; and, thereafter, the oil droplet continues to fall at a constant velocity known as the settling or terminal velocity.

The drag force is proportional to the droplet surface area perpendicular to the direction of gas flow and its kinetic energy per unit volume. Hence,

$$F_d = C_d \frac{\pi}{4} d^2 \frac{\rho_g u^2}{2} \tag{6.1}$$

where F_g is given by

$$F_g = \frac{\pi}{4} d^3 (\rho_o - \rho_g) g \tag{6.2}$$

where C_d is the drag coefficient, d is the diameter of the oil droplet (ft), u is the settling velocity of the oil droplet (ft/s), ρ_o and ρ_g are the oil and gas densities (lb/ft^3), respectively, and g is gravitational acceleration (ft/s^2).

The settling terminal velocity, u, is reached when $F_d = F_g$.

By equating Equations 6.1 and 6.2 and solving for u, the droplet settling velocity, we obtain

$$u^2 = \frac{8}{6} g \frac{(\rho_o - \rho_g)}{\rho_g} \left(\frac{d}{C_d} \right)$$

The droplet diameter, d, is normally expressed in microns, where 1 μm is equal to 3.2808×10^{-6} ft. Let d_m be the droplet diameter (in microns) and substitute 32.17 for g in the preceding equation to obtain the following expression for the settling velocity:

$$u = 0.01186 \left[\left(\frac{\rho_o - \rho_g}{\rho_g} \right) \frac{d_m}{C_d} \right]^{1/2} \text{ ft/s} \tag{6.3}$$

In designing gas–oil separators, the smallest oil droplet to be removed from the gas in the gravity settling section is normally taken as 100 μm. Under such a condition, the mist extractor will be capable of removing oil droplets smaller than 100 μm without getting flooded.

There are other special separators known as gas scrubbers that are normally used to remove liquid from gas streams, which have been through regular gas–liquid separators. Such gas streams are mostly gas with little liquid that may have formed due to condensation. Examples of such separators are the gas scrubbers used at the inlet to gas compressors and gas dehydration facilities. Scrubbers are also used on vents and flares. Since the amount of liquid is very small, the design of such gas scrubbers could be based on separation of liquid droplets up to 500 µm in the gravity settling section with no danger of flooding the mist extractor.

6.6.4 Gas Capacity of Separators

The volumetric flow rate of the gas processed by a gas–oil separator is directly related to the cross-sectional area of flow and the maximum allowable gas velocity at which the oil droplets will be suspended and not carried over. Mathematically,

$$Q_g = A_g u \text{ ft}^3/\text{s} \tag{6.4}$$

Equation 6.4 gives Q_g in cubic foot per second (ft³/s) under actual separator pressure and temperature. However, the volumetric flow rate of gas is normally reported at standard pressure and temperature of 14.7 psia and 520° R, respectively. Typically, Q_g is reported in units of millions of standard cubic feet per day (MMSCFD). Equation 6.4 could, therefore, be written as

$$\begin{aligned} Q_g &= (10^{-6} \times 60 \times 60 \times 24) A_g u \left(\frac{P}{14.7}\right)\left(\frac{520}{TZ}\right) \\ Q_g &= 3.056\left(\frac{P}{TZ}\right) u A_g \quad \text{MMSCFD} \end{aligned} \tag{6.5}$$

Solving for the gas velocity, u, gives

$$u = 0.327 Q_g \left(\frac{TZ}{P}\right)\left(\frac{1}{A_g}\right)\frac{\text{ft}}{\text{s}} \tag{6.6}$$

where Z is the gas compressibility at the operating pressure (P) and temperature (T) and A_g is the available area for gas flow (ft²).

6.6.5 Liquid Capacity of Separators

The basic relationship that combines the oil flow rate or oil capacity of a separator, Q_o, the volume of separator occupied by oil, V_o, and the retention time or residence time, t, is

$$Q_o\left(\frac{\text{ft}^3}{\text{min}}\right)=\frac{V_o\,(\text{ft}^3)}{t\,(\text{min})} \tag{6.7}$$

Equation 6.7 is rewritten in terms of barrels per day instead of cubic feet per minute:

$$Q_o\left(\frac{\text{bbl}}{\text{day}}\right)=257\frac{V_o}{t} \tag{6.8}$$

where 1 ft^3/min = 257 bbl/day.

6.6.6 Sizing Vertical Gas–Oil Separators

The size (diameter and height or length) of a separator is normally determined by consideration of its required capacity for gas and oil as discussed in the following sections.

6.6.6.1 Gas Capacity Constraint

For vertical separators, the upward average gas velocity should not exceed the downward terminal velocity of the smallest oil droplet to be separated. This condition is expressed mathematically by equating Equations 6.2 and 6.6:

$$0.327Q_g\left(\frac{TZ}{P}\right)\left(\frac{1}{A_g}\right)=0.01186\left[\left(\frac{\rho_o-\rho_g}{\rho_g}\right)\frac{d_m}{C_d}\right]^{1/2}$$

Substituting for A_g from

$$A_g=\frac{\pi}{4}\left(\frac{D}{12}\right)^2$$

where D is the internal diameter of the separator in inches, and then solving for D, we obtain

$$D^2=5.058Q_g\left(\frac{TZ}{P}\right)\left[\frac{\rho_g}{(\rho_o-\rho_g)}\frac{C_d}{d_m}\right]^{1/2}\ \text{in}^2 \tag{6.9}$$

Equation 6.9 provides the minimum acceptable diameter of the separator. Larger diameters yield lower gas velocities and, thus, better separation of the oil droplets from the gas. Smaller diameters, on the other hand, result in higher gas velocities and, therefore, the liquid droplets will be carried over with the gas.

A summary of the steps involved in the calculation of the gas capacity of a gas–oil separator is outlined in Figure 6.13.

In solving Equation 6.9, the value of the drag coefficient, C_d, must be first determined. C_d is related to the Reynolds number, Re, according to the following formula:

$$C_d = 0.34 + \frac{3}{\mathrm{Re}^{0.5}} + \frac{24}{\mathrm{Re}} \tag{6.10}$$

where the Reynolds number is given by

$$\mathrm{Re} = 0.0049 \frac{\rho_g d_m u}{\mu_g} \tag{6.11}$$

where μ_g is the gas viscosity.

The velocity, u, is given by Equation 6.2 and is a function of C_d. Therefore, C_d could only be determined by an iterative procedure as follows:

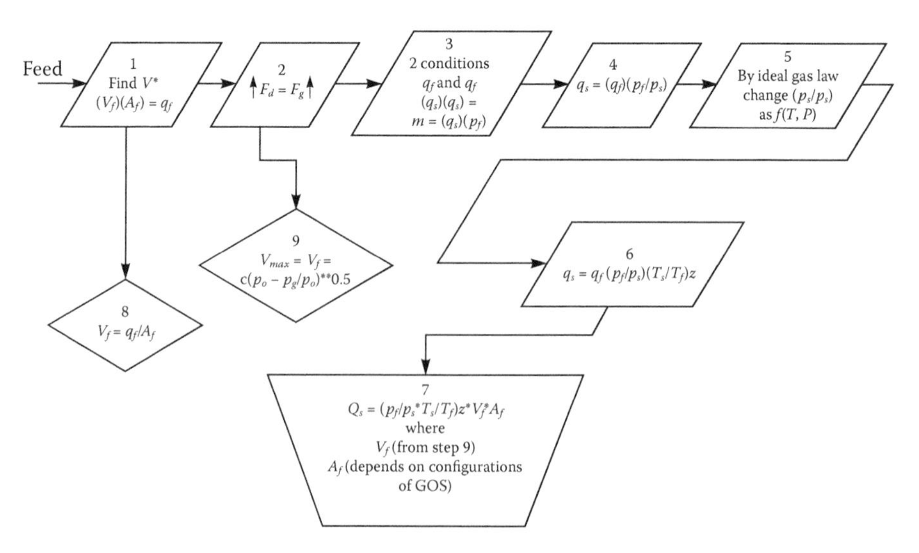

FIGURE 6.13
Systematic steps for sizing the gas capacity for a gas–oil separator.

1. Assume a value for C_d (a value of 0.34 could be used as a first assumption).
2. Calculate the velocity, u, from Equation 6.2.
3. Calculate Re from Equation 6.11.
4. Calculate C_d from Equation 6.10 and compare to the assumed value.
5. If no match is obtained, use the calculated value of C_d and repeat steps 2–4 until convergence is obtained.

6.6.6.2 *Oil Capacity Constraint*

The oil has to be retained within the separator for a specific retention time, t. The volume of separator occupied by oil, V_o, is obtained by multiplying the cross-sectional area by the height of the oil column, H (in). Equation 6.8 could, therefore, be rewritten as

$$Q_o = 257\left(\frac{\pi}{4}\right)\left(\frac{D}{12}\right)^2\left(\frac{H}{12}\right)\left(\frac{1}{t}\right)\left(\frac{\text{bbl}}{\text{day}}\right) \tag{6.12}$$

or

$$D^2H = 8.565Q_o t \text{ in}^3 \tag{6.13}$$

6.6.6.3 *Sizing Procedure*

In summary, the size (diameter and seam-to-seam length or height) of a vertical separator is determined as follows:

1. Equation 6.9 is used to determine the minimum allowable vessel diameter.
2. For diameters larger than the minimum, Equation 6.13 is used to determine combinations of D and H.
3. The seam-to-seam length, L_s, for each combination of D and H is determined using one of the following expressions as appropriate:

$$\text{If } D < 36 \text{ in, then } L_s = \frac{H+76}{12} \text{ ft} \tag{6.14}$$

$$\text{If } D > 36 \text{ in, then } L_s = \frac{H+D+40}{12} \text{ ft} \tag{6.15}$$

4. For each combination of D and L_s, the slenderness ratio, SR, defined as the ratio of length to diameter is determined. Separators with SR between 3 and 4 are commonly selected.

6.6.7 Sizing Horizontal Gas–Oil Separators

As with vertical separators, the size (diameter and length) of the horizontal separator is determined by its required capacity for gas and oil. It has been shown that the gas capacity constraint for vertical separators determines the minimum allowable vessel diameter. For horizontal separators, however, the gas capacity constraint yields, as shown in the following section, a relationship between the diameter and effective length of the separator. This along with a similar relationship derived from the liquid capacity constraint are used in determining the size of the separator. In reality, either the gas capacity constraint or the liquid capacity constraint governs the design and only one of the two constraints equations is used in determining the size.

In the following discussion, it is assumed that each of the gas and oil phases occupies 50% of the effective separator volume. Similar equations as those derived in the following could be obtained for other situations where either of the two phases occupies more or less than 50% of the separator effective volume.

6.6.7.1 Gas Capacity Constraint

Since the gas occupies the top half of the separator, its average flowing velocity within the separator, u_g, is obtained by dividing the volumetric flow rate, Q_g, by one-half of the separator cross-sectional area, A, that is,

$$u_g = \frac{Q_g}{0.5[(\pi/4)D^2]}$$

Q_g is usually reported in units of MMSCFD and should, therefore, be converted into actual cubic feet per second (ft^3/s); also D, which is usually given in inches, should be converted into feet in order to obtain the velocity in units of feet per second (ft/s). The preceding equation, therefore, becomes

$$u_g = 120\frac{Q_g}{D^2}\left(\frac{TZ}{P}\right) \frac{\text{ft}}{\text{s}} \tag{6.16}$$

The gas travels horizontally along the effective length of the separator, L (ft), in a time, t_g, that is given by

$$t_g = \frac{L}{u_g} \text{ s} \tag{6.17}$$

This time must, at least, be equal to the time it takes the smallest oil droplet to be removed from the gas to travel a distance of $D/2$ to reach the gas–oil interface. This settling time, t_s, is obtained by dividing the distance ($D/2$) by the settling velocity from Equation 6.2; therefore,

$$t_s = \left(\frac{D}{2 \times 12}\right)\left\{0.01186\left[\left(\frac{\rho_o - \rho_g}{\rho_g}\right)\frac{d_m}{C_d}\right]^{1/2}\right\}^{-1} \text{ s} \tag{6.18}$$

Equating Equation 6.18 and Equation 6.17, substituting for u_g from Equation 6.16, and solving for the product LD, we obtain

$$LD = 422\left(\frac{Q_g TZ}{P}\right)\left[\left(\frac{\rho_g}{\rho_o - \rho_g}\right)\left(\frac{C_d}{d_m}\right)\right]^{1/2} \text{ ft in.} \tag{6.19}$$

Equation 6.19 provides a relationship between the vessel diameter and effective length that satisfies the gas capacity constraint. Any combination of D and L satisfying Equation 6.19 ensures that all oil droplets having diameter d_m and larger will settle out of the gas flowing at a rate of Q_g MMSCFD into the separator that is operating at P psia and T°R.

6.6.7.2 Liquid Capacity Constraint

A gas–oil separator has to have a sufficient volume to retain the liquid for the specified retention time before it leaves the separator. For a horizontal separator that is half full of liquid, the volume occupied by the liquid is given by

$$V_o = 0.5\left(\frac{\pi}{4}\right)\left(\frac{D}{12}\right)^2 L \text{ ft}^3$$

Substituting in Equation 6.8, the following equation is obtained:

$$D^2L = 1.428Q_o t \text{ ft}^3 \quad (6.20)$$

Equation 6.20 provides another relationship between D and L that satisfies the liquid capacity (retention) time constraint.

6.6.7.3 Sizing Procedure

For a given set of operating conditions (pressure, temperature, gas and oil flow rates, gas and oil properties, and oil retention time), the size (diameter and seam-to-seam length) of a horizontal separator is determined as follows:

1. Assume various values for the separator diameter, D.
2. For each assumed value of D, determine the effective length, L_g, that satisfies the gas capacity constraint from Equation 6.19 and calculate the seam-to-seam length, L_s, from

$$L_s = L_g + \frac{D}{12} \text{ ft} \quad (6.21)$$

3. For each assumed value of D, determine the effective length, L_o, that satisfies the liquid capacity constraint from Equation 6.20 and calculate the seam-to-seam length, L_s, from

$$L_s = \frac{4}{3}L_o \text{ ft} \quad (6.22)$$

4. For each value of D used, compare the values of L_g and L_o to determine whether the gas capacity constraint or the oil capacity constraint governs the design of the separator. Of course, the larger required length governs the design.
5. Select reasonable combinations of D and L_s such that the slenderness ratio (SR) is in the range of 3 to 5. The cost and availability would then determine the final selection.

The preceding equations and sizing procedures are very sufficient for the determination of separator diameter and length as well as for the performance evaluation of existing separators. Students and practicing engineers should be familiar with such design equations and procedures before attempting the use of commercially available software. The retention time is an important parameter in designing gas–oil separators. It is best obtained from laboratory tests that simulate the field operating conditions. This, however, may not always be available. In such cases, experience and data from offset fields, if available, will be very valuable.

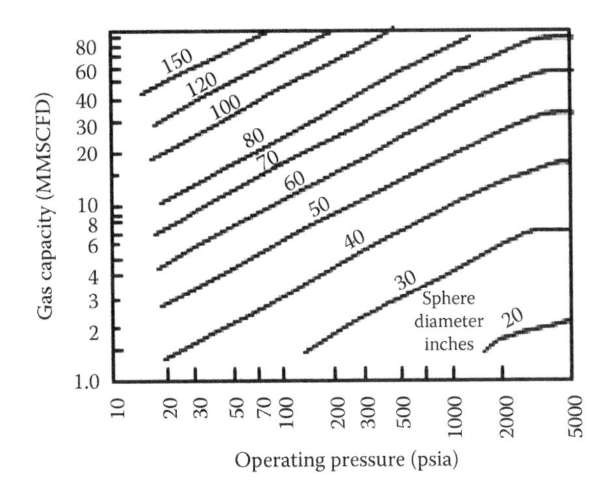

FIGURE 6.14
Gas capacities of spherical gas–oil separators.

On the commercial side, computer programs for sizing gas–oil separators have been developed by Ellis Engineering Inc. These curves are suitable for preliminary sizing for estimating gas and liquid capacities.

For spherical separators, curves have been developed for sizing gas capacity, as illustrated by the diagram given in Figure 6.14.

6.6.8 Solved Examples on Sizing Gas–Oil Separators

Example 6.1: Design Problem, Vertical Separator

Determine the diameter and height (seam-to-seam length) of a vertical separator for the following operating conditions:

Gas rate:	15 MMSCFD
Gas specific gravity:	0.6
Oil rate:	3000 bbl/day
Oil gravity:	35°API
Operating pressure:	985 psi
Operating temperature:	60°F
Retention time:	3 min

Solution

1. Determine gas and oil properties: For 0.6 specific gravity gas at 60°F and 985 psi (1000 psia),

$$Z = 0.84 \text{ and } \mu = 0.013 \text{ cP}$$

$$\rho_g = 2.7\gamma\frac{P}{TZ} = 2.7(0.6)\frac{1000}{520\times0.84} = \frac{3.708 \text{ lb}}{\text{ft}^3}$$

$$\rho_o = \rho_w\gamma_o = 62.4\frac{141.5}{131.5+35} = \frac{53.03 \text{ lb}}{\text{ft}^3}$$

2. Determine C_d: Assume $C_d = 0.34$. Using Equation 6.2,

$$u = 0.01186\left[\left(\frac{\rho_o-\rho_g}{\rho_g}\right)\frac{d_m}{C_d}\right]^{1/2}$$

$$= 0.01186\left[\left(\frac{53.03-3.708}{3.708}\right)\left(\frac{100}{0.34}\right)\right]^{1/2}$$

$$u = 0.7418 \ \frac{\text{ft}}{\text{s}}$$

$$\text{Re} = 0.0049\frac{\rho_g d_m u}{\mu_g} = 0.0049\frac{3.708\times100\times0.7418}{0.013} = 103.67$$

$$C_d = 0.34 + \frac{3}{\text{Re}^{0.5}} + \frac{24}{\text{Re}}$$

$$= 0.34 + \frac{3}{(103.67)^{0.5}} + \frac{24}{103.67} = 0.866$$

3. Repeat the preceding steps using the calculated value of C_d to obtain a new value of c_d and continue until convergence. The final value of c_d is 1.1709.
4. Check for gas capacity constraint, Equation 6.9:

$$D^2 = 5058Q_g\left(\frac{TZ}{P}\right)\left[\frac{\rho_g}{(\rho_o-\rho_g)}\frac{C_d}{d_m}\right]^{1/2}$$

$$= 5058\times15\left(\frac{520\times0.84}{1000}\right)\left(\frac{3.708}{53.03-3.708}\frac{1.1709}{100}\right)^{1/2}$$

$$D^2 = 983.246;$$

Therefore: $D_{\min} = 31.357$ in. This is the minimum allowable vessel diameter for separation of oil droplets down to 100 μm.

5. Check for liquid capacity (retention time constraint), Equation 6.13:

$$D^2H = 8.565Q_ot = 8.565 \times 3000 \times 3 = 77{,}805$$

6. Select values for D greater than 31.357 in and calculate the corresponding values of H from Equation 6.13 and the corresponding values of L_s from Equation 6.14 for $D < 36$ or Equation 6.15 for $D > 36$. The results are as follows:

D (in)	H (in)	L_s (ft)	SR
36	60.03	11.34	3.78
42	44.10	10.51	3.00
48	33.77	10.15	2.54
54	26.68	10.06	2.23
60	21.61	10.13	2.03
66	17.86	10.32	1.88
72	15.01	10.58	1.76

From the results in the table, the first two diameters provide slenderness ratios (SR) within the commonly used range; therefore, both are acceptable. A preferred selection will be a 36 in diameter by 12 ft separator as compared to a 42 in by 11 ft separator. This is because the 36 in separator is a standard unit and would probably be less costly than the larger diameter, shorter separator.

Example 6.2: Performance Problem, Vertical Separator

Determine the actual gas and oil capacity of a 36 in diameter by 12 ft seam-to-seam length vertical separator operating under the following conditions:

Gas specific gravity:	0.6
Oil gravity:	35°API
Operating pressure:	985 psi
Operating temperature:	60°F
Retention time:	3 min
C_d:	1.1709

Solution

1. To determine the actual gas capacity: Rearranging Equation 6.9 to solve for Q_g, we obtain

$$Q_g = 1.977 \times 10^{-4} D^2 \left(\frac{P}{TZ}\right)\left(\frac{(\rho_o - \rho_g)d_m}{\rho_g C_d}\right)^{1/2}$$

The same operating conditions of Example 6.1 are applicable in this case, can substitute the same values of ρ_g, ρ_o, and Z to calculate Q_g:

$$Q_g = 1.977 \times 10^{-4} (36)^2 \left(\frac{1000}{520 \times 0.84} \right) \left(\frac{(53.03 - 3.708)100}{3.708 \times 1.1709} \right)^{1/2}$$

Example 6.3

Determine the diameter and seam-to-seam length of a horizontal separator for the following operating conditions. Determine the actual gas and oil capacity of the designed separator.

Gas rate:	15 MMSCFD
Gas specific gravity:	0.6
Oil rate:	3000 bbl/day
Oil gravity:	35°API
Operating pressure:	985 psi
Operating temperature:	60°F
Retention time:	3 min

Solution

From Example 6.1, we have

$$\mu_g = 0.013 \text{ cP}$$

$$\rho_g = 3.708 \text{ lb/ft}^3$$

$$\rho_o = 53.03 \text{ lb/ft}^3$$

$$Z = 0.84$$

$$c_d = 1.1709$$

1. Check for gas capacity constraint, Equation 6.18:

$$LD = 422 \left(\frac{Q_g TZ}{P} \right) \left[\left(\frac{\rho_g}{\rho_o - \rho_g} \right) \left(\frac{c_d}{d_m} \right) \right]^{1/2}$$

$$= 422 \left(\frac{15 \times 520 \times 0.84}{1000} \right) \left[\left(\frac{3.708}{53.03 - 3.708} \right) \left(\frac{1.1709}{100} \right) \right]^{1/2}$$

$$LD = 82.04$$

2. Check for oil capacity (retention time), Equation 6.20:

$$D^2L = 1.428Q_o t = 1.428 \times 3000 \times 3$$

$$D^2L = 12{,}852$$

3. Assume values for D and determine corresponding effective length for gas capacity, L_g, and the corresponding effective length for oil capacity, L_o. The results are summarized as follows:

D (in)	L_g (ft)	L_s (gas)	L_o (ft)	L_s (oil)	SR = $12 l_s$ (oil)/D
30	2.73	5.23	14.28	19.04	7.62
36	2.28	5.28	9.92	13.22	4.41
42	1.95	5.45	7.29	9.71	2.78
48	1.71	5.71	5.58	7.44	1.86
54	1.52	6.02	4.41	5.88	1.31

Comparing the value of L_s for the oil capacity to those for the gas capacity shows that the gas capacity does not govern the design. Investigating the values of the slenderness ratio shows that the 36 in and 42 in separators are the only possible selections. The recommended size would be a 36 in diameter by 14 ft seam-to-seam length.

Recall that for the same conditions a vertical separator of the same diameter but shorter (12 ft) was suitable (Example 6.1). For such conditions, the vertical separator should be selected unless other operating conditions necessitate the selection of a horizontal separator.

Example 6.4: Performance Problem, Horizontal Separator

Determine the actual gas and oil capacity of a horizontal separator having a diameter of 36 in and a seam-to-seam length of 14 ft given the following operating conditions:

Gas specific gravity:	0.6
Oil gravity:	35°API
Operating pressure:	985 psi
Operating temperature:	60°F
Retention time:	3 min
C_d:	1.1709

Solution

From Example 6.1, because we have the same operating conditions, the fluid properties will be the same:

$$\rho_g = \frac{3.708 \text{ lb}}{\text{ft}^3}$$

$$\rho_o = \frac{53.03 \text{ lb}}{\text{ft}^3} \text{ and } Z = 0.84$$

1. To determine the actual gas capacity of the separator: Determine the separator effective length for gas capacity from Equation 6.21:

$$L_s = L_g + \frac{D}{12}$$
$$L_g = 14 - (12/36) = 14 - 3 = 11 \text{ ft}$$

Rearranging Equation 6.14 to solve for Q_g, we have

$$Q_g = 0.00237 L_g D\left(\frac{P}{TZ}\right)\left[\left(\frac{\rho_o - \rho_g}{\rho_g}\right)\left(\frac{d_m}{C_d}\right)\right]^{1/2}$$

Substitute $L_g = 11$ and $D = 36$ in the preceding equation and solve for Q_g:

$$Q_g = 0.00237(11 \times 36)\left(\frac{1000}{520 \times 0.84}\right)$$
$$\times\left[\left(\frac{53.03 - 3.708}{53.03}\right)\left(\frac{100}{1.1709}\right)\right]^{1/2}$$
$$Q_g = 72.4 \text{ MMSCFD}$$

2. To determine the actual oil capacity of the separator: Determine the separator effective length for oil capacity from Equation 6.22:

$$L_o = 34 L_s = 3 \times 414 = 10.5 \text{ ft}$$

Rearrange Equation 6.20 to solve for Q_o:

$$Q_o = 0.7\left(\frac{D^2 L}{t}\right) D^2 L$$

Substitute $L_o = 10.5$ and $D = 36$ in the preceding equation and solve for Q_o:

$$Q_o = 3176 \text{ bbl/day}$$

Comparing the preceding actual oil and gas capacities to the design production rates of Example 6.3 shows that the separator can handle almost five times the design gas rate of

Example 6.3. This is because the cross-sectional area for gas flow is much more than needed for settling. In such situations, it is evident that the assumption of allowing each phase to occupy one-half of the separator effective volume is not a good assumption. The size of the separator could certainly be reduced if the design was based on allowing the gas to flow through a smaller area than the 50% of the total area. It is recommended that the reader solve this problem assuming the gas occupies a smaller volume than 50% of the separator volume. Care must be taken in modifying the aforementioned design equations used.

Geometrical relationships relating the cross-sectional area occupied by the gas, the height of the gas column, and the vessel diameter could be derived. For the cases where the cross-sectional area of the gas flow, A_g, is not equal to one-half of the vessel cross-sectional area, A, that is, the height of the gas, H_g, is not equal to the radius, $D/2$, Lockhart presented the following simplified, but reasonably accurate, relationships.

a. To determine H_g/D as a function of A_g/A:

$0 \le A_g/A \le 0.2$:

$$\frac{H_g}{D} = 2.481\frac{A_g}{A} - 12.29\left(\frac{A_g}{A}\right)^2 + 31.133\left(\frac{A_g}{A}\right)^3$$

$0.2 \le A_g/A \le 0.8$:

$$\frac{H_g}{D} = 0.8123\frac{A_g}{A} + 0.0924$$

$0.8 \le A_g/A \le 1.0$:

$$\left(1-\frac{H_g}{D}\right) = 2.481\left(1-\frac{A_g}{A}\right) - 12.29\left(1-\frac{A_g}{A}\right)^2 + 31.133\left(1-\frac{A_g}{A}\right)^3$$

b. To determine A_g/A as a function of H_g/D:

$0 \le H_g/D \le 0.25$:

$$\frac{A_g}{A} = 0.21\left(\frac{H_g}{D}\right) + 3.52\left(\frac{H_g}{D}\right)^2 - 4.93\left(\frac{H_g}{D}\right)^3$$

$0.25 \le H_g/D \le 0.75$:

$$\frac{A_g}{A} = 1.231\left(\frac{H_g}{D}\right) - 0.1138$$

$0.75 \le H_g/D \le 1.0$:

$$\left(1-\frac{A_g}{A}\right) = 0.21\left(1-\frac{H_g}{D}\right) + 3.52\left(1-\frac{H_g}{D}\right)^2 - 4.93\left(1-\frac{H_g}{D}\right)^3$$

6.7 Optimum Pressure for Gas–Oil Separators

6.7.1 Introduction

In order to study the effect of operating pressure in gas–oil separation in general, we will consider first the case of a single-stage separation plus a storage tank. Now, what is the effect of the operating pressure on the recovery of stock tank-oil for the following two extreme cases?

1. High-pressure operation—This will diminish the opportunity of light hydrocarbons in the feed to vaporize and separate. However, once this liquid stream is directed to the storage tank (normally operating at or close to atmospheric pressure), violent flashing occurs due to the high-pressure drop, with subsequent severe losses of the heavier hydrocarbons into the gas phase.
2. Low-pressure operation—Here, large quantities of light hydrocarbons will separate from the gas–oil separator, carrying along with them heavier hydrocarbons, causing a loss in the recovered oil. Upon directing this liquid stream to the storage tank, it suffers very little loss in heavy components, because the bulk of light gases were separated in the separator.

It may be concluded that a proper operating pressure has to be selected and its value has to be between the two extreme cases as described in order to maximize the oil yield. This conclusion is illustrated as shown in Figure 6.15, in which an optimum pressure of 45 psig is selected to give 0.75 bbl of oil yield.

The same concept is discussed quantitatively using Rault's and Dalton's laws. For a multicomponent mixture of hydrocarbons, the following thermodynamic relationships are applicable:

- In the liquid phase, the partial pressure exerted by each component $i\left(P_i^l\right)$ is related to its vapor pressure $\left(P_i^0\right)$ and mole fraction (X_i) by

$$P_i^l = P_i^0 X_i \tag{6.23}$$

- In the vapor phase, the partial pressure of component $i\left(P_i^v\right)$ in the gas mixture is related to its mole fraction (Y_i) and total pressure (P_t) by

$$P_i^v = P_t Y_i \tag{6.24}$$

At given conditions of T and P, thermodynamic equilibrium exists, because the vapor and the liquid are in intimate contact. Consequently, the partial

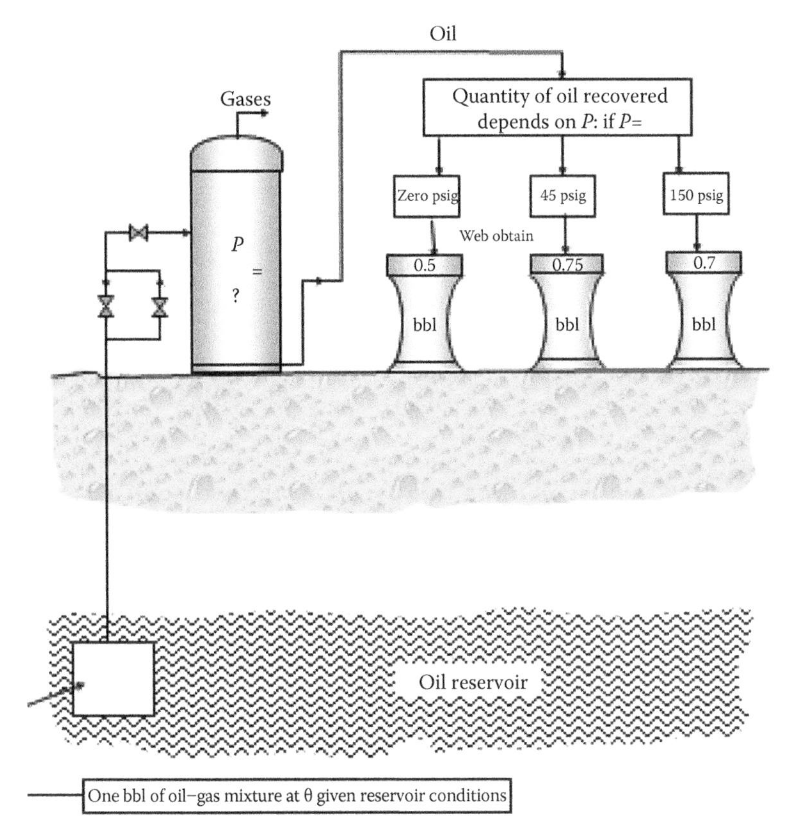

FIGURE 6.15
Effect of operating pressure in gas–oil separators on crude yield.

pressure of a component in the vapor is equal to the partial vapor pressure of the component in the liquid. Therefore, equating the two equations and rearranging, the following relationship is obtained:

$$\frac{P_i^0}{P_t} = \frac{Y_i}{X_i} = K_i = \text{Equilibrium constant} \tag{6.25}$$

where P_i^v is the partial pressure of component i in the mixture, P_i^0 is the vapor pressure of a pure component i, X_i is the mole fraction of component i in the liquid phase, Y_i is the mole fraction of component i in the gas phase, and P_t is the total pressure on the system (inside the separator).

Now, it can be concluded that if the total pressure on the system is increased, the mole fraction in the gas phase has to decrease in accordance with Equation 6.21. In other words, the tendency of vaporization diminishes as the pressure inside the gas–oil separator increases.

In considering the preceding relationship based on Rault's law, there are some shortcomings in using it, especially for gas mixtures containing methane (its critical temperature is –116°F). However, for mixture of C_3, C_4, and C_5, Rault's law could be applied fairly well at temperatures up to 150°F and pressures up to 100 psia.

6.7.2 Pressure Profile of a Three-Stage Gas–Oil Separator Plant

In the determination of the optimum operating pressure for a GOSP consisting of three stages (high-, intermediate-, and low-pressure separators) it is the second-stage pressure that could be freely changed, hence optimized.

The pressure in the first stage (high pressure) is normally fixed under one of the following conditions:

- Matching certain requirements to supply high-pressure gas for gas-injection facilities existing in the field
- Selling the gas through pipelines
- Flow conditions of the producing wells

Similarly, the pressure in the third stage (low pressure) is fixed for the following cases:

- The last stage is a storage tank.
- An existing gas-gathering or vapor recovery facilities utilizing the gas.
- The last stage operates at the relatively low pressure.

Figure 6.16 illustrates a three-stage GOSP in which the pressure in the first and the third stages are fixed for the aforementioned reasons.

The optimum pressure of operating a GOSP is defined as the second-stage pressure that provides the desired gas–oil separation with maximum oil recovery in the stock tank and minimum gas–oil ratio (GOR).

If R designates the recovery of oil and is defined as

$$R = \text{O/G (bbl of oil/SCF of gas)} \tag{6.26}$$

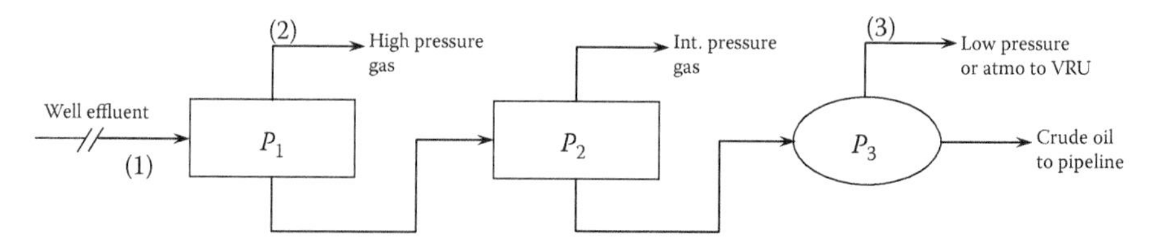

FIGURE 6.16
Optimization of the second stage in a three-stage gas–oil separator.

then the optimum pressure in the second-stage separator is the value that makes R maximum or $1/R$ minimum.

Apart from obtaining a high recovery of oil, operating pressures have other important considerations in the processing of the separated streams. A minimum pressure has to be maintained in order for the oil to be delivered to the next processing stage. In addition, using high pressure will deliver the gas stream for sales at higher output pressure, thus reducing the compressor horsepower used for gas pumping.

In general, pressure of 50–100 psia is considered an optimum value of the second-stage operation, whereas a minimum pressure for the third stage will be in the range of 25–50 psig.

6.7.3 Determination of the Optimum Second-Stage Operating Pressure

Three methods can be used to determine the optimum operating pressure for the second stage.

6.7.3.1 *Experimental Measurements*

In this method, experimental runs are carried out in which the composition of the gas leaving the separator is analyzed and the content of some key component (e.g., C_5^+) is determined. Now, while increasing the pressure in the second stage, we should calculate the ratio of gas to oil (G/O) for both second and third stages. A graphical plot for P_2 versus G/O is given in Figure 6.17. It is observed that as the pressure increases, $(G/O)_2$ decreases because condensation of C_5^+ is enhanced at higher pressure, whereas $(G/O)_3$ increases because the pressure difference between stages 2 and 3 becomes higher, causing more hydrocarbons to vaporize from the third stage. The optimum operating pressure is the value that makes $(G/O)_T$ minimum.

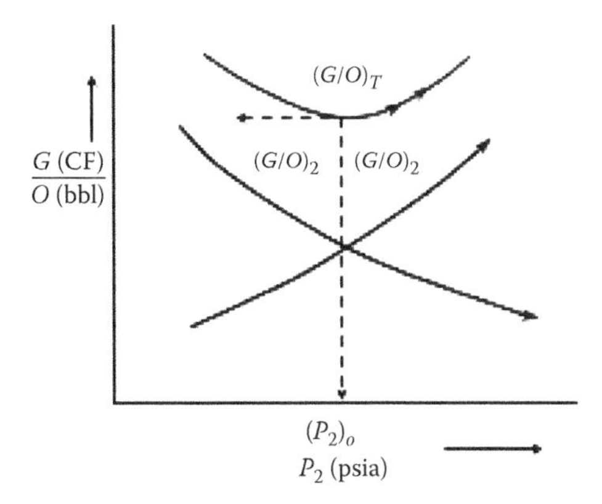

FIGURE 6.17
Variation of G/O with P_2 for three-stage gas oil separator.

Further verification of the value of $(P_2)_o$ is done as follows:

1. Determine experimentally the C_5^+ content in the gas samples leaving the top of separators 2 and 3, in gallons/MCF of the gas (gal C_5^+/MCF).
2. Calculate the (gal C_5^+/bbl oil) by multiplying results of step 1 by G/O, the gas/oil ratio (MCF/bbl oil). Sum the values for both stages 1 and 2 to give the total gal C_5^+ lost per barrel of oil.
3. Repeat for different values of operating pressure, P_2.
4. Plot the value of gal C_5^+/MCF for both stages 2 and 3 versus P_2, as shown in Figure 6.18. It is to be noted that the C_5^+ content per MCF in the gas streams decreases with the increase in P_2 for both stages. However, it is higher for stage 3 than stage 2.

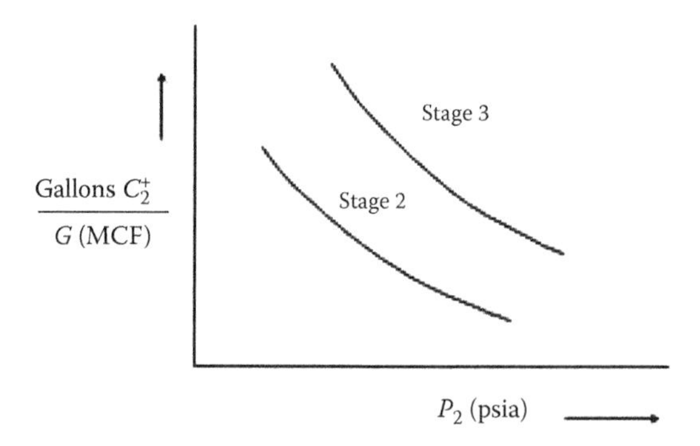

FIGURE 6.18
Variation of gasoline content with P_2.

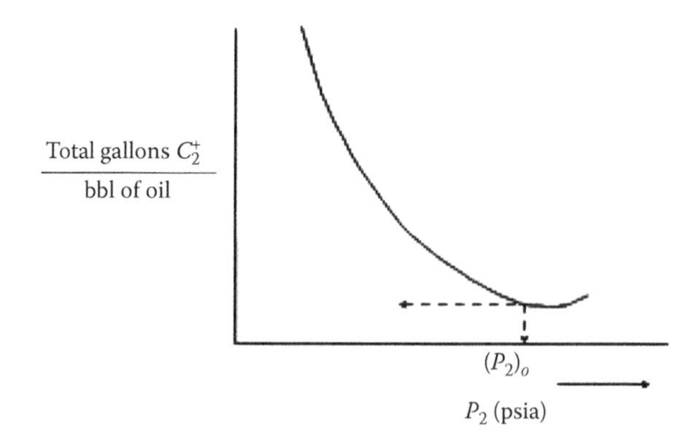

FIGURE 6.19
Determination of $(P_2)_o$ experimentally.

5. Plot the total gal C_5^+ lost per barrel of oil versus P_2, as given in Figure 6.19. An optimum value, $(P_2)_o$ is determined, which corresponds to a minimum loss of C_5^+ in gallons per barrel of oil.
6. This is a double check on the value obtained before, from Figure 6.17.

6.7.3.2 Approximate Formula

The following simplified formula can be used for determining $(P_2)_o$ as a function of P_1 and P_3:

$$(P_2)_o = (P_1 P_3)^{0.5} \tag{6.27}$$

Example 6.5

A well fluid with the composition (in mole fraction) $C_1 = 0.4$, $C_2 = 0.2$, $C_3 = 0.1$, $C_4 = 0.1$, $C_5 = 0.1$, $C_6 = 0.05$, and $C_7 = 0.05$ is to undergo a gas–oil separation process using a three-stage GOSP. The last stage is operated at atmospheric pressure. The first stage is fixed at a pressure of 500 psia. Calculate the optimum operating pressure for the second stage by using Equation 6.27.

Solution

Direct substitution of the values of $P_1 = 500$ psia and $P_3 = 14.7$ psia in Equation 6.27 yields $(P_2)_o = 85.5$ psia.

6.7.3.3 Equilibrium Flash Vaporization Calculation

Equations describing flash vaporization, as given next, can be applied to calculate the value of optimum operating pressure for the second-stage separator for a three-stage GOSP. An outline of a trial-and-error calculation procedure is presented in Figure 6.20.

6.8 Selections and Performance of Gas–Oil Separators

For the selection of a particular separator, a preliminary survey of all conditions and factors prevailing on site in an oil field should be considered. The main factors underlying this survey as well as the basic steps included in the selection of a separator are embodied in one integrated scheme outlined in Figure 6.21.

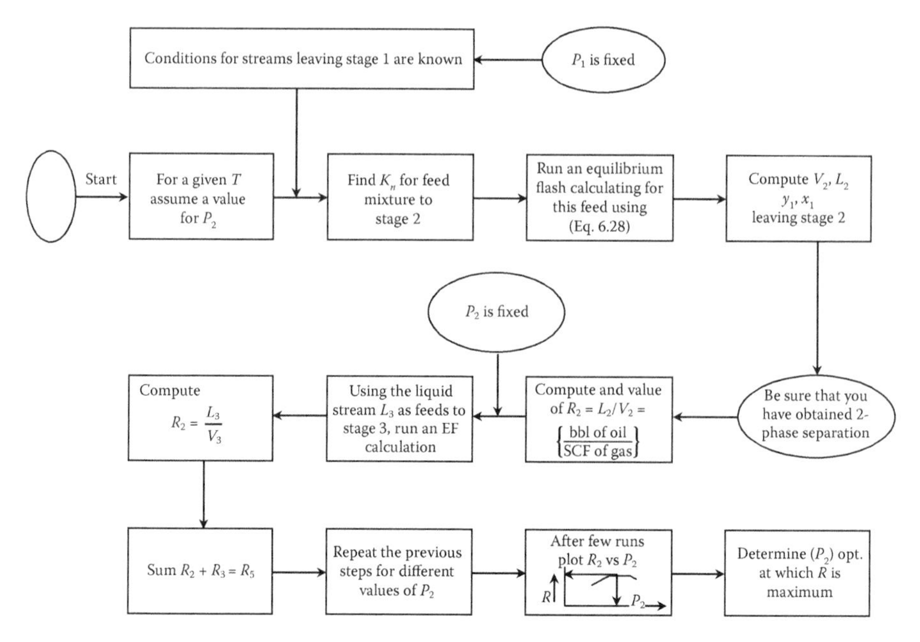

FIGURE 6.20
How to compute $(P_2)_o$ for a GOSP.

The performance of gas–oil separators, on the other hand, is controlled by four factors:

1. Operating temperature—A higher temperature will cause more evaporation of the hydrocarbons, diminishing the recovery of the liquid portion.
2. Operating pressure—A higher pressure will allow more hydrocarbons to condense, increasing liquid recovery. However, after reaching a certain peak, a higher pressure causes liquid to decrease. This is in accordance with the well-known retrograde phenomena.
3. Number of stages—Increasing the number of stages in general will increase the efficiency of separation, resulting in a higher yield of the stable stock tank oil. This is true for a number of stages in the range of two to three. However, if the number increases beyond three, improvement in the recovery diminishes. A four-stage GOSP is not economically attractive, as the recovery of stable stock tank oil increases by 8% by adding one more stage.
4. In addition to the preceding factors, the composition of the well streams has to be considered in evaluating the performance of a gas–oil separator.

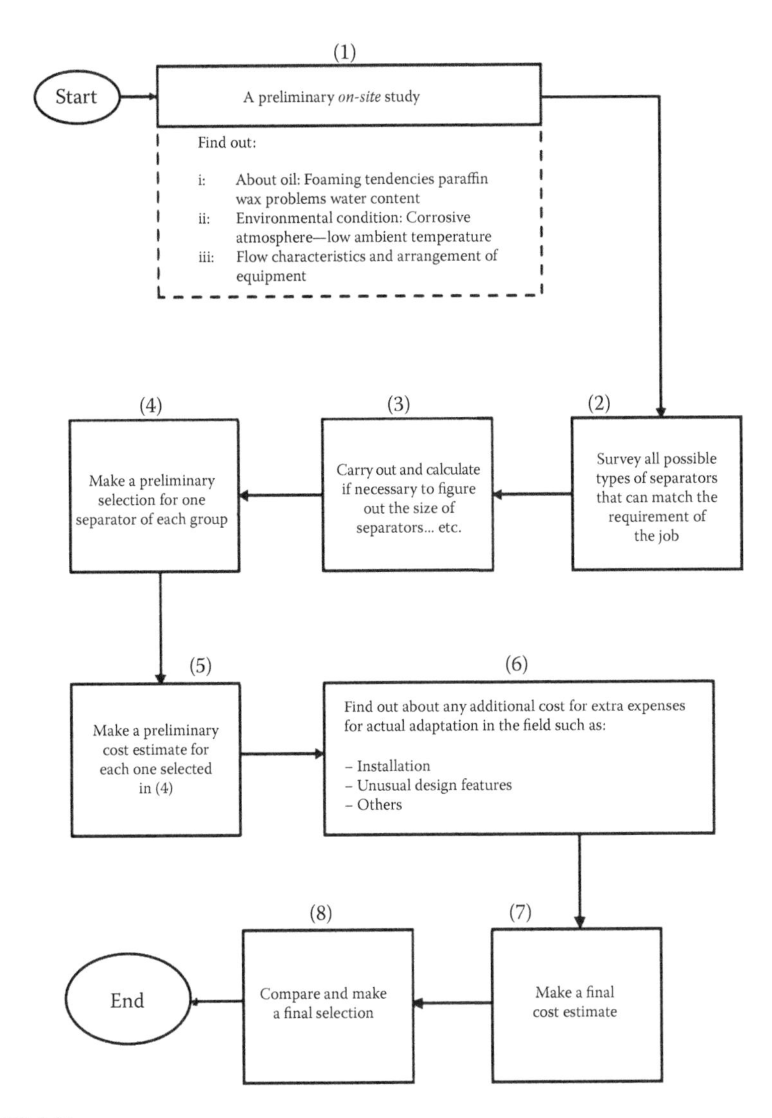

FIGURE 6.21
How to select a gas–oil separator for a given job.

In order to judge the performance of a GOSP, a number of tests are commercially carried out to evaluate the efficiency of operation. Most important are the following:

Evaluation of particle size—The method requires determining the size of liquid particles entrained by the gas stream. The efficiency of the gas–oil separator is thus evaluated based on this size. It is accepted that liquid particles with sizes larger than 10 μm coming in the gas stream is an indication of poor performance.

Determination of the quantity of liquid carryover—The method requires determining the volume of liquid entrained or *carried over* by the gas stream. A limit of 0.10 gal/MCF in the exit gas is usually specified (where M = 1000).

Stain test—It is a rather old test (formerly named the *handkerchief* test). It simply consists of holding and exposing a white cloth in the gas stream leaving the separator. The performance of the separator is considered acceptable if a brown strain does not form on the cloth in 1 min.

6.9 Flash Calculations

6.9.1 Introduction

Under the assumption of equilibrium conditions, and knowing the composition of the fluid stream coming into the separator and the working pressure and temperature conditions, we could apply our current knowledge of vapor/liquid/equilibrium (flash calculations) and calculate the vapor and liquid fractions at each stage.

The problem of separating the gas from crude oil for well fluids (crude oil mixtures) breaks down to the well-known problem of flashing a partially vaporized feed mixture into two streams: vapor and liquid. In the first case, we use a gas–oil separator. In the second case, we use what we call a flashing column, as shown in Figure 6.22.

A *flash* is a single-stage distillation in which a feed is partially vaporized to give a vapor that is richer in the more volatile components. This is the case of a feed heated under pressure and flashed adiabatically across a valve to a lower pressure, the vapor being separated from the liquid residue in a flash drum. This is the case of *light liquids*. Apart from the gas–oil separation problem addressed here, methods used in practice to produce and hence separate two-phase mixtures are as follows:

Initial Phase	Process and Conditions
Higher-pressure liquids (light)	Heat under pressure, then flash adiabatically using valve
Low-pressure liquids (heavy)	Partial vaporization by heating, flash isothermally (no valve)
Gas	Cool after initial compression
Gas	Expand through a valve or engine

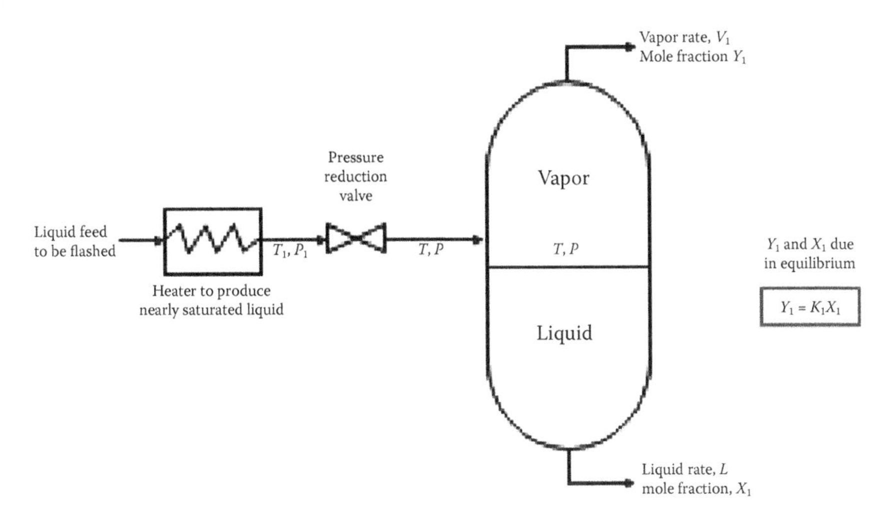

FIGURE 6.22
Schematic presentation of adiabatic flash vaporization.

6.9.2 Conditions Necessary for Flashing

For flashing to take place, the feed has to be two-phase mixture, that is, it satisfies the following: $T_{BP} < T_f < T_{DP}$ (as indicated in Figure 6.23) or the sum of z_iK_i for all components is greater than 1 and the sum of z_i/K_i for all components is less than 1, where T_{BP}, T_f, and T_{DP} are the bubble point of the feed mixture, flash temperature, and dew point of the feed mixture, respectively; Z_i and K_i are the feed composition and equilibrium constant, respectively, for component i.

6.9.3 The Flash Equation

The need to discuss flash calculation arises from the fact that it provides a tool to determine the relative amounts of the separation products V (gas) and L (oil) and their composition Y_i and X_i, respectively.

The flash equation is derived by material balance calculations, using the unit operation's concept, as presented in Figure 6.24.

Two forms are presented:

$$\text{First simple form: } X_i = \frac{Z_i}{1 - \frac{V}{F}(1 - K_i)} \qquad (6.28)$$

For given conditions of P and T, the solution of the equation to find the value of X_i is obtained by trial and error assuming a value for V/F (take F unity), until the sum of $X_i = 1$ is satisfied.

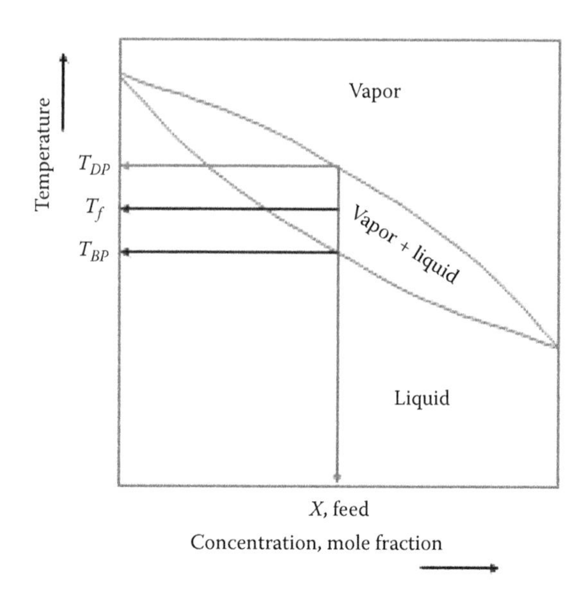

FIGURE 6.23
Conditions for the flashing of a binary system.

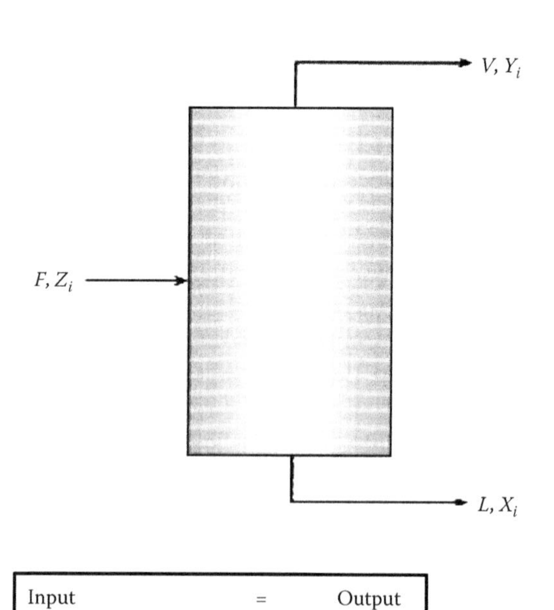

Input	=	Output

Total MB: $F = L + V$

Component MB: $FZ_i = LX_i + VY_i$

FIGURE 6.24
Material balance (MB) for flash column.

$$\text{Second functional form: } f(g) = \sum_{i=1}^{i=c} \left[\frac{Z_i}{1 - g(1 - K_i)} \right] - 1 \tag{6.29}$$

where $g = V/F$. The desired root to this function, $gr = (V/F)r$, is the value that makes the function $f(g)$ goes to zero or the sum of $X_i = 1$. The solution of this flash equation is carried out using computers. Details on using functional technique are fully described in many references.

The vapor–liquid equilibrium constant, defined as

$$K_i = Y_i/X_i$$

where Y_i is the mole fraction of component i in the vapor phase, X_i is the mole fraction of component i in the liquid phase, and K_i is the equilibrium constant of component i for a given T and P is considered the key concept used in the computation of phase behavior of hydrocarbon mixtures in oil and gas streams. K is called the distribution constant because it predicts the distribution of a component in each phase: vapor and liquid.

K is a function of T, P, and the composition of a given system. The K values most widely used are those developed by the National Gas Processors Association (NGPA). It is important that the value of the operating pressure of the system should be below the convergence pressure used in predicting the K values, since the K values are designated for a specific *convergence pressure* (defined as the pressure at a given T, where the values of K for all components in a system become or tend to become equal to unity).

6.9.4 Some Important Applications for the Flash Equation

Other applications for the flash equation are as follows:

- To find the number of stages needed for a given separation—This simply utilizes the stage-to-stage concept in which another stage is added to each trial. Convergence is established when the liquid composition leaving the last stage equals the specified feed composition.
- To determine the optimum pressure for the second stage—This involves a trial-and-error procedure.

REVIEW QUESTIONS AND EXERCISE PROBLEMS

1. What is the purpose of each of the following internal components of a separator?
   a. Inlet diverter
   b. Mist extractor
   c. Vortex breaker
   d. Wave breaker

2. What are the forces acting on a liquid droplet in the gravity settling section of a separator?
3. What is the significance of retention time in separator design?
4. Complete the following statements:
   a. In the gravity settling of a gas–oil separator, oil droplets down to a size of __________ should be removed in order to prevent __________.
   b. In a gas scrubber, removal of 500 μm droplets in the gravity settling section is sufficient because __________.
5. Which type of separator (vertical or horizontal) is more suitable for the following cases (justify your selection)?
   a. Handling solids in the fluid
   b. Handling large volumes of gas
   c. Handling liquid surges
   d. Offshore operations
6. State whether each of the following statements is true (T) or false (F):
   a. Carryover is the escape of free liquid with the gas due to vortexes.
   b. In the separator design unless laboratory data are available, the smallest water droplet size that should be removed from oil is 100 μm.
   c. Blowby is the escape of free gas with the liquid that may result from excessive foaming.
   d. In long horizontal separators, waves may develop at the interface. This results in level control problems, which could be eliminated by the installation of coalescing plates.
7. Circle the correct answers (note that more than one answer could be correct):
   a. Foamy crude oil may result in:
      i. Improper gas–oil interface level control
      ii. Carryover
      iii. Better separation in the gravity settling section
      iv. Entrapment of foam in the oil
   b. Accumulation of produced sand in the separator may:
      i. Result in incomplete separation of oil and gas
      ii. Be better handled in horizontal separators
      iii. Reduce the gas capacity of the separator
      iv. Reduce the liquid (oil) capacity of the separator

8. Derive the equations for gas capacity and oil capacity constraints for a horizontal separator for the following two cases:
   a. The gas occupies 75% of the separator effective volume.
   b. The gas occupies 25% of the separator effective volume.
9. The following data are given for a small oil field:

Oil rate:	8000 bbl/day
Gas–oil ratio:	1200 SCF/bbl
Operating pressure:	500 psia
Operating temperature:	80°F
Oil gravity:	35°API
Gas specific gravity:	0.6
Oil viscosity:	10 cP
Oil retention time:	180 s

   a. Design a horizontal separator to handle the field production.
   b. If the field is located offshore, design the appropriate separator to handle the field production.
   c. Determine the actual capacity of a horizontal two-phase separator having a 60 in diameter and 15 ft seam-to-seam length for the preceding operating conditions.
10. Determine the actual gas and oil capacities for vertical and horizontal separators having a 48 in. diameter and 16 ft seam-to-seam length, operating at 585 psi and 80°F, given the following data:

Oil specific gravity:	0.875
Gas specific gravity:	0.6
Oil viscosity:	5.0 cP
Operating pressure:	700 psia
Operating temperature:	80°F
Oil retention time:	120 s
C_d:	0.86
Z:	0.89

11. Design a vertical and horizontal gas scrubber for the following conditions:

Gas rate:	80 MMSCFD
Gas specific gravity:	0.6
Operating pressure:	1200 psia
Operating temperature:	80°F
Oil rate:	Negligible
Oil gravity:	40°API

7

Three-Phase Oil–Water–Gas Separators

The materials presented in Chapter 6 for two-phase separators apply, in general, to the separation of any gas–liquid system such as gas–oil, gas–water, and gas–condensate systems. However, in almost all production operations the produced fluid stream consists of three phases: oil, water, and gas.

The term *separator* in oil field terminology designates a pressure vessel used for separating well fluids produced from oil and gas wells into gaseous and liquid components. Separators work on the principle that the three components have different densities, which allows them to stratify when moving slowly with gas on top, water on the bottom, and oil in the middle. Any solids such as sand will also settle in the bottom of the separator.

In this chapter, the two types of three-phase separators, horizontal and vertical, are described and the basic design equations are developed for each.

7.1 Introduction

To meet process requirements, the oil–gas separators are normally designed in stages, in which the first stage separator is used for preliminary phase separation, while the second and third stage separator are applied for further treatment of each individual phase (gas, oil, and water).

Generally, water produced with the oil exists partly as free water and partly as water-in-oil emulsion. In some cases, however, when the water–oil ratio is very high, oil-in-water rather than water-in-oil emulsion will form. Free water produced with the oil is defined as the water that will settle and separate from the oil by gravity. To separate the emulsified water, however, heat treatment, chemical treatment, electrostatic treatment, or a combination of these treatments would be necessary in addition to gravity settling.

Along with the water and oil, gas will always be present and, therefore, must be separated from the liquid. The volume of gas depends largely on the producing and separation conditions. When the volume of gas is relatively small compared to the volume of liquid, the method used to separate free water, oil, and gas is called a *free-water knockout*. In such a case, the separation of the water from oil will govern the design of the vessel. When there is a large volume of gas to be separated from the liquid (oil and water), the vessel is called a *three-phase separator* and either the gas capacity requirements or the water–oil separation constraints may govern the vessel design. Free-water knockout and three-phase separators are basically similar in shape and components. Further, the same design concepts and procedures are used for both types of vessel. Therefore, the term *three-phase separator* will be used for both types of vessel throughout this chapter.

Three-phase separators may be either horizontal or vertical pressure vessels similar to the two-phase separators described in Chapter 6. However, three-phase separators will have additional control devices and may have additional internal components.

7.2 Three-Phase Horizontal Separators

Three-phase separators differ from two-phase separators in that the liquid collection section of the three-phase separator handles two immiscible liquids (oil and water) rather than one. This section should, therefore, be designed to separate the two liquids to provide means for controlling the level of each liquid and to provide separate outlets for each liquid.

Figures 7.1 and 7.2 show schematics of two common types of horizontal three-phase separators. The difference between the two types is mainly in the method of controlling the levels of the oil and water phases. For the separator in Figure 7.1, an interface controller and a weir provide the control. The design of the second type (Figure 7.2), normally known as the *bucket and weir design*, eliminates the need for an interface controller.

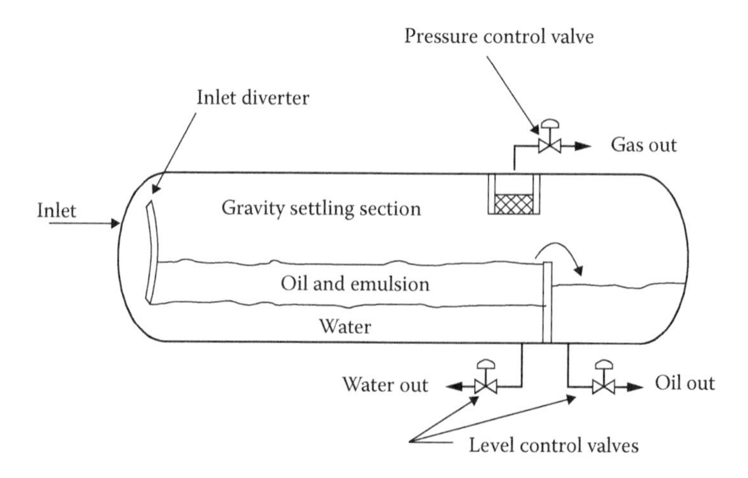

FIGURE 7.1
Three-phase separator horizontal (with an interface controller and weir).

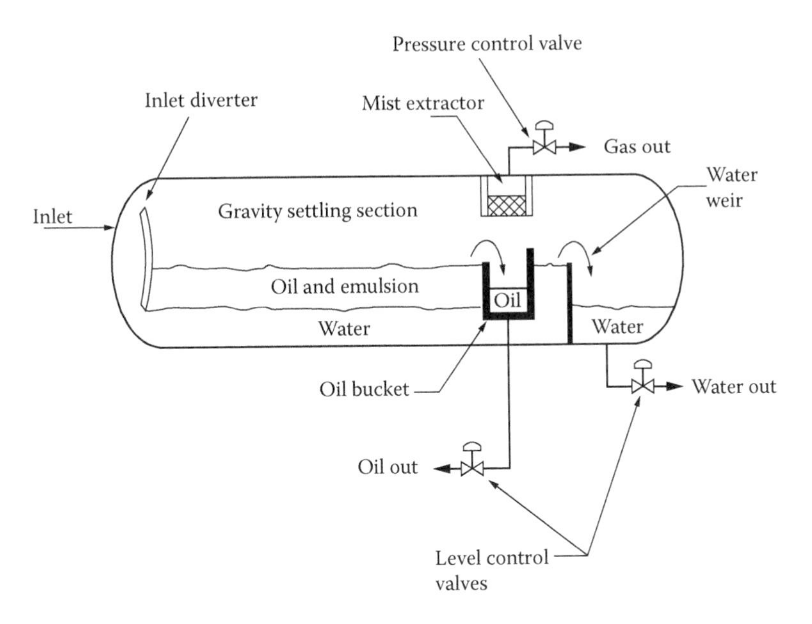

FIGURE 7.2
Three-phase horizontal separator (with bucket and weir design).

The operation of the separator is, in general, similar to that of the two-phase separator. The produced fluid stream coming either directly from the producing wells or from a free-water knockout vessel enters the separator and hits the inlet diverter, where the initial bulk separation of the gas and liquid takes place due to the change in momentum and difference in fluid densities. The gas flows horizontally through the gravity settling section (the top part of the separator) where the entrained liquid droplets, down to a certain minimum size (normally 100 μm), are separated by gravity. The gas then flows through the mist extractor, where smaller entrained liquid droplets are separated, and out of the separator through the pressure control valve, which controls the operating pressure of the separator and maintains it at a constant value. The bulk of liquid, separated at the inlet diverter, flows downward, normally through a downcomer that directs the flow below the oil–water interface. The flow of the liquid through the water layer, called *water washing*, helps in the coalescence and separation of the water droplets suspended in the continuous oil phase. The liquid collection section should have sufficient volume to allow enough time for the separation of the oil and emulsion from the water. The oil and emulsion layer forming on top of the water is called the *oil pad*. The weir controls the level of the oil pad and an interface controller controls the level of the water and operates the water outlet valve. The oil and emulsion flow over the weir and collect in a separate compartment, where its level is controlled by a level controller that operates the oil outlet valve.

The relative volumes occupied by the gas and liquid within the separator depend on the relative volumes of gas and liquid produced. It is a common practice, however, to assume that each of the two phases occupies 50% of the separator volume. In such cases, however, where the produced volume of one phase is much smaller or much larger than the other phase, the volume of the separator should be split accordingly between the phases. For example, if the gas–liquid ratio is relatively low, we may design the separator such that the liquid occupies 75% of the separator volume and the gas occupies the remaining 25% of the volume.

The operation of the other type of horizontal separator (Figure 7.2) differs only in the method of controlling the levels of the fluids. The oil and emulsion flow over the oil weir into the oil bucket, where its level is controlled by a simple level controller that operates the oil outlet valve. The water flows through the space below the oil bucket, then over the water weir into the water collection section, where its level is controlled by a level controller that operates the water outlet valve. The level of the liquid in the separator, normally at the center, is controlled by the height of the oil weir. The thickness of the oil pad must be sufficient to provide adequate oil retention time. This is controlled by the height of the water weir relative to that of the oil weir. A simple pressure balance at the bottom of the separator between the water side and the water and oil side can be used to approximately determine the thickness of the oil pad as follows:

$$H_o = \frac{H_{ow} - H_{ww}}{1 - (\rho_o/\rho_w)} \tag{7.1}$$

where H_o is the thickness of the oil pad, H_{ow} is the height of the oil weir, H_{ww} is the height of the water weir, and ρ_o and ρ_w are the oil and water densities, respectively.

Equation 7.1 gives only an approximate value for the thickness of the oil pad. A more accurate value could be obtained if the density of the oil in Equation 7.1 is replaced by the average value of the density of oil and density of emulsion, which depends on the thickness of the oil and emulsion layers within the oil pad. The height of the water weir should not be so small as to avoid the downward growth of the oil pad and the possibility of the oil flowing below the oil bucket, over the water weir, and out with the water. It is advisable to have the oil bucket as deep as possible and to have either the oil weir, or the water weir, or both to be adjustable to accommodate any unexpected changes in flow rates and/or liquids properties. Such problems are easily accommodated in the interface controller and weir design of Figure 7.1, as the interface controller could be easily adjusted. In some cases, however, when the difference in density between the water and oil, or the water and emulsion are small (e.g., in heavy oil operations), the operation of the interface controller becomes unreliable and the bucket design (Figure 7.2) will be preferred.

7.3 Three-Phase Vertical Separators

As discussed in Chapter 6, the horizontal separators are normally preferred over vertical separators due to the flow geometry that promotes phase separation. However, in certain applications, the engineer may be forced to select a vertical separator instead of a horizontal separator despite the process-related advantages of the later. An example of such applications is found in offshore operations, where the space limitations on the production platform may necessitate the use of a vertical separator.

Figure 7.3 shows a schematic of a typical three-phase vertical separator. The produced fluid stream enters the separator from the side and hits the inlet diverter, where the bulk separation of the gas from the liquid takes place. The gas flows upward through the gravity settling sections, which are designed to allow separation of liquid droplets down to a certain minimum size (normally 100 μm) from the gas. The gas then flows through the mist extractor, where the smaller liquid droplets are removed. The gas leaves the separator at the top through a pressure control valve that controls the separator pressure and maintains it at a constant value.

The liquid flows downward through a downcomer and a flow spreader that is located at the oil–water interface. As the liquid comes out of the spreader, the oil rises to the oil pad and the water droplets entrapped in the oil settle down and flow, countercurrent to the rising oil phase, to collect in the water collection section at the bottom of the separator. The oil flows over a weir into an oil chamber and out of the separator through

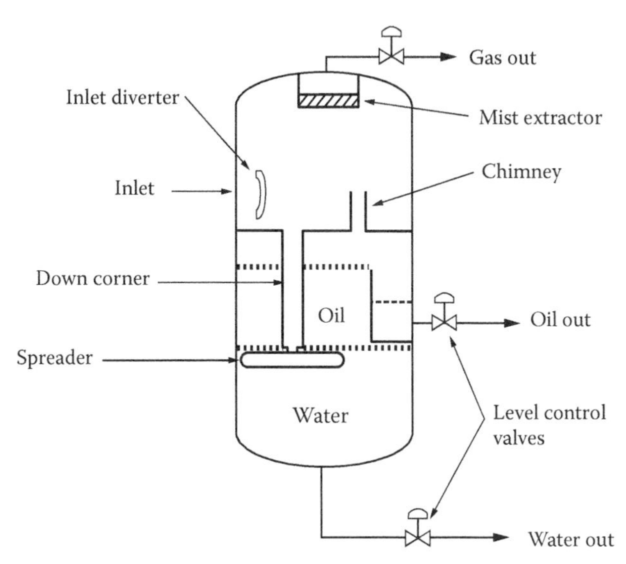

FIGURE 7.3
Schematic of three-phase vertical separator.

the oil outlet valve. A level controller controls the oil level in the chamber and operates the oil outlet valve. Similarly, the water out of the spreader flows downward into the water collection section, whereas the oil droplets entrapped in the water rise, countercurrent to the water flow, into the oil pad. An interface controller that operates the water outlet valve controls the water level. In the design shown in Figure 7.3, a chimney must be provided, as shown in the figure, to allow the gas liberated from the oil to rise and join the rest of the separated gas and, thus, avoid overpressurizing the liquid section of the separator. The use of the oil weir and chamber in this design provides good separation of water from oil, as the oil has to rise to the full height of the weir before leaving the separator. The oil chamber, however, presents some problems. First, it takes up space and reduces the separator volume needed for the retention times of oil and water. It also provides a place for sediments and solids to collect, which creates cleaning problems and may hinder the flow of oil out of the vessel. In addition, it adds to the cost of the separator.

Liquid–liquid interface controllers will function effectively as long as there is an appreciable difference between the densities of the two liquids. In most three-phase separator applications, water–oil emulsion forms and a water–emulsion interface will be present in the separator instead of a water–oil interface. The density of the emulsion is higher than that of the oil and may be too close to that of the water. Therefore, the smaller density difference at the water–emulsion interface will adversely affect the operation of the interface controller. The presence of emulsion in the separator takes up space that otherwise would be available for the oil and the water. This reduces the retention time of the oil and water and thus results in a less efficient oil–water separation. In most operations where the presence of emulsion is problematic, chemicals known as de-emulsifying agents are injected into the fluid stream to mix with the liquid phase. These chemicals help in breaking the emulsion, as will be described next in Chapter 8. Another method that is also used for the same purpose is the addition of heat to the liquid within the separator. In both cases, however, the economics of the operations have to be weighted against the technical constraints.

7.4 Separation Theory

The basic separation concepts and settling equations developed for two-phase separators in Chapter 6 are, in general, valid for three-phase separators. In particular, the equations developed for separation of liquid droplets from the gas phase, which determined the gas capacity constraint, are exactly the same for three-phase separators.

Treatment of the liquid phase for three-phase separators is, however, different from that used for two-phase separators. The liquid retention time constraint was the only criterion used for determining the liquid capacity of two-phase separators. For three-phase separators, however, the settling and separation of the oil droplets from water and of the water droplets from oil must be considered in addition to the retention time constraint. Further, the retention time for both water and oil, which might be different, must also be considered.

In separating oil droplets from water, or water droplets from oil, a relative motion exists between the droplet and the surrounding continuous phase. An oil droplet, being smaller in density than the water, tends to move vertically upward under the gravitational or buoyant force, F_g. The continuous phase (water), on the other hand, exerts a drag force, F_d, on the oil droplet in the opposite direction. The oil droplet will accelerate until the fractional resistance of the fluid drag force, F_d, approaches and balances F_g; thereafter, the oil droplet continues to rise at a constant velocity known as the *settling* or *terminal velocity*. Similarly, a water droplet, being higher in density than the oil, tends to move vertically downward under the gravitational or buoyant force, F_g. The continuous phase (oil), on the other hand, exerts a drag force, F_d, on the water droplet in the opposite direction. The water droplet will accelerate until the frictional resistance of the fluid drag force, F_d, approaches and balances F_g; thereafter, the water droplet continues to rise at a constant velocity known as the *settling* or *terminal velocity*. Upward settling of oil droplets in water and downward settling of water droplets in oil follow Stokes' law, and the terminal settling velocity can be obtained as follows. The drag force, F_d, is proportional to the droplet surface area perpendicular to the direction of flow, and its kinetic energy per unit volume. Hence,

$$F_d = C_d \frac{\pi}{4} d^2 \frac{\rho_c u^2}{2g} \tag{7.2}$$

whereas F_g is given by

$$F_g = \frac{\pi}{6} d^3 (\Delta\rho) \tag{7.3}$$

where d is the diameter of the droplet (ft), u is the settling velocity of the droplet (ft/s), ρ_c is the density of the continuous phase (lb/ft^3), g is gravitational acceleration (ft/s), and C_d is the drag coefficient. For a low Reynolds number, Re, flow, the drag coefficient is given by

$$C_d = \frac{24}{\text{Re}} = \frac{24\mu' g}{\rho d u} \tag{7.4}$$

where μ' is the viscosity of the continuous phase (lb-s/ft^2).

Substituting for C_d from Equation 7.4 into Equation 7.2 yields

$$F_d = 3\pi\mu' d u \tag{7.5}$$

The settling terminal velocity, u, is reached when $F_d = F_g$. Therefore, equating Equations 7.3 and 7.5 and solving for u, the droplet settling velocity, we obtain

$$u = \frac{(\Delta\rho) d^2}{18\mu'}$$

The typical units used for droplet diameter and viscosity are the micrometers and centipoise, respectively. Letting μ be the viscosity in centipoise and d_m be the droplet diameter in micrometers, the preceding equation becomes

$$u = \frac{(\Delta\rho)(3.281 \times 10^{-6} d_m)^2}{18(2.088 \times 10^{-5} \mu)}$$

$$u = 2.864 \times 10^{-8} \frac{(\Delta\rho) d_m^2}{\mu} \frac{\text{ft}}{\text{s}} \tag{7.6}$$

or

$$u = 1.787 \times 10^{-6} \frac{(\Delta\gamma) d_m^2}{\mu} \frac{\text{ft}}{\text{s}} \tag{7.7}$$

where $\Delta\gamma = \gamma_w - \gamma_o$. γ_o and γ_w are the specific gravity of oil and water, respectively, u is the terminal settling velocity of the droplet (ft/s), d_m is the diameter of the droplet (μm), and μ is the viscosity of the continuous phase (cP).

Equation 7.6 (or Equation 7.7) shows that the droplet settling velocity is inversely proportional to the viscosity of the continuous phase. Oil viscosity is several magnitudes higher than the water viscosity. Therefore, the settling velocity of water droplets in oil is much smaller than the settling velocity of oil droplets in water. The time needed for a droplet to settle out of one continuous phase and reach the interface between the two phases depends on the settling velocity and the distance traveled by the droplet. In operations where the thickness of the oil pad is larger than the thickness of the water layer, water droplets would travel a longer distance to reach the water–oil interface than that traveled by the oil droplets. This combined with the much slower settling velocity of the water droplets, makes the time needed for separation of water from oil longer than the time needed for separation of oil from water. Even in operations with a very high water–oil ratio, which might result in having a water layer that is thicker than the oil pad, the ratio of the thickness of the water layer to that of the oil pad would not offset the effect of viscosity. Therefore, the separation of water droplets from the continuous oil phase would always be taken as the design criterion for three-phase separators.

The minimum size of the water droplet that must be removed from the oil and the minimum size of the oil droplet that must be removed from the water to achieve a certain oil and water quality at the separator exit depend largely on the operating conditions and fluid properties. Results obtained from laboratory tests conducted under simulated field conditions provide the best data for design. The next best source of data could be obtained from nearby fields. If such data are not available, the minimum water droplet size to be removed from the oil is taken as 500 μm. Separators design with this criterion have produced oil and emulsion containing between 5% and 10% water. Such produced oil and emulsion could be treated easily in the oil dehydration facility. Experience has also shown that three-phase separators designed based on the 500 μm water droplet removal produces water with a suspended oil content that is below 2000 mg/L. This produced water must be treated before it is disposed of, as will be described in Chapter 15.

Another important aspect of separator design is the retention time, which determines the required liquid volumes within the separator. The oil phase needs to be retained within the separator for a period of time that is sufficient for the oil to reach equilibrium and liberates the dissolved gas. The retention time should also be sufficient for appreciable coalescence of the water droplets suspended in the oil to promote effective settling and separation. Similarly, the water phase needs to be retained within the separator for a period of time that is sufficient for coalescence of the suspended oil droplets. The retention times for oil and water are best determined from laboratory tests; they usually range from 3 to 30 min, based on operating conditions and fluid properties. If such laboratory data are not available, it is a common practice to use a retention time of 10 min for both oil and water.

7.5 Separator Sizing Equations and Rules

In this section, the equations and rules used for determining the dimensions of horizontal and vertical three-phase separators are developed and presented. It should be realized that these equations are generally used for preliminary sizing of the separators. Other important aspects of the design should not be ignored. The changes in operating conditions, such as production rates, gas–liquid ratio, water–oil ratio, fluid properties, pressure, and temperature, over the life of the field should be incorporated in the design. For new field development, there is always some degree of uncertainty in the available data and information. This should be an integral part of the facility design. Also, cost, availability, and space limitation could affect the design and selection of equipment.

The sizing procedure is generally similar to that for two-phase separators with the exception that the separation of water from oil, and oil from water are additional constraints for three-phase separators.

7.5.1 Sizing Equations for Horizontal Separators

As with two-phase separators, consideration of the gas capacity constraint and the liquid retention time constraint results in developing two equations; each relates the vessel diameter to its length. Analysis of the two equations determines the equation that governs the design and that should be used to determine possible combinations of diameters and lengths. For three-phase horizontal separators, consideration of the settling of water droplets in oil results in a third equation that determines the maximum diameter of the separator. Therefore, in determining the vessel's diameter–length combinations, the diameters selected must be equal to or less than the determined maximum diameter.

7.5.1.1 Water Droplet Settling Constraint

In comparison to two-phase separators, the additional constraint in the design of three-phase horizontal separators is that the oil retention time should be sufficient for the water droplets of certain minimum size to settle out of the oil. To be on the conservative side, we shall assume that the water droplets to be separated are at the top of the oil pad. Therefore, such droplets have to travel a distance equal to the thickness of the oil pad before they reach the water–oil interface. This constraint can be translated into a useful relationship by equating the time needed for the water droplets to travel through the oil pad thickness to the oil retention time.

The time needed for the water droplets to travel through the oil pad, t_{wd} min, is obtained by dividing the oil pad thickness, H_o (in), by the water settling velocity given by Equation 7.7; therefore,

$$t_{wd} = \left(\frac{1}{60}\right)\frac{(H_o/12)}{1.787\times10^{-6}(\Delta\gamma)d_m^2/\mu_o} \text{ min} \tag{7.8}$$

Equating Equation 7.8 to the oil retention time, t_o, and solving for H_o, we obtain the maximum allowable oil pad thickness, $H_{o,\max}$ expressed as follows:

$$H_{o,\max} = \frac{1.28 \times 10^{-3} t_o (\Delta\gamma) d_m^2}{\mu_o} \text{ in} \tag{7.9}$$

The minimum water droplet diameter to be removed, d_m, is determined, as discussed in Section 6.1, from laboratory tests. In case such data are not available, d_m may be assigned the value of 500 μm.

The oil and water flow rates and retention times and the vessel diameter control the height of the oil pad. Considering a separator that is half full of liquid, the following geometrical relation is easily derived:

$$\frac{A}{A_w} = \left(\frac{1}{\pi}\right)\left[\cos^{-1}\left(\frac{2H_o}{D}\right) - \left(\frac{2H_o}{D}\right)\left(1 - \frac{4H_o^2}{D^2}\right)^{-0.5}\right] \tag{7.10}$$

where A_w and A are the cross-sectional area of the separator occupied by water and the total cross-sectional area of the separator, respectively, and D is the diameter of the vessel. For given oil and water flow rates and retention times, the ratio A_w/A can be determined as follows. For a separator that is half full of liquid, the total cross-sectional area of the separator, A, is equal to twice the area occupied by the liquid, which is equal to the area occupied by water, A_w, and the area occupied by oil, A_o; therefore, $A = 2(A_o + A_w)$.

It follows that

$$\frac{A_w}{A} = 0.5 \frac{A_w}{A_o + A_w}$$

Since the volume occupied by each phase is the product of the cross-sectional area and the effective length, the cross-sectional area is directly proportional to the volume. Further, the volume occupied by any phase is also determined as the product of the flow rate and retention time. Therefore,

$$\frac{A_w}{A} = 0.5 \frac{Q_w t_w}{Q_o t_o + Q_w t_w} \tag{7.11}$$

Therefore, once the ratio A_w/A is determined from Equation 7.11, Equation 7.10 can be solved to determine the ratio H_o/D. This is then used with the value of $H_{o,\max}$ determined from Equation 7.9 to determine the maximum

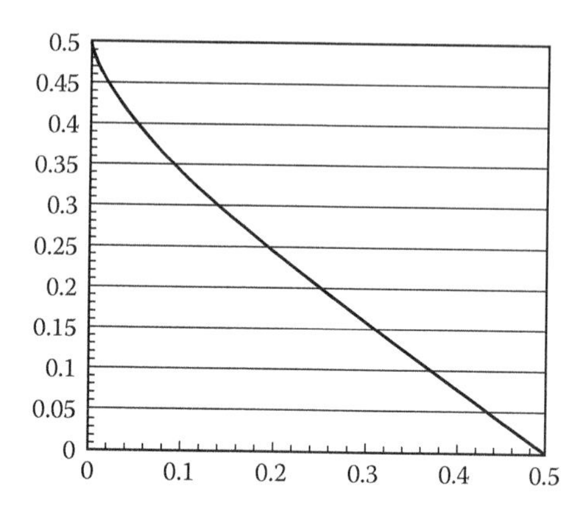

FIGURE 7.4
H_o/D as a function of A_w/A.

vessel diameter associated with the maximum oil pad height according to Equation 7.12:

$$D_{\max} = \frac{H_{o,\max}}{H_o/D} \tag{7.12}$$

This, therefore, sets the upper limit for the separator diameter. To obtain the value of H_o/D from Equation 7.9 it is convenient to use the graphical solution of Equation 7.10 (given in Figure 7.4).

Equations similar to Equations 7.10 and 7.11 could be derived for other cases where the liquid may occupy more or less than half the volume of the separator.

7.5.1.2 Gas Capacity Constraint

The gas capacity constraint equation developed for two-phase horizontal separators in Chapter 3 is also valid for three-phase horizontal separators. The equation provides a relationship between the separator diameter and effective length as follows:

$$LD = 422\left(\frac{Q_g TZ}{P}\right)\left[\left(\frac{\rho_g}{\rho_o - \rho_g}\right)\left(\frac{C_d}{d_m}\right)\right]^{1/2} \tag{7.13}$$

where D is the separator internal diameter (in), L is the effective length of the separator (ft), T is the operating temperature (°R), Z is the gas compressibility at operating pressure and temperature, P is the operating pressure, ρ_g is the gas density (lb/ft^3), ρ_o is the oil density (lb/ft^3), C_d is the drag coefficient, and d_m is the minimum oil droplet size to be separated from gas (μm). As discussed in Chapter 3, d_m is normally taken as 100 μm and C_d is determined by the iterative procedure described there.

Using diameters smaller than the maximum diameter determined from the aforementioned water droplet settling constraint, Equation 7.13 is used to determine possible diameter and length combinations that satisfy the gas capacity constraint.

7.5.1.3 Retention Time Constraint

The separator size must provide sufficient space for the oil and water such that each phase is retained within the separator for the desired retention time. The assumption that the liquid will occupy half of the separator volume, which was used for two-phase separators, is also used here. However, in the present case, both oil and water occupy that volume. Therefore, the volume occupied by the liquid phase (both oil and water), V_l, in a separator having a diameter, D (in), and effective length, L (ft), is given by

$$V_l = 0.5\left(\frac{\pi}{4}\right)\left(\frac{D}{12}\right)^2 L \text{ ft}^3$$

Because 1 barrel (bbl) = 5.61 ft^3,

$$V_l = 4.859 \times 10^{-4} D^2 L \text{ bbl} \tag{7.14}$$

The volume of separator occupied by oil, V_o, is the product of the oil flow rate, Q_o, and the oil retention time, t_o. If Q_o is in barrels per day (BPD) and t_o is in minutes, then

$$V_o = \frac{t_o Q_o}{24 \times 60} \text{ bbl} \tag{7.15}$$

Similarly, the volume of separator occupied by water, V_w, is the product of the water flow rate, Q_w, and water retention time, t_w:

$$V_w = \frac{t_w Q_w}{24 \times 60} \text{ bbl} \tag{7.16}$$

Because $V_l = V_o + V_w$,

$$4.859 \times 10^{-4} D^2 L = \frac{Q_o t_o + Q_w t_w}{24 \times 60}$$

Therefore,

$$D^2 L = 1.429(Q_o t_o + Q_w t_w) \text{ in}^2 \text{ ft} \tag{7.17}$$

Again, using diameters smaller than the maximum diameter determined from the water droplet settling constraint, Equation 7.12 is used to determine possible diameter and length combinations that satisfy the retention time constraint.

The procedure for determining the diameter and length of a three-phase horizontal separator can, therefore, be summarized in the following steps:

1. Determine the value of Aw/A from Equation 7.11.
2. Use Figure 7.4 to determine the value of H_o/D for the calculated value of A_w/A.
3. Determine the maximum oil pad thickness, $H_{o,max}$ from Equation 7.9 with d_m equal to 500 μm.
4. Determine D_{max} from Equation 7.12.
5. For diameters smaller than D_{max}, determine the combinations of D and L that satisfy the gas capacity constraint from Equation 7.13, substituting 100 μm for d_m.
6. For diameters smaller than D_{max}, determine the combinations of D and L that satisfy the retention time constraint from Equation 7.17.
7. Compare the results obtained in steps 5 and 6 and determine whether the gas capacity or retention time (liquid capacity) governs the separator design.
8. If the gas capacity governs the design, determine the seam-to-seam length of the separator, L_s, from

$$L_s = L + \frac{D}{12} \tag{7.18}$$

   If the liquid retention time (liquid capacity) governs the design, determine L_s from

$$L_s = 4\frac{L}{3} \tag{7.19}$$

9. Recommend a reasonable diameter and length with a slenderness ratio in the range of 3–5. In making the final selection, considerations such as cost and availability will be important. It should be mentioned that, in some cases, the slenderness ratio might be different from the range of 3–5. In such cases, especially when the slenderness ratio is larger than 5, internal baffles should be installed to act as wave breakers in order to stabilize the gas–liquid interface.

Example 7.1

Determine the diameter and seam-to-seam length of a three-phase horizontal separator for the following operating conditions:

Oil production rate:	8000 BPD
Water production rate:	3000 BPD
Gas–oil ratio:	1000 SCF/bbl
Oil viscosity:	20 cP
Oil specific gravity:	0.89
Water specific gravity:	1.04
Gas specific gravity:	0.65
Gas compressibility:	0.89
Operating pressure:	250 psia
Operating temperature:	95°F
Oil retention time:	15 min
Water retention time:	10 min

Solution

Using Equation 7.9, determine $H_{o,\max}$:

$$H_{o,\max} = \frac{1.28\times10^{-3}\, t_o(\gamma_o - \gamma_w)d_m^2}{\mu_o}\ \text{in}$$

$$H_{o,\max} = \frac{1.28\times10^{-3}(15)(1.04-0.89)(500)^2}{20} = 36\ \text{in}$$

Use Equation 7.11 to determine the ratio A_w/A:

$$\frac{A_w}{A} = \frac{0.5Q_w t_w}{Q_o t_o + Q_w t_w}$$

$$\frac{A_w}{A} = \frac{0.5(3000\times10)}{(3000\times10)+(8000\times15)} = 0.1$$

From Figure 7.4, determine the ratio H_o/D for $A_w/A = 0.1$:

$$\frac{H_o}{D} = 0.338$$

Therefore,

$$D_{\max} = \frac{H_{o,\max}}{H_o D} = \frac{36}{0.338} = 106.5\ \text{in}$$

This is the maximum allowable vessel diameter.

The gas capacity constraint, Equation 7.13, yields

$$DL = 420\left(\frac{TZQ_g}{P}\right)\left(\frac{\rho_g C_d}{d_m(\rho_o - \rho_g)}\right)^{0.5}$$

$$DL = 420\left(555 \times 0.89 \times \frac{8}{250}\right)\left(\frac{\rho_g C_d}{100(\rho_l - \rho_g)^1}\right)^{0.5}$$

Determine the gas and oil densities and substitute $C_d = 0.65$ in the preceding equation:

$$\rho_g = \frac{2.7\gamma_g p}{TZ} = \frac{2.7 \times 0.65 \times 250}{555 \times 0.89} = 0.888\ \text{lb/ft}^3$$

$$\rho_l = \rho_w \gamma_1 = 62.4 \times 0.89 = 55.54\ \text{lb/ft}^3$$

Therefore, the gas capacity constraint is expressed by

$$DL = 68.22 \qquad (7.20)$$

Use Equation 7.17 to check the liquid capacity (retention time) constraint:

$$D^2L = 1.429(3000 \times 10 + 8000 \times 15) = 214{,}350 \qquad (7.21)$$

Select diameters smaller than the determined maximum diameter and find the corresponding effective length from Equations 7.20 and 7.21 for the gas capacity and liquid capacity constraints, respectively. Investigation of Equation 7.20, however, shows that for any selected diameter, the effective length is too small compared to that calculated from Equation 7.21. Therefore, the gas capacity does not govern the design. For the liquid capacity constraints, the results are tabulated as follows:

D (in)	L (ft) (Equation 7.21)	$L_s(= 4L/3)$ (ft)	$L/(d/12)$
66	49.21	65.61	11.93
72	41.35	55.13	9.19
78	35.23	46.98	7.23
84	30.38	40.50	5.786
90	26.46	35.28	4.71
96	23.26	31.01	3.88
102	20.65	27.47	3.23

The last three diameters and length combinations in the above table will be suitable selections, because the most common slenderness ratio is between 3 and 5. Therefore, the recommended separator size can be either 90 in by 36 ft, or 96 in by 31 ft, or 102 in by 28 ft based on cost and availability. Normally, the smaller diameter and longer separator is less expensive than the larger diameter and shorter separator.

The selected separator will be able to handle a much higher gas flow rate. the actual separator gas capacity can be calculated from Equation 7.8 by substituting the values of d and L and calculating the value of Q_g. For a 96 in by 31 ft separator ($L = 3L_s/4 = 23.26$), the gas capacity is 263 MMSCFD. This is much larger than the production rate of 8 MMSCFD. This indicates that designing the separator on the basis of being half full of liquid is not efficient. The size of the separator could be made smaller by allowing the liquid to occupy more than half the volume of the separator.

7.5.2 Sizing Equation for Vertical Separators

Sizing of a vertical three-phase separator is done in a similar manner to that used in sizing vertical two-phase separators (see Chapter 6); that is, the gas capacity constraint is used to determine the minimum diameter of the vessel and the liquid retention time constraint is used to determine the height of the vessel. For three-phase separators, however, a third constraint is added. This is the requirement to settle water droplets of a certain minimum size out of the oil pad. This results, as shown in the following section, in a second value for the minimum diameter of the separator. Therefore, in selecting the diameter of the vessel, the larger of the minimum diameters determined from the gas capacity constraint and water settling constraint should be considered as the minimum acceptable vessel diameter.

7.5.2.1 Water Droplets Settling Constraint

The condition for the settling and separation of water droplets from the oil is established by equating the average upward velocity of the oil phase, u_o, to the downward settling velocity of the water droplets of a given size, u_w. The average velocity of the oil is obtained by dividing the oil flow rate by the cross-sectional area of flow. If Q_o is the oil rate (in BPD), then

$$u_o = \frac{Q_o \times 5.61}{24 \times 3600}\frac{\text{ft}^3}{\text{s}}\left(\frac{4 \times 144}{\pi D^2}\right)\text{ft}^2$$

$$u_o = 0.0119\frac{Q_o}{D^2}\frac{\text{ft}}{\text{s}} \tag{7.22}$$

Equation 7.7 gives the water droplet settling velocity:

$$u_w = 1.787 \times 10^{-6} \frac{(\Delta\gamma)d_m^2}{\mu_o} \frac{\text{ft}}{\text{s}} \tag{7.23}$$

For water droplets to settle out of the oil, u_w must be larger than u_o. Equating u_w to u_o would result, therefore, in determining the minimum diameter of the separator, D_{min}, that satisfies the water settling constraint. From Equations 7.22 and 7.23, it follows that

$$D_{\min}^2 = 6686 \frac{Q_o \mu_o}{(\Delta\gamma)d_m^2} \text{in}^2 \tag{7.24}$$

where D is the separator internal diameter (in), Q_o is the oil flow rate (BPD), μ_o is the oil viscosity (cP), $\Delta\gamma$ is the difference in specific gravity of oil and water, and d_m is the minimum water droplet size to be separated from gas (μm).

Any diameter larger than the minimum diameter determined from Equation 7.24 yields a lower average oil velocity and, thus, ensures water separation.

7.5.2.2 Gas Capacity Constraint

As described in Chapter 6, the gas capacity constraint for a vertical separator yields an expression for the minimum vessel diameter as follows:

$$D_{\min}^2 = 5058 Q_g \left(\frac{TZ}{P}\right) \left(\frac{\rho_g}{\rho_o - \rho_g} \frac{C_d}{d_m}\right)^{1/2} \text{in}^2 \tag{7.25}$$

where D is the separator internal diameter (in), T is the operating temperature (°R), Z is the gas compressibility at operating pressure and temperature, P is the operating pressure, ρ_g and ρ_o are the gas and oil densities (lb/ft^3), C_d is the drag coefficient, and d_m is the minimum oil droplet size to be separated from gas (μm).

Any diameter that is larger than the minimum diameter determined from Equation 7.25 results in a lower gas velocity and therefore ensures settling and separation of liquid droplets of diameters equal to and larger than d_m out of the gas.

7.5.2.3 Liquid Retention Time (Capacity) Constraint

The separator must provide sufficient volume for the oil and water to be retained within the separator for the required retention times. The retention times are determined to allow separation of the entrained water droplets from the oil, separation of the entrained oil droplets from the water, and for the oil to reach equilibrium with the gas. As explained in Section 7.2, retention times are best determined from laboratory tests and they normally range from 3 to 30 min based on fluid properties and operating conditions. In the absence of laboratory data, a retention time of 10 min may be used for both oil and water.

Let H_o and H_w be the heights of the oil and water (in inches), respectively. Therefore, the volume of each phase within the separator is given by

$$V_o = \left(\frac{1}{12}\right)^3 \left(\frac{\pi}{4}\right) D^2 H_o \text{ ft}^3$$

and

$$V_w = \left(\frac{1}{12}\right)^3 \left(\frac{\pi}{4}\right) D^2 H_w \text{ ft}^3$$

Therefore,

$$V_o + V_w = 4.543 \times 10^{-4} D^2 (H_o + H_w) \text{ ft}^3 \quad (7.26)$$

The volume is also calculated by multiplying the volumetric flow rate by the retention time; therefore,

$$V_o = Q_o \frac{5.61}{24 \times 60} \frac{\text{ft}}{\text{min}} \times t_o \text{ min ft}^3$$

$$V_w = Q_w \frac{5.61}{24 \times 60} \frac{\text{ft}^3}{\text{min}} \times t_w \text{ min ft}^3$$

Then,

$$V_o + V_w = 3.896 \times 10^{-13} (Q_o t_o + Q_w t_w) \quad (7.27)$$

From Equations 7.26 and 7.27, we obtain

$$(H_o + H_w) D^2 = 8.576 (Q_o t_o + Q_w t_w) \text{ in}^3 \quad (7.28)$$

In summary then, the diameter and seam-to-seam length of vertical three-phase separators are determined as follows:

1. Determine the minimum diameter that satisfies the water droplets settling constraint from Equation 7.24.
2. Determine the minimum diameter that satisfies the gas capacity constraint from Equation 7.25.
3. The larger of the two minimum diameters determined in steps 1 and 2 is then considered as the minimum allowable vessel diameter.
4. For various values of diameter larger than the minimum allowable vessel diameter, use Equation 7.28 to determine combinations of diameters and liquid heights.
5. For each combination, determine the seam-to-seam length from the following:

For $D > 36$ in

$$L_s = \frac{1}{12}(H_o + H_w + D + 40) \text{ ft} \tag{7.29}$$

For $D < 36$ in

$$L_s = \frac{1}{12}(H_o + H_w + 76) \text{ ft} \tag{7.30}$$

Example 7.2

Determine the diameter and seam-to-seam length of a three-phase vertical separator for the following operating conditions:

Oil production rate:	6000 BPD
Water production rate:	3000 BPD
Gas production rate:	8.0 MMSCFD
Oil viscosity:	10 cP
Oil specific gravity:	0.87
Water specific gravity:	1.07
Gas specific gravity:	0.6
Gas compressibility:	0.88
Operating pressure:	500 psia
Operating temperature:	90°F
Oil retention time:	10 min
Water retention time:	10 min
c_d:	0.64

Solution

Determine ρ_g and ρ_o:

$$\rho_g = \frac{2.7 \times 0.6 \times 500}{550 \times 0.88} = 1.674 \text{ lb/ft}^3 4$$

$$\rho_o = 0.87 \times 62.4 = 54.288 \text{ lb/ft}^3$$

Use Equation 7.24 with d_m equal to 500 m to determine the minimum diameter for water droplet settling:

$$D_{\min}^2 = \frac{6686 Q_o \mu_o}{(\gamma_o - \gamma_w) d_m^2} \text{ in}^2$$

$$D_{\min}^2 = \frac{6686 \times 6000 \times 10}{0.2 \times 500^2} = 8023.2 \text{ in}^2$$

$$D_{\min}^2 = 89.57 \text{ in}$$

Use Equation 7.25 with d_m equal to 100 m to determine the minimum diameter for gas capacity constraint:

$$D_{\min}^2 = 5058 Q_g \left(\frac{TZ}{P}\right)\left(\frac{\rho_g}{\rho_o - \rho_g} \frac{C_d}{d_m}\right)^{1/2}$$

$$D_{\min}^2 = 5058 \times 8 \left(\frac{550 \times 0.88}{500}\right)\left(\frac{1.674}{52.614} \frac{0.64}{100}\right)^{1/2}$$

$$= 835.86 \text{ in}^2$$

$$D_{\min} = 28.911 \text{ in}$$

Therefore, the minimum allowable vessel diameter is taken as the larger of the two calculated; that is 89.57 in.

Use Equation 7.28 to determine the relation between the diameter and liquid height for retention time constraint:

$$\begin{aligned}(H_o + H_w)D^2 &= 8.576(Q_o t_o + Q_w t_w) \text{ in}^3 \\ &= 8.576 \times (6000 \times 10 + 3000 \times 10) \\ &= 771{,}840 \text{ in}^3\end{aligned} \tag{7.31}$$

For diameters larger than 89.57 in, use Equation 7.31 to determine the value of $H_o + H_w$; then, calculate the corresponding values of the seam-to-seam length from Equation 7.29, and, finally, calculate the corresponding slenderness ratio. The results are summarized in this table.

D (in)	$H_o + H_w$ (in)	L_s (ft)	$12L_s/d$
90	95.289	18.774	2.50
96	83.750	18.312	2.29
102	74.187	18.016	2.12
108	66.173	17.848	1.98
114	59.391	17.783	1.87
120	53.600	17.800	1.78
132	44.298	18.025	1.64
144	37.222	18.4351	1.54

All combinations of diameter and seam-to-seam length in the above table are acceptable because the slenderness ratio falls between 1.5 and 3 for all of them. The final selection would, therefore, depend on cost and availability. A 96-in by 19-ft separator is probably the best choice.

REVIEW QUESTIONS AND EXERCISE PROBLEMS

1. State whether each of the following statements is true (T) or false (F):
   a. The knockout drum is used to separate the bulk of free water and free gas from the produced stream.
   b. Separation of oil droplets from water is easier and faster than separation of water droplets from oil because the density of oil is lower than that of water.
   c. In separator design, unless laboratory data are available, the smallest water droplet size that should be removed from oil is 100 μm.
2. Circle the correct answers (note that more than one answer could be correct):
   a. In three-phase separators, the produced fluid is separated into the following:
      i. Gas, oil, and water + emulsion
      ii. Gas oil + emulsion, and water
      iii. Gas, oil, emulsion, and water
   b. The formulation of emulsion in the separators may cause:
      i. Improper oil–gas interface level control
      ii. Reduction of oil and water effective retention times
      iii. Blowby
      iv. Plugging of mist extractor
   c. Separation of water droplets from oil compared to separation of oil droplets from water is:

i. More difficult because oil density is less than that of water
ii. Easier because water viscosity is lower than oil viscosity
iii. More difficult because oil viscosity is higher than water viscosity
iv. A more important criterion is designing the separator

3. The following data are given for a small oil field:

Oil rate:	8000 BPD
Water rate:	3000 BPD
Gas–oil ratio:	2000 SCF/bbl
Operating pressure:	500 psia
Operating temperature:	80°F
Oil gravity:	35°API
Water specific gravity:	1.07
Gas specific gravity:	0.6
Oil viscosity:	10 cP
Oil retention time:	600 s
Water retention time:	360 s

a. Design a horizontal separator to handle the field production.
b. Design a vertical separator to handle the field production.

4. Determine the actual gas, oil, and water capacities of a horizontal three-phase separator having a 60 in diameter and 15 ft seam-to-seam length for the preceding operating conditions.
5. Determine the actual gas, oil, and water capacities for a three-phase horizontal separator having a 48 in diameter and a 16 ft seam-to-seam length, operating at 585 psi and 80°F, given the following data:

Gas specific gravity:	0.6
Water specific gravity:	1.07
Oil specific gravity:	0.875
Water viscosity:	1.12 cP
Oil viscosity:	5.0 cP
Water–oil ratio:	1:1.5
Oil retention time:	600 s
Water retention time:	480 s
K:	0.26
Z:	0.89

6. Determine the actual gas, oil, and water capacities of a vertical three-phase separator for the same conditions given in Problem 5. Compare the results for the two separators.

Section III

Crude Oil Treatment

Crude oil treatment represents the second most important process in surface oil field operations, next to gas–oil separation. Many oil production processes present a significant challenge to oil and water treating equipment design and operations. The nature of crude oil emulsions continuously changes as the producing field depletes and conditions change with time.

In today's market, the high value of crude oil and hydrocarbon liquids, in general, makes it essential to install reliable, high-performing systems for treatment to maximize the economic recovery of crude oil.

Crude oil is seldom produced alone because it comes with water in different forms. The water creates several problems, in particular emulsion formation. The produced water, once separated from the oil, has to be treated as discussed in Chapter 15. These treatment steps increase costs. Furthermore, sellable crude oil must comply with certain product specifications, including the amount of basic sediment and water (BS&W) and salt content in terms pounds of salt per thousand barrels of oil (PTB).

Section III is concerned with emulsion treatment and dehydration, desalting, and stabilization and sweetening of crude oil as presented in Chapters 8 through 10, respectively. A fairly new chapter is introduced in this section to discuss other treatment options geared toward quality upgrading of crude oil (Chapter 11).

8

Emulsion Treatment and Dehydration of Crude Oil

Generally what comes out of a well is a mixture of oil, water, gas, and even sand. Once crude oil is separated from the gas, it undergoes further treatment steps. An important aspect during oil field development is the design and operation of wet crude handling facilities. Foreign materials such as water and sand must be separated from the oil and the gas before they can be sold. This process is known as oil treating or oil dehydration.

This chapter deals with the dehydration stage of treatment. The objective of this treatment is first to remove free water and then break the oil emulsions to reduce the remaining emulsified water in the oil. Depending on the original water content of the oil as well as its salinity and the process of dehydration used, oil field treatment can produce oil with a remnant water content of between 0.2 and 0.5 of 1%. The remnant water is normally called the bottom sediments and water (BS&W).

To summarize, the dehydration step is a dual function process: to ensure that the remaining free water is totally removed from the bulk of oil and to apply whatever tools in order to break the oil emulsion. A dehydration system, in general, comprises various types of equipment according to the type of treatment involved: water removal or emulsion breaking. Basic types of equipment include the separator, free-water knockout drum, and heater-treater.

8.1 Introduction

Many oil production processes present a significant challenge to the oil and gas field processing facilities. This applies to the design and operations of the processing equipment. A typical example is the fact that the nature of crude oil emulsions changes continuously as the producing field depletes. Therefore, conditions change as well.

The fluid produced at the wellhead consists usually of gas, oil, free water, and emulsified water (water–oil emulsion). Before oil treatment begins, we must first remove the gas and free water from the well stream. This is essential in order to reduce the size of the oil–treating equipment.

As presented in Chapters 6 and 7, the gas and most of the free water in the well stream are removed using separators. Gas, which leaves the separator, is known as *primary gas*. Additional gas will be liberated during the oil treatment processes because of the reduction in pressure and the application of heat. Again, this gas, which is known as *secondary gas*, has to be removed. The free water removed in separators is limited normally to water droplets of 500 μm and larger. Therefore, the oil stream leaving the separator would normally contain free water droplets that are 500 μm and smaller in addition to water emulsified in the oil. This oil has yet to go through various treatment processes (dehydration, desalting, and stabilization) before it can be sent to refineries or shipping facilities.

The treatment process and facilities should be carefully selected and designed to meet the contract requirement for BS&W. Care should be taken not to exceed the target oil dryness. Removal of more remnant water than allowed by contract costs more money while generating less income because the volume of oil sold will be based on the contract value of the BS&W.

The basic principles for the treating process are as follows:

1. Breaking the emulsion, which could be achieved by either any or a combination of the addition of heat, the addition of chemicals, and the application of electrostatic field
2. Coalescence of smaller water droplets into larger droplets
3. Settling, by gravity, and removal of free water

The economic impact of these treating processes is emphasized by Abdel-Aal and Alsahlawi (2014).

8.2 Oil Emulsions

Rarely does oil production take place without water accompanying the oil. Salt water is thus produced with oil in different forms as illustrated in Figure 8.1. Apart from free water, emulsified water (water-in-oil emulsion) is the one form that poses all of the concerns in the dehydration of crude oil.

Oil emulsions are mixtures of oil and water. In general, an emulsion can be defined as a mixture of two immiscible liquids, one of which is dispersed as droplets in the other (the continuous phase) and is stabilized by an emulsifying agent. In the oil field, crude oil and water are encountered as the two immiscible phases together. They normally form water-in-oil emulsion (W/O emulsion), in which water is dispersed as fine droplets in the bulk of oil. This is identified as part (c) in Figure 8.2. However, as the water cut increases, the possibility of forming reverse emulsions (oil-in-water, or O/W emulsion) increases, as shown by part (b) in Figure 8.2.

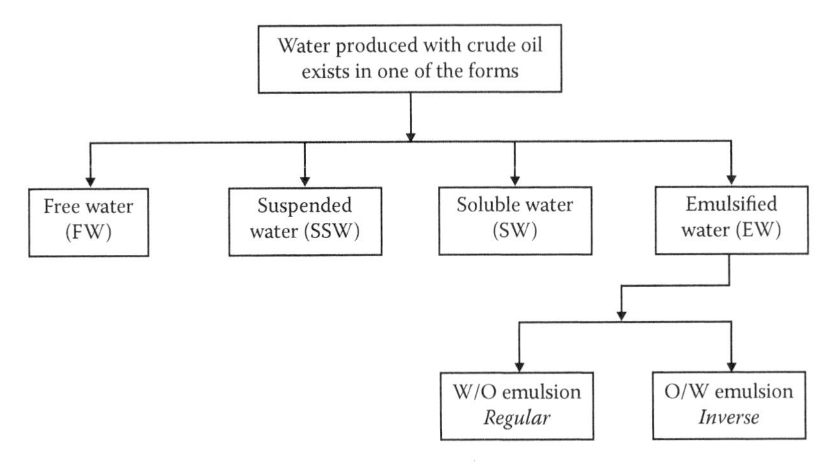

FIGURE 8.1
Different forms of water produced with crude oil.

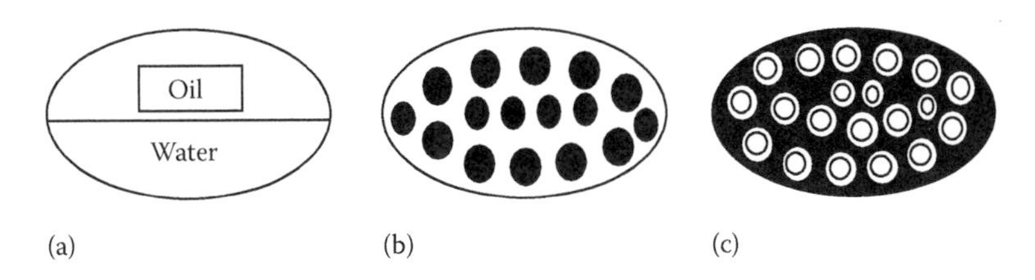

FIGURE 8.2
Schematic representations of different forms of emulsions. (a) Two immiscible fluids, (b) O/W inverse emulsion, and (c) W/O normal emulsion.

For two liquids to form a stable emulsion, three conditions are to be fulfilled:

1. The two liquids must be immiscible.
2. There must be sufficient energy of agitation to disperse one phase into the other.
3. An emulsifying agent must be present.

8.2.1 Energy of Agitation

Emulsions normally do not exist in the producing formation but are formed because of the agitation that occurs throughout the oil production system. Starting within the producing formation, the oil and water migrate through the porous rock formation, making their way into the wellbore, up the well tubing, through the wellhead choke, and through the manifold into the surface separators. Throughout this journey, the fluids are subjected to agitation due to the turbulent flow. This energy of agitation, which forces the water drops into the bulk of oil, functions in the following pattern:

- Energy is spent first to overcome the viscous force between the liquid layers leading to their separation into thin sheets or parts. This is what we call *shearing energy* and is mathematically approximated by the formula

$$SE = \tau A D_o \tag{8.1}$$

where SE is the shearing energy, A is the shear surface area, D_o is the characteristic length, and τ is the shearing force per unit area, defined next by Equation 8.2:

$$\tau = \frac{C_d \rho v^2}{2 g_c} \tag{8.2}$$

where C_d is the drag coefficient, ρ is the density of the fluid, v is the velocity of flow, and g_c is a conversion factor = 32.174[(ft)(lbm)]/[(sec^2)(lbf)].

- Energy is used in the formation of *surface energy*, which occurs as a result of the separation of the molecules at the plane of cleavage. This surface energy is related to the surface tension, which involves the creation of an enormous area of interface with attendant free-surface energy. Energy contained per unit area is referred to as *surface tension*, having the units of dynes/cm.

The drops attain the spherical shape, which involves the least energy contained for a given volume. This is in accordance with the fact that all energetic systems tend to seek the lowest level of free energy. The fact that surface tension is defined as *the physical property due to molecular forces existing in the surface film of the liquid*, this will cause the volume of a liquid to be contracted or reduced to a shape or a form with the least surface area. This is the same force that causes raindrops to assume a spherical shape.

A crucial question that can be asked now is the following: Can the plant designer prevent emulsion formation? Well, the best he can do is to reduce its extent of formation based on the fact that the liquids initially are not emulsified. From the design point of view, primarily reducing the flowing velocity of the fluid and minimizing the restrictions and sudden changes in flow direction could minimize formation of emulsion.

8.2.2 Emulsifying Agents

If an oil emulsion is viewed through a microscope, many tiny spheres or droplets of water will be seen dispersed through the bulk of oil, as depicted in Figure 8.3. A tough film surrounds these droplets; this is called a stabilizing film. Emulsifying agents, which are commonly found in crude oil or water in the natural state or introduced in the system as contaminants during drilling and maintenance operations, create this type of film.

Some of the common emulsifiers are as follows:

- Asphaltic materials
- Resinous substances
- Oil-soluble organic acids
- Finely dispersed solid materials such as sand, carbon, calcium, silica, iron, zinc, aluminum sulfate, and iron sulfide

These emulsifying agents support the film formation encasing the water droplets, hence the stability of an emulsion.

The stability of oil–water emulsions could be viewed through the following analysis. The relative difficulty of separating an emulsion into two phases is a measure of its stability. A very stable emulsion is known as a *tight* emulsion and its degree of stability is influenced by many factors. Accordingly, we can best understand the resolution problem and, hence, the treatment procedure if we consider the following factors:

- Viscosity of oil—Separation is easier for a less viscous oil phase.
- Density or gravity difference between oil and water phases—Better separation is obtained for a larger difference.

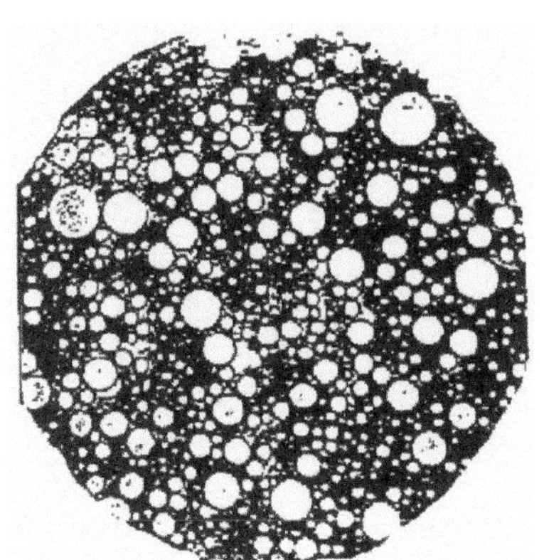

FIGURE 8.3
Photomicrograph of loose emulsion containing about 30% emulsified water droplets (size about 60 μm and less).

- Interfacial tension between the two phases (which is related to the type of emulsifying agent)—Separation is promoted if this force is lowered (i.e., decreasing the interfacial tension).
- Size of dispersed water droplets—The larger the size of water drops, the faster is the separation. The size of dispersed water droplets is an important factor in emulsion stability. A typical droplet size distribution for emulsion samples was determined by using a special computer scanning program. Results reported in Figure 8.4 indicate that most of the droplets found in oil emulsions are below 50 μm.
- Percentage of dispersed water—The presence of a small percentage of water in oil under turbulence conditions could lead to a highly emulsified mixture. Water droplets are finely divided and scattered with very little chance of agglomerating to larger particles.
- Salinity of emulsified water—Highly saline water will lead to a faster separation because of a higher density difference between the oil and the water phases.
- Percentage of dispersed water—The presence of a small percentage of water in oil under turbulence conditions could lead to highly emulsified mixture. Water droplets are finely divided and scattered with very little chance of agglomerating to larger particles.
- Salinity of emulsified water—Highly saline water will lead to a faster separation because of a higher density difference between the oil and the water phases.

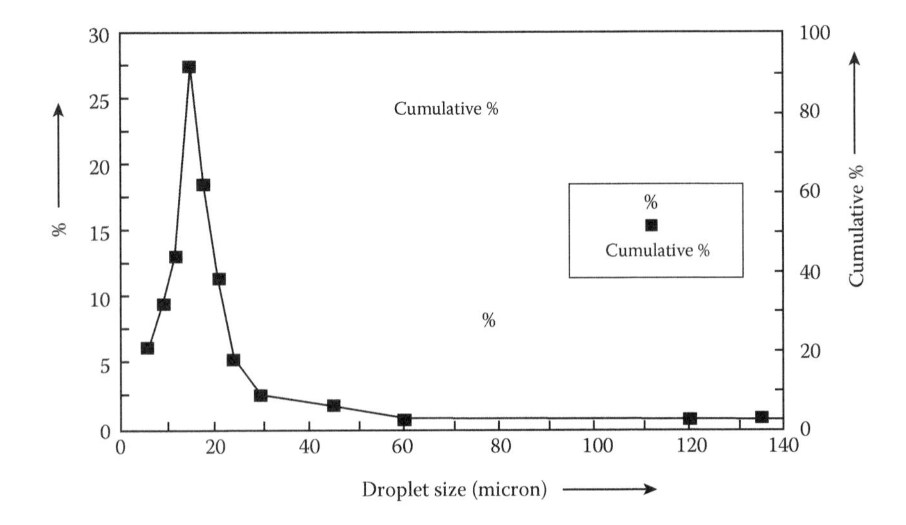

FIGURE 8.4
Droplet size distributions for an emulsion sample. (After Al-Tahini, A., Crude oil emulsions, Co-op Report, Department of Chemical Engineering, KFUPM, Dhahran, Saudi Arabia, 1996.)

8.3 Dehydration/Treating Processes

The method of treating *wet* crude oil for the separation of water associated with it varies according to the form(s) in which water is found with the crude. Free-water removal comes first in the treating process, followed by the separation of *combined* or emulsified water along with any foreign matter such as sand and other sediments. The basic approaches of handling wet crude oils are illustrated in Figure 8.5.

Again, from an economic point of view, removal of free water at the beginning will reduce the size of the treating system, hence its cost. The same applies for the separation of associated natural gas from oil in the gas–oil separator plant (GOSP).

A dehydration system in general comprises various types of equipment. Most common are the following:

- Free-water knockout vessel
- Wash tank
- Gun barrel
- Flow treater (heater-treater)
- Chemical injector
- Electrostatic dehydrator

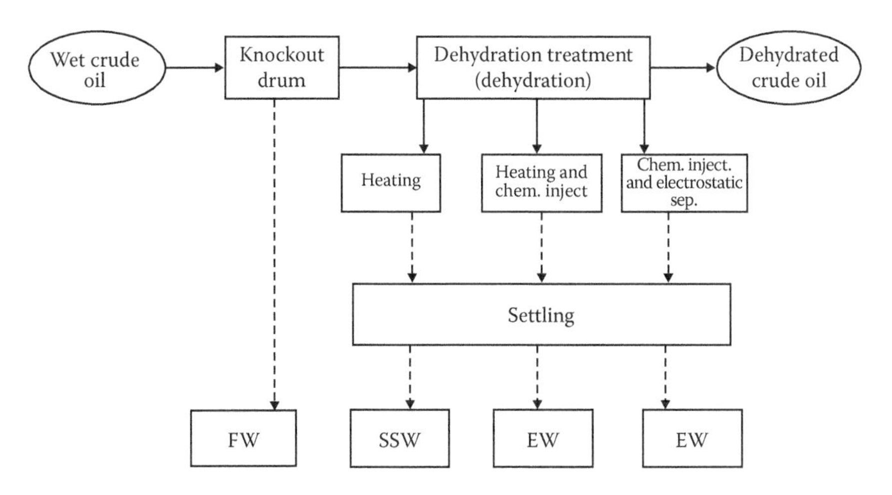

FIGURE 8.5
Basic approach of handling wet crude oil. (EW, emulsified water; FW, free water; SSW, suspended water.)

8.3.1 Removal of Free Water

Free water is simply defined as that water produced with crude oil and will settle out of the oil phase if given little time. There are several good reasons for separating the free water first: reduction of the size of flow pipes and treating equipment; reduction of heat input when heating the emulsion (water takes about twice as much heat as oil); and minimization of corrosion, because free water comes into direct contact with the metal surface, whereas emulsified water does not.

Free water has its distinctive benefits. Free water found in the reservoir fluid will carry twice as much heat as oil and take it up the tubing to the surface. Eventually, it will help in breaking oil emulsions. It is to be observed that an oil well producing salt water (free water) will be much warmer than the one producing oil only. Further, free water contributes to what is called *water wash*, which is the action of the salt water to break the oil emulsions.

Free water removal takes place using a knockout vessel, which could be an individual piece of equipment or incorporated in a flow treater.

8.3.2 Resolution of Emulsified Oil

Resolution of emulsified oil is the heart of the dehydration process, which consists of three consecutive steps:

1. Breaking the emulsion—This requires weakening and rupturing the stabilizing film surrounding the dispersed water droplets. This is a destabilization process and is affected by using what is called an *aid*, such as chemicals and heat.
2. Coalescence—This involves the combination of water particles that became free after breaking the emulsion, forming larger drops. Coalescence is a strong function of time and is enhanced by applying an electrostatic field, impingement on a solid surface area, and water washing.
3. Gravitational settling and separation of water drops—The larger water droplets resulting from the coalescence step will settle out of the oil by gravity and be collected and removed.

The resolution process is exhibited by the above steps in series. The slowest one is the most controlling. Out of these, coalescence is the slowest step. In other words, using either heat or chemicals followed by gravitational settling can break some emulsions, but the process is dependent on the time spent in coalescence. This time is the element that determines the equipment size, hence its capital cost.

8.3.3 Emulsion Treatment

As explained in Section 8.2.1, using chemicals followed by settling can break some emulsions. Other emulsions require heating and allowing the water to settle out of the bulk of oil. More difficult (tight) emulsions require, however, both chemicals and heat, followed by coalescence and gravitational settling.

Basically, a dehydration process that utilizes any or a combination of two or more of the treatment aids (heating, adding chemicals, and an applying electrical field) is used to resolve water–oil emulsions. The role of each of these aids is discussed next in detail.

8.4 Heating

Heating is the most common way of treating water–oil emulsions. To understand how heating aids in the resolution of water–oil emulsions and separation of the water droplets from the bulk of oil, reference is made to the droplet settling velocity equation derived in Chapter 7 (Equation 7.7):

$$u = 1.787 \times 10^{-6} \frac{(\Delta\gamma)d_m}{\mu_o} \text{ ft/sec} \tag{8.3}$$

where u is the water droplet settling velocity, $\Delta\gamma$ is the difference between water and oil specific gravities, d_m is the diameter of the water droplet (in µm), and μ_o is the oil viscosity.

8.4.1 Benefits and Drawbacks of Heating

Heating of water–oil emulsions aids in the resolution of the emulsion and the separation of the emulsified water in several ways. The most significant effect is the reduction of oil viscosity with temperature. The viscosity of all types of crude oil drops rapidly with temperature. From Equation 8.3, such reduction in viscosity results in increasing the water droplet settling velocity and, thus, speeds and promotes the separation of water from the oil.

As the water and oil mixture is heated, the density (specific gravity) of both water and oil is reduced. However, the effect of temperature on oil density is more pronounced than on water density. The result is that the difference in density (or specific gravity) increases as the emulsion is heated. For example, if oil and water are heated from 60°F to 150°F, the following changes in their relative specific gravity take place:

	At 60°F	At 150°F
Oil specific gravity	0.83	0.79
Water specific gravity	1.05	1.03
Difference in specific gravity	0.22	0.24

With reference to Equation 8.3 an increase in $\Delta\gamma$ increases the settling velocity and, therefore, promotes the separation of water droplets from the bulk of oil. The change in the specific gravity difference is, however, small. Therefore, this effect is not as significant as the effect of viscosity. In fact, we may completely ignore the effect of specific gravity on the process up to a temperature of 200°F. For some specific crude oils, increased temperature may cause a reverse effect on the difference in specific gravity. For some heavy oils, the specific gravity of the oil and water will be equal at certain temperature. This situation must be avoided, as it will stop the separation process completely. Therefore, care should be exercised when determining the treating temperature for a specific crude oil.

Another beneficial effect of heating is that the increased temperature promotes movements of the small water droplets, which upon collision with one another may form larger size droplets. The increased droplet size significantly speeds the settling process. Heat will also help to destabilize (weakening) the emulsifying film, hence breaking the emulsion. Further, heating will dissolve the small paraffin and asphaltenes crystals and, thus, neutralize their potential effect as emulsifiers.

Despite of all of the aforementioned benefits of heating, there are some drawbacks associated with this method of treatment. Heating of the oil can result in significant loss of the lighter hydrocarbon ends and thus results in loss of the oil volume. For example, on heating 35°API oil from 100°F to 150°F results in losing more than 1% of the oil volume by evaporation. Of course, the light ends could be collected and sold with the gas. This, however, will not make up for the loss of revenue resulting from oil losses. In addition to oil losses, evaporation of the light ends leaves the treated oil with lower API gravity (i.e., lower quality), which will be sold at a lower price. Finally, heating requires additional investment for heating equipment and additional operating cost for the fuel gas and maintenance. In conclusion, it is generally recommended to avoid using heating as a treatment process if at all possible. Otherwise, some of the benefits of heating may be realized with the minimum amount of heating.

8.4.2 Methods of Heating Oil Emulsions

The fuel used to supply heat in oil-treating operations is practically natural gas. Under some special conditions, crude oil may be used.

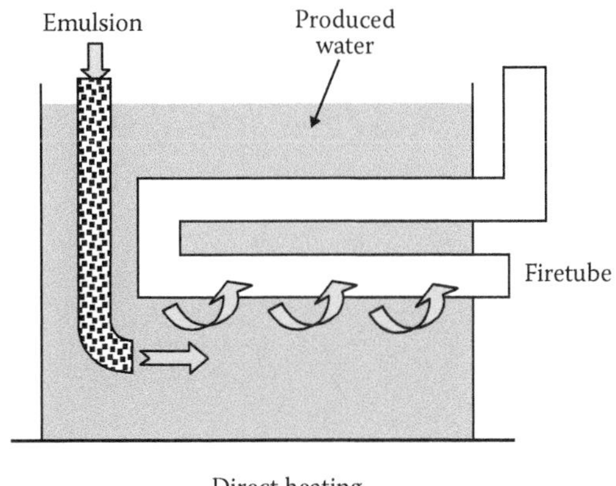

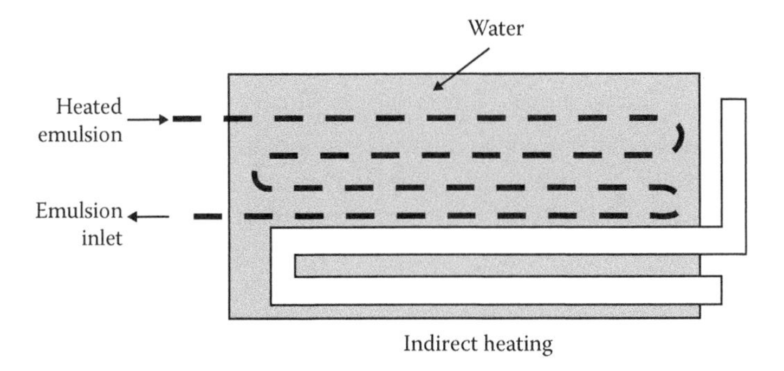

FIGURE 8.6
Two basic methods of heating emulsions.

Heaters are generally of two basic types: *Direct heaters*, in which oil is passed through a coil exposed to the hot flue gases of the burned fuel or to introduce the emulsion into a vessel heated using a fire tube heater. *Indirect heaters*, in which heat is transferred from the hot flue gases to the emulsion via water as a transfer medium. The emulsion passes through tubes immersed in a hot water bath.

In general, the amount of free water in the oil emulsion will be a factor in determining which method is to be used. If free water is found to be 1%–2%, then use an indirect heater. If the free-water content is enough to hold a level around the fire tube, then use a direct heater. Both types are shown in Figure 8.6.

8.4.3 Heat Requirement

The amount of heat transferred to, or gained by, a fluid is generally given by

$$q = mc\Delta T \text{ Btu/h} \tag{8.4}$$

where q is the rate of heat transferred/gained (Btu/h), m is the mass flow rate of fluid (lb/h), c is the specific heat of fluid (Btu/lb °F), and ΔT is the increase in temperature due to heat transfer (°F).

The mass flow rate, m, could be calculated from the volumetric flow rate Q that is normally given in BPD (bbl/day) as follows:

$$m = Q\left(\frac{\text{bbl}}{\text{day}}\right)\left(\frac{1}{24}\right)\left(\frac{\text{day}}{\text{h}}\right)5.61\left(\frac{\text{ft}^3}{\text{bbl}}\right)62.4\gamma\left(\frac{\text{ft}^3}{\text{lb}}\right)\frac{\text{lb}}{\text{h}}$$

Therefore,

$$m = 14.59\gamma Q \cong 15\gamma Q \text{ lb/h} \tag{8.5}$$

where γ is the specific gravity of the liquid.

When heating is used for emulsion treatment, the bulk of the free water must be removed prior to the application of heat (as described in Chapter 4). The remaining water would be mostly emulsified water and some free water that was not separated (water droplets of 500 μm and smaller). In estimating the total amount of heat required for the treatment (q), we should account for the heat transferred to the water (q_w) along with that transferred to the oil (q_o). Further, we should also account for the heat losses (q_l). These could be estimated using Equations 8.2 and 8.3 as follows:

$$q_o = 15\gamma_o Q_o C_o(\Delta T) \tag{8.6}$$

Normally, the amount of water is given as a percentage of the oil volume (i.e., $Q_w = wQ_o$, where w is the water percentage). Therefore,

$$q_w = 15\gamma_w w Q_o C_w(\Delta T) \tag{8.7}$$

The heat loss is normally expressed as a percentage of the total heat input. Let l be the percent of total heat lost; then,

$$q_l = lq \tag{8.8}$$

Because $q = q_o + q_w + q_l$, adding Equations 8.6 to 8.8 and solving for q, we obtain

$$q = \frac{1}{1-l} 15Q_o(\Delta T)(\gamma_o c_o + w\gamma_w c_w)\frac{\text{Btu}}{\text{h}} \tag{8.9}$$

where l is the percent of heat loss (fraction); Q_o is the flow rate of oil (bbl/day); ΔT is the increase in temperature due to heating (°F); γ_o and γ_w are the specific gravity of oil and water, respectively; c_o and c_w are the specific heats of oil and water, respectively; and w is the percent, by volume, of water in the oil (fraction). Equation 8.9 is used to estimate the required heat duty of the burner.

8.4.4 Types of Heater-Treaters

The various types of field heaters and heater-treaters are presented in Figure 8.7. They include vertical treaters, horizontal treaters, and gunbarrel settling tanks.

Vertical treaters are commonly used as single-well treaters. Oil plus the emulsion stream enters the treater from the side at the top section of the vessel where gas, if any, separates and leaves the vessel at the top through a mist extractor. The liquid flows downward through a pipe called the downcomer and exits through a flow spreader located slightly below the water–oil interface to water wash the oil–emulsion stream. Water washing helps in coalescing the small water droplets suspended in the oil. The oil and emulsion flow upward, exchanging heat with the heater fire tubes, then through the coalescing section. The coalescing section, normally packed with porous material such as hay, is sized to provide sufficient time for the coalescence of the water droplets and their settling out of the oil. The treated oil is then collected from the treater.

Horizontal treaters are normally used in centralized multiwell-treating facilities GOSP. The oil and emulsion stream is introduced to heating section

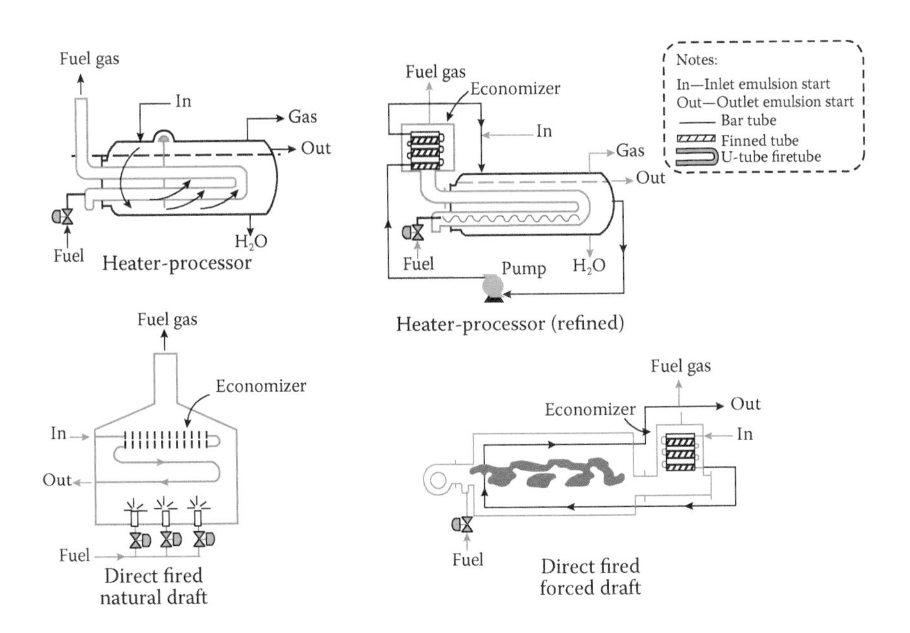

FIGURE 8.7
Types of field heaters.

of the treater near the top where gas is flashed, separated, and exits the vessel at the top through a mist extractor. The liquid is made to flow tangent to the inside surface of the vessel and falls below the water–oil interface, where it is water washed. Water washing causes coalescence and separation of free water. The oil plus emulsion rises up, exchanging heat with the fire tubes, and flows over a weir into an oil surge chamber. The hot oil plus emulsion leaves the oil surge chamber near the bottom of the vessel and enters the coalescing section of the treater through a flow spreader, which ensures that the oil flows evenly throughout the length of the coalescing section. The oil flows upward, where it is withdrawn from the vessel through a collector. The spreader–collector system allows the oil flow to be vertical. This section of the treater is sized to allow sufficient retention time for the coalescence and settling of the water out of the oil. The separated water is removed from the treater at two locations: one at the bottom of the heating section and the other at the bottom of the coalescing section. Interface level controllers control both outlet valves.

Gunbarrel settling tanks are large-diameter vertical tanks operating mostly at atmospheric pressure. They are generally used for small fields where no or minimum heating is required for separation of the emulsion. When heating is needed, the most common way is to preheat the oil and emulsion stream before it enters the tank. The oil plus emulsion stream enters the tank at the top (where gas is flashed and separated) into a downcomer. It leaves the downcomer through a spreader located below the water–oil interface and rises vertically upward, flowing through the large cross-sectional area of the tank. As the oil + emulsion rises, it is first water washed to coalesce the water droplets. Then, it is retained for a sufficient time in the settling section to allow for the separation of the water droplets, which flow countercurrent to the oil flow and collects at the bottom section of the tank.

8.5 Chemical Treatment

As mentioned in Section 8.3, some oil emulsions will readily break upon heating with no chemicals added; others will respond to chemical treatment without heat. A combination of both *aids* will certainly expedite the emulsion-breaking process. Chemical additives, recognized as the second aid, are special surface-active agents comprising relatively high-molecular-weight polymers. These chemicals (de-emulsifiers), once adsorbed to the water–oil interface, can rupture the stabilizing film and displace the stabilizing agent due to the reduction in surface tension on the inside of the film (i.e., on the water side of the droplet). In other words, when the de-emulsifiers are added to the oil, they tend to migrate to the oil–water interface and rupture the stabilizing

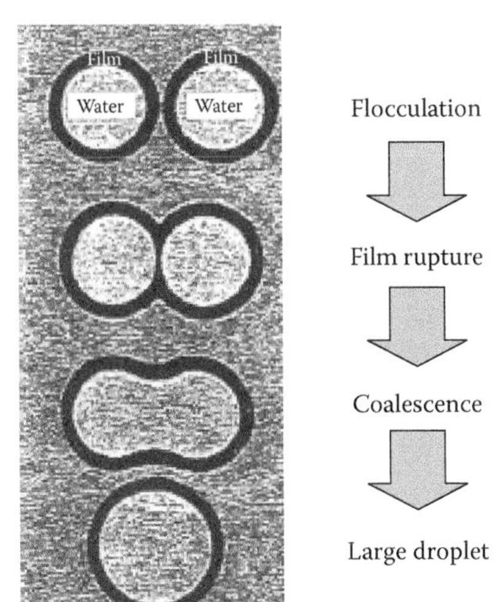

FIGURE 8.8
How de-emulsifiers lead to larger oil drops.

film. The de-emulsifier, as it reaches the oil–water interface, functions in the following pattern: flocculation, then film rupture, followed by coalescence. The faster the de-emulsifier reaches the oil–water interface, the better results it achieves. Figure 8.8 illustrates these steps.

8.5.1 Selection and Injection of Chemicals (De-Emulsifiers)

The very first step for selecting the proper chemical for oil treating is testing an oil sample. The representative sample is measured into a number of bottles (12 or more). To each bottle, a few drops of different chemicals are added, followed by shaking to ensure good mixing between the emulsion and the chemical. Heat could be applied if needed. Final selection of the right chemical will be based on testing a sample of the oil to find out how complete the water removal was.

From the practical point of view, most oil de-emulsifiers are oil soluble rather than water soluble. It is therefore recommended to dilute the chemical with a solvent to have a larger volume of the solution to inject to ensure thorough mixing.

The point of injection of de-emulsifiers will depend largely on the type used. For the case of water-soluble de-emulsifiers, injection is carried out after free water has been removed; otherwise, most of the chemical is lost down the drain. Three points of injection are recommended:

1. Upstream of the choke, where violent agitation takes place in the choke as the pressure is lowered from wellhead to that corresponding to the gas–oil separator. It is considered the ideal injection point. This is illustrated in Figure 8.9.

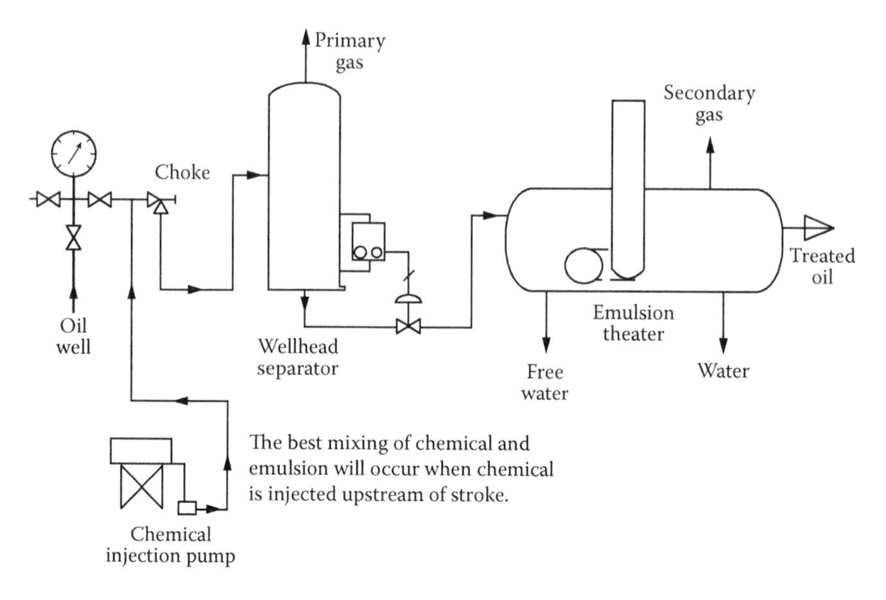

FIGURE 8.9
Chemical injection: upstream of the choke.

2. Upstream of the level control valve on the separator, where agitation occurs in the valve as the pressure is lowered, as shown in Figure 8.10.
3. For the case in which the treating system does not include a gas–oil separator, the injection point is placed 200–250 ft from the emulsion treater.

Chemicals are applied and injected using a small chemical pump. The pump is of the displacement plunger type. The chemical pump should be able to deliver small quantities of the de-emulsifier into the oil line. At normal treating conditions, 1 L of chemical is used for each 15–20 m^3 of oil (or about 1 qt per 100 bbl of oil). Dilution of the chemical with proper solvents is necessary. Based on the type of oil, the required concentration of the chemical ranges between 10 and 60 ppm (parts per million). Chemical de-emulsifiers (emulsion breakers) are complex organic compounds with surface-active characteristics. A combination of nonionic, cationic, and anionic materials contribute to the surface-active properties. Some of the de-emulsifiers are sulfonates, polyglycol esters, polyamine compounds, and many others.

A final and important word is that excessive amounts of chemicals can do harm. Too many chemicals is usually called overtreating. In addition to the unnecessary additional operating cost, excessive treatment would lead to what is known as *burning of the emulsion* (i.e., unbreakable or tight emulsion).

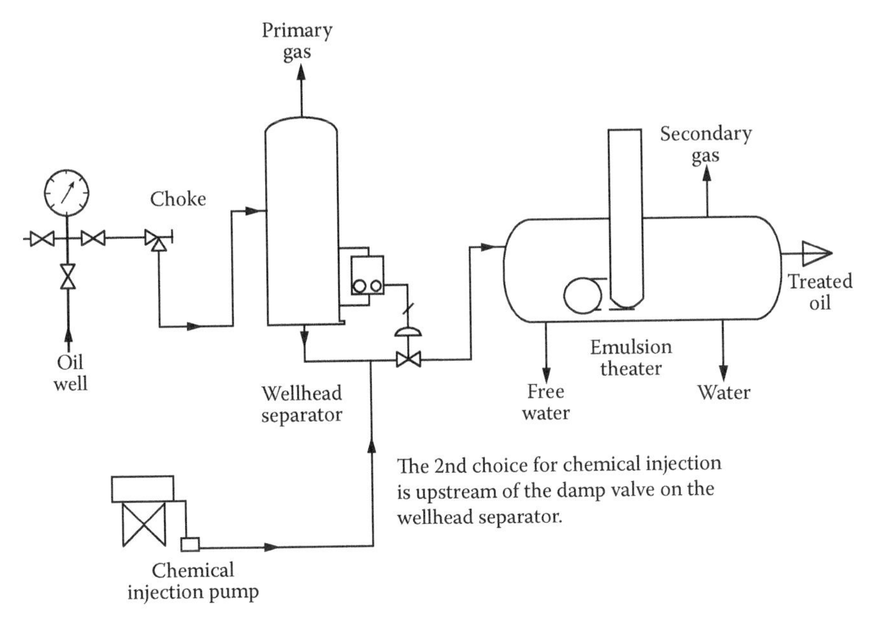

FIGURE 8.10
Chemical injection upstream of the control valve of the gas–oil separator.

8.6 Electrical Aid

Electrical is the third aid of emulsion treating in crude oil dehydration. However, it should be realized that both heating and chemical treating work in order to *break the emulsion*, whereas electrical emulsion treating is aimed at speeding up *coalescence*, hence settling. In other words, electric dehydration does not break the emulsion electrically.

Looking at the three consecutive steps involved in the dehydration of emulsified crude oils (breaking the emulsion, coalescence of water droplets, and settling and separation) and assuming that the first and third steps are fast compared to the second step, it can be concluded that coalescence is the controlling step. In other words, coalescence, which is a function of time, influences settling.

Consequently, in the design of dehydrators, some means should be implemented to reduce the coalescence time, hence the settling time. Some of these means are installing a coalescing medium in the settling section to speed up the build up and the formation of water drops, applying centrifugal force to the emulsion that can promote separation, and applying an electrical field in the settling section of the treater.

The principle underlying the breaking oil–water emulsions using electrical current is known as electrostatic separation. Ionization of these emulsions with the aid of an electric field was introduced in 1930 for crude oil desalting in oil refineries. A high-voltage field (10,000 to 15,000 v) is used to help dehydration according to the following steps:

1. The water droplet is made up of polar molecules, because the oxygen atom has a negative end, and the hydrogen atoms have positive charges. These polar forces are magnetized and respond to an external electrical force field. Therefore, a dipole attraction between the water droplets in the emulsion is established, leading to coalescence, hence settling and separation.
2. As a result of the high-voltage field, the water droplets vibrate rapidly, causing the stabilizing film to weaken and break.
3. The surface of the water droplets expand (their shapes change into ellipsoids); thus attracted to each other, they collide and then coalesce, as depicted in Figure 8.11.
4. As the water droplets combine, they grow in size until they become heavy enough to separate by settling to the bottom of the treater.

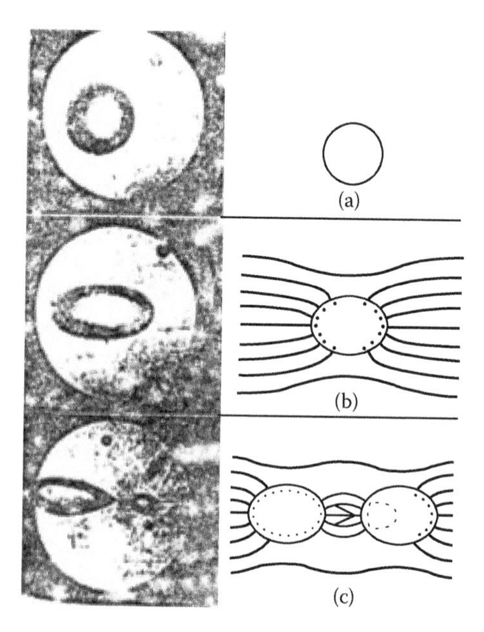

FIGURE 8.11
Emulsion breaking by electric current. (a) Breaking of stabilized film, (b) expansion of water surface into ellipsoid, and (c) attraction of two water drops.

8.7 Chemielectric Dehydrators (Emulsion Treaters)

It is normal practice to call emulsion treaters *heater-treaters*. However, when other or additional treating aids are used, the name of the treater would be made to reflect such aids of treatment. Consequently, a name such as *chemielectric dehydrator* is used to indicate that both chemical and electrical aids are used (in addition to heating) in the treatment.

Figure 8.12 is a diagram for a typical chemielectrical treater. Once the oil is heated, it flows to the settling section. Free water separates from the emulsion (under the effect of both heat and chemicals) and settles to the bottom. The oil on the other hand moves slowly upward, passing across the electric grid in the settling section, where remnant emulsified water is separated as explained in Section 8.3. Finally, clean oil flows to the top of the treater.

It should be made clear that most of the emulsified water is removed by the dual action of both heat and chemicals before the oil passes to the electric grid. The water content in the oil could be reduced to 1%–0.5% before it gets to the grid.

The application of the electrical field has a significant impact on the treater's performance. This is exemplified by Figure 8.13, which clearly illustrates the improvement of a treater when an electric field is applied.

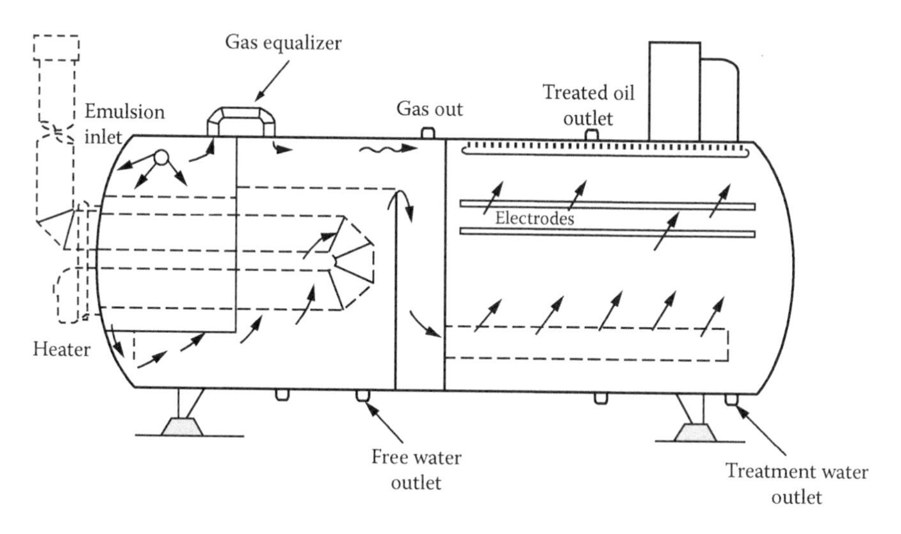

FIGURE 8.12
Chemielectrical dehydrator.

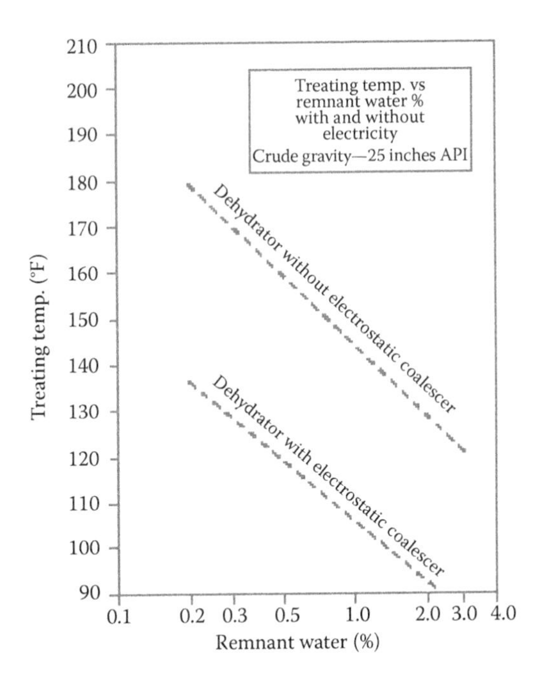

FIGURE 8.13
Effect of electrical field on coalescence.

8.8 Design of Treaters

In the present discussion, we shall limit our design consideration to determination of the dimensions of the coalescing section and the heat input requirement.

In order to achieve a certain quality for the treated oil (i.e., certain maximum water content), we must determine the smallest water droplet size that must be removed. The temperature of the emulsion, the retention time, and the viscosity of the oil affect the water droplet size that must be removed.

Heating the emulsion results in a state of excitation where water droplets collide and coalesce into larger droplets. Therefore, by increasing the temperature of the emulsion, the water droplet size that must be removed increases.

Increasing the retention time allows water droplets to grow larger. However, after a certain initial retention time, increasing the time does not significantly affect the water droplet size. For practically sized treaters, the retention time is normally kept between 10 and 30 min. The higher retention time is usually associated with heavier oils.

The viscosity of the oil is the most influencing factor on the water droplet size that can be removed. The more viscous the oil, the larger the droplet size that can be removed (settled) within a reasonable time. In an article by Thro and Arnold (1994), the following equations relating the water droplet diameter, d_m (that must be removed to achieve 1% water content in the treated oil), to oil viscosity were developed using field data:

$$\text{For } \mu_o < 80 \text{ cP}; d_m = 200\mu_o \text{ } \mu\text{m} \tag{8.10}$$

$$\text{For electrostatic treaters, } 3 \text{ cP} < \mu_o < 80 \text{ cP}; d_m = 170\mu_o \text{ } \mu\text{m} \tag{8.11}$$

The preceding equations are useful in the absence of actual laboratory measurements, which are very difficult to obtain.

With the water droplet size that must be removed determined from the preceding equations, the dimensions of the coalescing/settling section of the treater must be sufficient to allow settling of such water droplets and also allow for the required retention time. These two conditions could be used to develop two equations that govern the dimensions of the treater as detailed in the following subsections for horizontal and vertical treaters.

8.8.1 Sizing Horizontal Treaters

8.8.1.1 Water Droplets Settling Constraint

We start with the settling equation (Equation 8.3):

$$u = 1.787 \times 10^{-6} \frac{(\Delta\gamma)d_m}{\mu_o} \text{ ft/S}$$

This equation gives the terminal settling velocity of the water droplet. For settling to occur, the upward average velocity of the oil must not exceed the water settling velocity. The average oil velocity, u_o, is obtained by dividing the oil volumetric flow rate, Q_o, by the flow cross-sectional area, A.

Let D be the treater inside diameter in inches and L be the effective length of the settling/coalescing section in feet. Therefore,

$$\begin{aligned} u_o &= \frac{Q_o(\text{bbl/day}) \times 5.61(\text{ft}^3/\text{bbl}) \times 1.1574 \times 10^{-5}(\text{day/s})}{(D/12)L} \\ u_o &= 7.792 \times 10^{-4} \frac{Q_o}{DL} \frac{\text{ft}}{\text{s}}, \end{aligned} \tag{8.12}$$

Equating u_o (Equation 8.12) to u (Equation 8.3), we obtain

$$DL = 436 \frac{Q_o\mu_o}{(\Delta\gamma)d_m} \text{ in ft} \tag{8.13}$$

8.8.1.2 Retention Time Constraint

The retention time, t, can be obtained by dividing the volume of the settling/coalescing section occupied by oil, V_o, by the oil volumetric flow rate, Q_o. Assuming that the oil occupies only 75% of the coalescing/settling section,

$$t = \frac{0.75(\pi D^2 L/4 \times 144)}{5.61 Q_o \times 6.944 \times 10^{-4} (\text{day}/\text{min})} \text{ min.}$$

Solving for $D^2 L$, we get

$$D^2 L = \frac{Q_o t}{1.05} \text{ in}^2 \text{ ft} \tag{8.14}$$

8.8.1.3 Sizing Procedure

The following procedure is mostly aimed at determining the minimum size of the coalescing/settling section of the treater and the rating of the burner. Such information will be very useful in preparing equipment specifications for vendors and for evaluating the quotations received from the vendors. The vendors would provide the detailed design and dimensions of the treater.

1. The first step is to decide on a treating temperature. This is best determined from laboratory tests. The optimum treating temperature must provide a minimum loss of oil volume and quality along with a practical treater size. If laboratory data are not available, the treating temperature may be determined based on experience. In such cases, however, the design (following steps) may be executed for different assumed treating temperature and a final decision is made based on analysis of the design results.
2. Determine the diameter of the water droplet that must be removed (from Equation 8.10 or Equation 8.11).
3. Use Equation 8.13 to obtain the relation between D and L that satisfies the settling constraint. Assume various values of D and determine the corresponding values of L from this relation.
4. Use Equation 8.14 to obtain another relation between D and L that satisfies the retention time constraint. For the same values of D assumed in step 3, determine corresponding values of L from this relation.
5. Compare the results obtained from the previous two steps and select a combination of D and L that satisfies both settling and retention time constraints.

6. Use Equation 8.9 to determine the heat requirement for the selected treating temperature.

Example 8.1

Determine the heat requirement and the size of the settling/coalescing section of a horizontal heater-treater for the following conditions:

Oil flow rate:	7000 BPD
Inlet BS&W:	15%
Outlet BS&W:	1%
Oil specific gravity:	0.86
Oil viscosity:	45 cP at 85°F
	20 cP at 105°F
	10 cP at 125°F
Water specific gravity:	1.06
Specific heat of oil:	0.5 Btu/lb °F
Specific heat of water:	1.1 Btu/lb °F
Inlet temperature:	85°F
Retention time:	20 min
Treating temperature:	Examine 105°F, 125°F, and no heating

Solution

Use Equation 8.10 to determine the water droplet diameter for each treating temperature:

$$\text{For T} = 125°\text{F}: \quad d_m = 200\mu_o = 200(10)^{0.25} = 356\ \mu\text{m}$$
$$\text{For T} = 105°\text{F}: \quad d_m = 200\mu_o = 200(20)^{0.25} = 423\ \mu\text{m}$$
$$\text{For T} = 85°\text{F}: \quad d_m = 200\mu_o = 200(45)^{0.25} = 518\ \mu\text{m}$$

Ignoring the effect of temperature on specific gravity, use Equation 8.13 to determine the settling constraint for each treating temperature:

$$\text{For } T = 125°\text{F};\ DL = 436\frac{Q_o\mu_o}{(\Delta\gamma)d_m} = 436\frac{7000\times 10}{0.2\times 356^2} = 1204 \text{ in ft} \tag{8.15}$$

$$\text{For } T = 105°\text{F};\ DL = 436\frac{Q_o\mu_o}{(\Delta\gamma)d_m} = 436\frac{7000\times 20}{0.2\times 423^2} = 1706 \text{ in ft} \tag{8.16}$$

$$\text{For } T = 85°\text{F};\ DL = 436\frac{Q_o\mu_o}{(\Delta\gamma)d_m} = 436\frac{7000\times 45}{0.2\times 518^2} = 2559 \text{ in ft} \tag{8.17}$$

Use Equation 8.14 to determine the relationship for retention time constraints:

$$D^2 L = \frac{Q_o t}{1.05} = \frac{7000 \times 20}{1.05} = 133{,}333 \text{ in}^2 \text{ ft} \tag{8.18}$$

Assume different values for D and determine the corresponding values of L from Equations 8.15 through 8.18. The results are summarized in the following table and are plotted for comparison.

D (in)	L (ft) (Equation 8.18)	L (ft) (Equation 8.17)	L (ft) (Equation 8.16)	L (ft) (Equation 8.15)
60	37.04	**42.65**	28.43	20.07
72	25.72	**35.54**	23.69	16.72
84	18.90	**30.46**	**20.31**	14.33
96	14.47	**26.66**	**17.77**	12.54
108	11.43	**23.69**	**15.80**	11.15
120	9.26	**21.33**	**14.22**	**10.03**
132	7.65	**19.39**	**12.92**	**9.12**
144	6.43	**17.77**	**11.85**	**8.36**
156	5.48	**16.40**	**10.94**	**7.72**

Analyzing the tabulated/plotted results yields the following conclusions:

- Any combination of D and L that exists in the plot area below the retention time curve is not acceptable.
- For the treater diameters selected in the table, only the values of L shown in bold are acceptable, as they satisfy both settling and retention time constraints.
- As the treating temperature increases, the size of the coalescing/settling section decreases.
- There is no need to treat the emulsion at 125°F, as the reduction in treater size is not significant, and the increased temperature would negatively affect the volume and quality of the treated oil.
- There is a good potential of treating this oil without any heating aid, as the treater size required seems to be practical.
- A practical and economical selection would be an 84 in. diameter by 21 ft long coalescing section with a burner that can provide a treating temperature of 105°F.

Now use Equation 8.9 to calculate the heat requirement, assuming 10% heat losses:

$$q = \frac{1}{1-l} 15Q_o(\Delta T)(\gamma_o c_o + w\gamma_w c_w)$$

$$q = \frac{1}{1-0.1} 15 \times 7000(105-85)(0.86 \times 0.5 + 0.15 \times 1.06 \times 1.1)$$
$$= 1,411,433 \text{ Btu/h}$$

Therefore, a burner rated at 1.5 MM Btu/h would be a good selection.

It should be noted that after the installation of the treater in the field, the operator will run the treater at various settings while inspecting samples of the treated oil until he determines the optimum operating conditions that provide the required treatment with minimum heating. Optimization of the operating conditions is an important activity and should be conducted as often as possible to adapt to changing field conditions.

8.8.2 Sizing Vertical Treaters

8.8.2.1 Water Droplets Settling Constraint

Similar to the analysis performed for the horizontal treater, we shall start with the settling equation (Equation 8.3):

$$u = 1.787 \times 10^{-6} \frac{(\Delta\gamma) d_m}{\mu_o} \text{ ft/sec}$$

This equation gives the terminal settling velocity of the water droplet. For settling to occur, the upward average velocity of the oil must not exceed the water settling velocity. The average oil velocity, u_o, is obtained by dividing the oil volumetric flow rate, Q_o, by the flow cross-sectional area, A, which is the cross-sectional area of the theater: $A = (\pi/4)(D/12)^2$. Therefore,

$$u_o = \frac{Q_o(\text{bbl/day}) \times 5.61(\text{ft}^3/\text{bbl}) \times 1.1574 \times 10^{-5}(\text{day/s})}{(\pi/4)(D/12)^2}$$
$$u_o = 1.19 \times 10^{-2} \frac{Q_o}{D^2} \frac{\text{ft}}{\text{s}} \tag{8.19}$$

Applying equation u_o (Equation 8.19) to u (Equation 8.3), we obtain

$$D^2 = 6665\frac{Q_o\mu_o}{(\Delta\gamma)d_m^2} \text{ in ft} \tag{8.20}$$

8.8.2.2 Retention Time Constraint

The reaction time, t, can be obtained by dividing the volume of the settling/coalescing section occupied by oil, V_o, by the oil volumetric flow rate, Q_o. Let H be the height of the coalescing/settling section (in inches); then,

$$t = \frac{(\pi D^2/4\times 144)(H/12)}{5.61Q_o\times 6.944\times 10^{-4}(\text{day/min})}\text{min}$$

Solving for D^2L, we get

$$D^2H = 8.575Q_o t \text{ in}^2\text{ ft} \tag{8.21}$$

Equations 8.20 and 8.21 can be used for gunbarrel tanks as well. However, when the diameter of the tank is larger than 48 in, the equations must be multiplied by a factor (greater than 1.0) to account for short circuiting.

8.8.2.3 Sizing Procedure

Similar to horizontal treaters, the following producer is primarily aimed at determining the minimum size of the coalescing/settling section of the treater and the rating of the burner.

1. Determine the optimum treating temperature that provides the minimum lose of oil volume and quality along with a practical treater size. If this is not available, the design (following steps) may be executed for different assumed treating temperatures and a final decision is made based on analysis of the design results.
2. Determine the diameter of the water droplet that must be removed (from Equation 8.10 or Equation 8.11).
3. Use Equation 8.20 to obtain the minimum treater diameter D that satisfied the settling constraint.
4. Repeat the previous steps for different assumed treating temperatures and determine the values of D for each treating temperature.
5. Use Equation 8.21 to obtain a relation between D and H that satisfies the retention time constraint. Then, assume different values of D and determine corresponding value of H from this relation.

6. Analyze the results to determine the combinations of D and H for each treating temperature that satisfy both settling and retention time constraints.
7. Use Equation 8.9 to determine the heat requirement for the selected treating temperature.

Example 8.2

Determine the heat requirement and the size of the settling/coalescing section of a single-well vertical heater-treater for the same conditions of Example 8.1 given that the well flow rate is 1200 BPD.

Solution

Use Equation 8.20 to determine the minimum diameter at the three treating temperatures:

For $T = 85°F$;

$$D^2 = 6665\frac{Q_o\mu_o}{(\Delta\gamma)d_m} = 6665\frac{1200\times 45}{0.2(518)^2} = 6707 \text{ in}^2$$

$$D_{85} = 81.89 \text{ in.}$$

For $T = 105°F$;

$$D^2 = 6665\frac{Q_o\mu_o}{(\Delta\gamma)d_m} = 6665\frac{1200\times 20}{0.2(423)^2} = 4476 \text{ in}^2$$

$$D_{105} = 66.86 \text{ in.}$$

For $T = 85°F$;

$$D^2 = 6665\frac{Q_o\mu_o}{(\Delta\gamma)d_m} = 6665\frac{1200\times 10}{0.2(356)^2} = 3155 \text{ in}^2$$

$$D_{125} = 56.17 \text{ in.}$$

Now, use Equation 8.21 for the retention time constraint:

$$D^2H = 8.575Q_o t = 8.575 \times 1200 \times 20 = 205{,}800 \text{ in}^3$$

Assume different values for D and determine corresponding values of H from the preceding relation. The results are plotted as follows: From the figure, all diameters and heights that fall below the retention time curve are not acceptable. For the three treating temperatures, a coalescing section height equal to the value at the intersection with the retention time curve, or larger, will satisfy both retention time and settling constraints. A reasonable selection will be a treater with a 66 in diameter and a 60 in coalescing section height. The treating temperature will be 105°F, with the possibility of treatment being at lower temperatures in reality. The burner rating in determined from Equation 8.9.

REVIEW QUESTIONS AND EXERCISE PROBLEMS

1. Circle the correct answers for the following:
   a. A normal goal in dehydration of crude oil is to reduce the water content to below:
      i. 0.5% by volume
      ii. 5% by volume
      iii. 10% by volume
      iv. None of the above
   b. The objective of oil dehydration is to:
      i. Separate free water only
      ii. Reduce the emulsified water in the oil
      iii. Separate free water (if found) and reduce the emulsified water

9

Desalting of Crude Oil

The next stage in the treatment process of crude oil is desalting. Salt in crude oil is in most cases found dissolved in the remnant water within the oil. It is evident that the amount of salt found in crude oil is attributed to two factors:

1. The quantity of remnant water that is left in oil after normal dehydration
2. The salinity or the initial concentration of salt in the source of this water

In this chapter, a detailed description of the desalting process is presented for single-stage and two-stage types along with some basic design considerations. Emphasis is placed on electrostatic desalting of crude oil. The effect of operating parameters in the desalting operation is discussed along with some tips on troubleshooting problems.

9.1 Introduction

The removal of salts found in the form of what we may call *remnant brine* is carried out in the desalting process. This will reduce the salt content in the crude oil to the acceptable limits of 15 to 20 PTB (pounds of salt, expressed as equivalent sodium chloride, per thousand barrels of oil). After treating the oil by the dehydration and the desalting process, the possibility of stabilizing the crude oil and sweetening exists in the case of sour oil, as will be discussed in Chapter 10. The removal of salt from crude oil for refinery feed stocks has been and still is a mandatory step. This is particularly true if the salt content exceeds 20 PTB. The most economical place for the desalting process is usually in the refinery. However, when marketing or pipeline requirements are imposed, field plants are needed for processing the salty oil prior to shipping. The principles involved are the same whether desalting takes place at the refinery or in the field. Salt in crude oil is, in most cases, found dissolved in the remnant brine within the oil.

The remnant brine is that part of the salty water that cannot be further reduced by any of the dehydration methods described in Chapter 8. It is commonly reported as basic sediments and water (BS&W). It is understood that this remnant water exists in the crude oil as a dispersion of very fine droplets highly

TABLE 9.1

Properties of Crude Oils Shipped to Refineries

	Range	Average
Water in crude, % by volume of crude	0.1–2.0	0.3–0.5
Salt content in crude, PTB	10–250	60–130
Salt concentration in brine, wt%	0.4–25	–
Salt concentration in brine, ppm	4000–50,000	–

TABLE 9.2

Average Values for the PTB for Some Typical Crude Oils

Source of Oil	Average Salt Content (PTB)
Middle East	8
Venezuela	11
Pennsylvania	1
Wyoming	5
East Texas	28
Gulf Coast	35
Oklahoma and Kansas	78
West Texas	261
Canada	200

emulsified in the bulk of oil. The mineral salts of this brine consist mainly of chlorides of sodium, calcium, and magnesium. A summary of the properties of crude oil as received at the refinery is given in Table 9.1, along with the average PTB values for some typical crude oils, as presented in Table 9.2.

9.2 Salt Content of Crude Oil

The amount of salt in the crude oil, defined as pounds of salt per thousand barrels of oil (PTB) is a function of the amount of the brine (remnant water) that remains in the oil W_R (% BS&W) and of its salinity S_R in parts per million (ppm). Mathematically, we can put this in a functional relationship:

$$\text{PTB (Salt content)} = f\,(W_R, S_R)$$

According to Manning and Thomson (1995), the following relationship is reported:

$$\text{PTB} = 350\gamma_{\text{Brine}} \frac{1000\,W_R}{100 - W_R}\left(\frac{S_R}{10^6}\right) \tag{9.1}$$

where the specific gravity is given by γ_{Brine}.

The method of reducing the PTB by lowering the quantity of remnant water (W_R) is usually referred to as the treating process of oil dehydration. This was the main theme of Chapter 8. The other alternative of reducing the PTB is to substantially decrease the dissolved salt content of the remnant water (i.e., its concentration, S_R). This practice is the one we are dealing with in this chapter and is known as desalting, and is achieved by adding water of dilution, followed by separation using one of the aids shown in Figure 9.1.

Desalting of crude oil will eliminate or minimize problems resulting from the presence of mineral salts in crude oil. These salts often deposit chlorides on the heat transfer equipment of the distillation units and cause fouling effects. In addition, some chlorides will decompose under high temperature, forming corrosive hydrochloric acid:

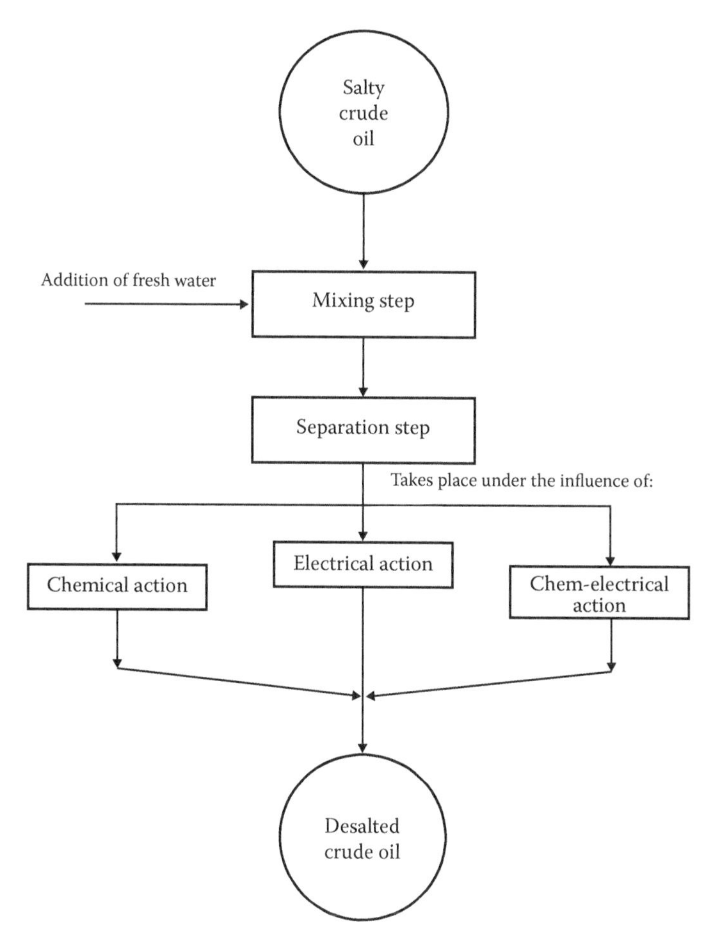

FIGURE 9.1
Illustration of the basic concept in the desalting operation of crude oil.

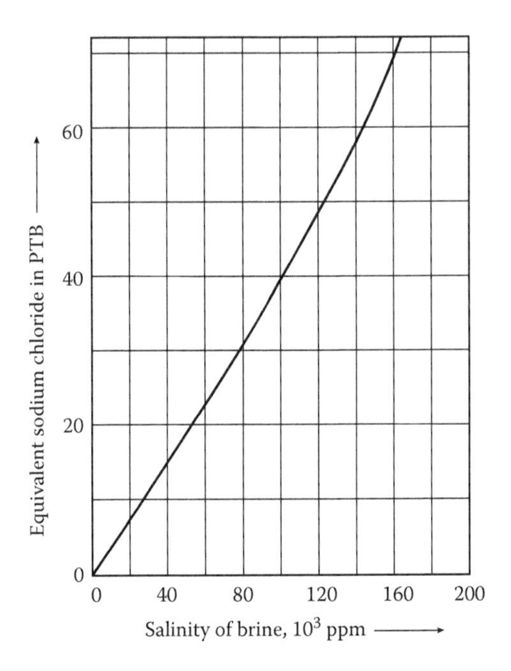

FIGURE 9.2
Salt content for 0.1% by volume remnant brine (water).

$$MgCl_2(aq.) \xrightarrow[H_2O]{\text{High temp.}} Mg(OH)_2 + HCl \tag{9.2}$$

The removal of these salts is aimed at providing an economical operating cycle in the refining process of crude oil. The reduction of salt content to 5 PTB is feasible. Even with this low salt content, it has been reported that the processing of 25,000 bbl/day of crude oil could result in an amount of hydrogen chloride (HCl) equal to 65 lb/day.

The salt content of crude oil could be determined using the following approach. Taking, for example, 1/10 of 1% by volume of water to remain in the crude as a basis for our calculation (remnant brine), the relationship between the salt content of this remnant water expressed in pounds of salt per thousand barrels of oil (PTB) and its concentration or salinity expressed in parts per million is presented graphically in Figure 9.2. For other volumes, simple multiples of the numbers given by this graph are used. The following example illustrates the use of this relationship.

Example 9.1

Find the PTB of a crude oil having 10% by volume remnant water if its concentration is estimated to be 40,000 ppm at 25°C.

Solution

The example is solved using two approaches: One is based on Figure 9.2 and the other approach utilizes basic calculations.

1. Using Figure 9.2, the PTB of crude oil having 0.1% remnant water with 40,000 ppm salinity is found to be 14 PTB. For crude oil containing 10% remnant water, the value of PTB obtained from Figure 9.2 should be multiplied by 100; therefore, the given crude contains 1400 PTB.
2. Take a basis of 1000 bbl of wet oil; the BS&W = 10%, and the saline water concentration = 40,000 ppm = 4%. Then,

Quantity of water in oil = (1000)0.1 = (100 bbl)(5.6 ft^3/bbl) = 560 ft^3

Now, the density of the saline water is estimated using Table 9.3. For 4% concentration and at 25°C, the density is 1.0253 g/cm^3, or 63.3787 lb/ft^3. Hence,

Mass of water = (560 ft^3)63.3787 lb/ft^3 = 35,828 lb

The quantity of NaCl salt found in this mass of water is (35,828)(40,000)/10^6 = 1433 lb. Since our basis is 1000 bbl of oil, the salt content is 1433 PTB.

3. Using Equation 9.1 we get

$$PTB = 350\gamma_{Brine} \frac{1000\,W_R}{100 - W_R}\left(\frac{S_R}{10^6}\right)$$

$$PTB = (350)(1.0253)\frac{1000(10)}{100-10}\left(\frac{40,000}{10^6}\right)$$

$$PTB = 1595$$

TABLE 9.3

Densities of Aqueous Inorganic Solutions (Sodium Chloride [NaCl])

%	0°C	10°C	25°C	40°C	60°C	80°C	100°C
1	1.00747	1.00707	1.00409	0.99908	0.9900	0.9785	0.9651
2	1.01509	1.01442	1.01112	1.00593	0.9967	0.9852	0.9719
4	1.03038	1.02920	1.02530	1.01977	1.0103	0.9988	0.9855
8	1.06121	1.05907	1.05412	1.04798	1.0381	1.0264	1.0134
12	1.09244	1.08946	1.08365	1.07699	1.0667	1.0549	1.0420
16	1.12419	1.12056	1.11401	1.10688	1.0962	1.0842	1.0713
20	1.15663	1.15254	1.14533	1.13774	1.1268	1.1146	1.1017
24	1.18999	1.18557	1.17776	1.16971	1.1584	1.1463	1.1331

Example 9.2

Rework Example 8.2 of Chapter 8, assuming that the water salinity is 40,000 ppm instead of 20,000 ppm and dehydration is done to 2/10 of 1%. Calculate the PTB of the oil.

Solution

With reference to Figure 9.2, the PTB corresponding to 1/10 of 1% is found to be 14. Hence, the PTB of 2/10 of 1% is (14)(0.2/0.1) = 28. This salt content is above the maximum limit of 20 PTB. Therefore, desalting is recommended as a second stage following dehydration in order to reduce the PTB.

9.3 Description of the Desalting Process

It is clear from Example 9.2 that we cannot economically achieve a satisfactory salt content in oil by using dehydration only (single stage). This is particularly true if the salinity of the water produced with oil is much greater than 20,000 ppm (formation water has a concentration of 50,000–250,000 mg/L). Accordingly, a two-stage system (a dehydration stage and a desalting stage) as shown in Figure 9.3a is used. Under certain conditions, however, a three-stage system may be used that consists of a dehydration stage and two consecutive desalting units as shown in Figure 9.3b.

As shown in Figure 9.3, *wash water*, also called *dilution water*, is mixed with the crude oil coming from the dehydration stage. The wash water, which could be either fresh water, or water with lower salinity than the remnant water, mixes with the remnant water, thus diluting its salt concentration. The mixing results in the formation of a water–oil emulsion. The oil (and emulsion) is then dehydrated in a manner similar to that described in Chapter 8. The separated water is disposed of through the field-produced water treatment and disposal system. In the two-stage desalting system, dilution water is added in the second stage and all, or part, of the disposed water in the second stage is recycled and used as the dilution water for the first desalting stage. Two-stage desalting systems are normally used to minimize the wash water requirements.

The mixing step in the desalting of crude oil is normally accomplished by pumping the crude oil (which is the continuous phase) and wash water (which is the dispersed phase) separately through a mixing device. The usual mixing device is simply a throttling valve. The degree of mixing can be enhanced if the interfacial area generated upon mixing is increased. A useful device for such a purpose is the application of multiple-orifice-plate

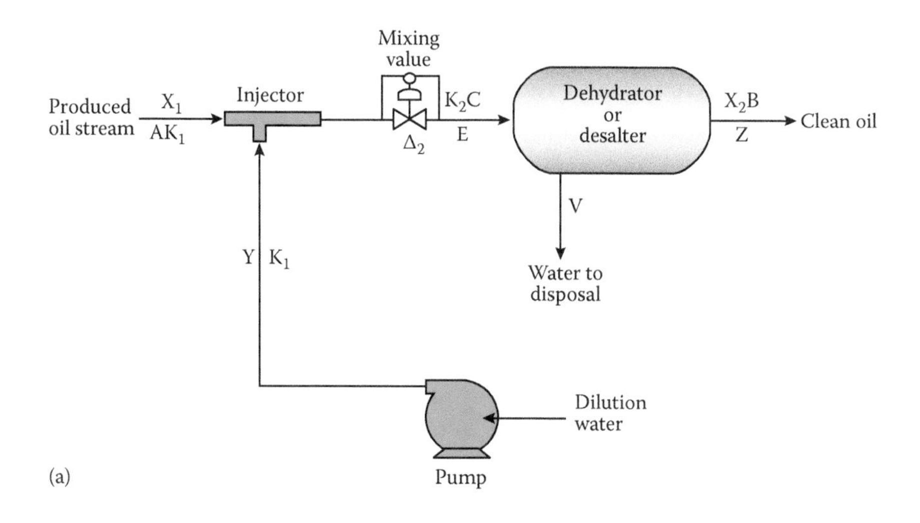

(a)

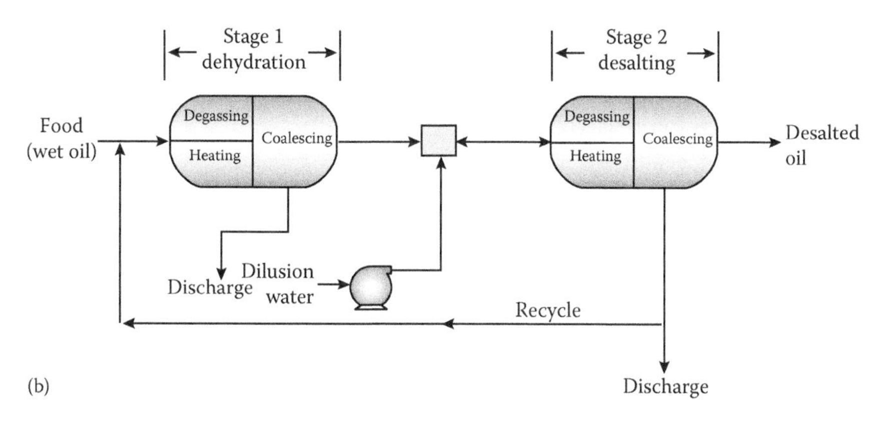

(b)

FIGURE 9.3
(a) Single-stage desalting. (b) Two-stage desalting.

mixers. It is of importance to point out that although the theory of dilution of remnant water with fresh water is sound in principle, it can become impossible to implement in actual application. It all depends on the intimate mixing of remnant water with dilution water.

In the emulsion-treating step, a heating, chemical, or electrical demulsifying aid (or a combination of them) is commonly used (as shown in Figure 9.1). The chemical desalting process involves adding chemical agents and wash water to the preheated oil, followed by settling. The settling time varies from a few minutes to 2 hours. Some of the commonly used chemical agents are sulfonates, long-chain alcohols, and fatty acids.

9.4 Electrostatic Desalting

In this case, an external electric field is applied to coalesce the small water droplets and thus promote settling of the water droplets out of the oil. The electric field may be applied in any of the following modes:

- Alternating current (ac) field devices for water-rich emulsions—Alternating current (ac) is applied, which alternates the polar water molecule arrangements leading to better coalescence. A schematic diagram of ac electrostatic coalescence is shown in Figure 9.4.
- ac/dc field for maximum dehydration—A combination of ac and dc (direct current) is used in this case. The basic configuration of this process is shown in Figure 9.5. The ac is produced in the zone beneath the electrodes, whereas the dc field is produced between adjacent electrodes. This arrangement achieves maximum water removal.
- Variable gradient field for maximum salt reduction—If the field gradient is increased beyond a certain limit (E_c), this will shatter the drops; it is expressed as

$$E_c \leq k\left(\frac{\gamma}{d}\right)^{1/2}$$

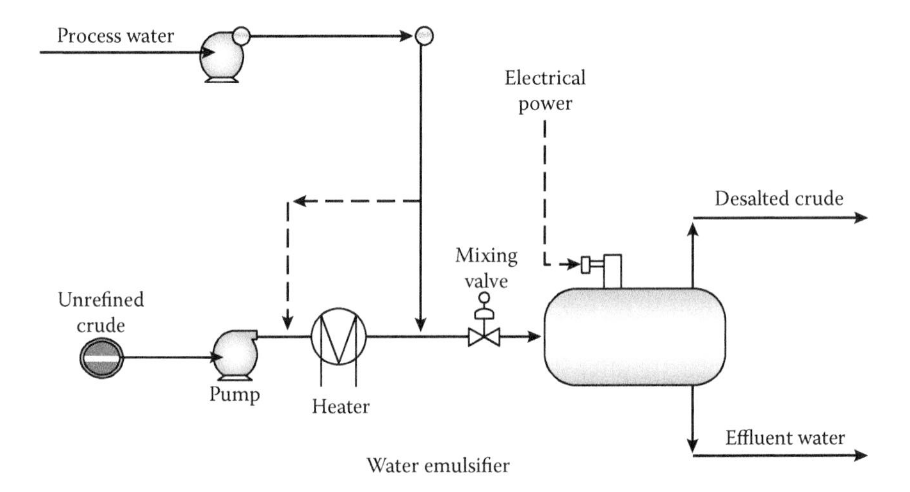

FIGURE 9.4
Electrical desalting.

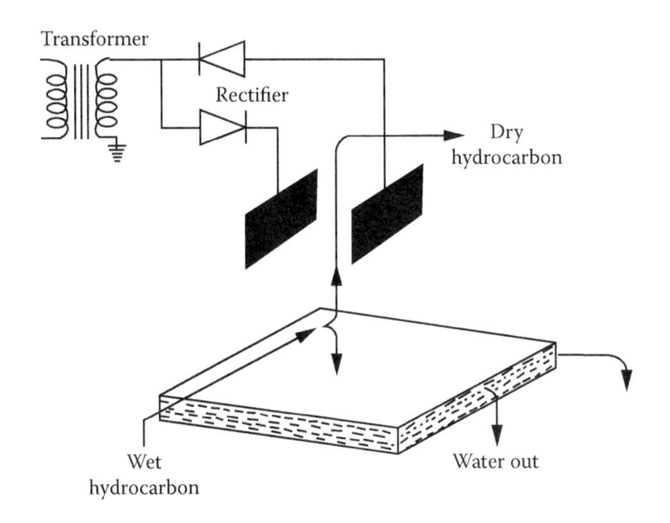

FIGURE 9.5
Dual polarity ac/dc field.

where k is the dielectric constant, γ is the interfacial tension, and d is the drop diameter. Thus, the drop size can be controlled by the field gradient. The electric field can be used both to mix and separate the drops.

By cycling the field strength, the process can be repeated many times during the retention time of the drops within the electric field. The electrostatic field causes the dipolar water molecules to align with the field.

In the electrical desalting process, a high potential field (16,500–33,000 V) is applied across the settling vessels to help coalescence. The chemielectric concept utilizing both chemical agents and an electrical field is a very effective practice.

In the desalting process, it is a common practice to apply enough pressure to suppress any loss of hydrocarbon due to vaporization of the oil. The pressure normally used in a desalting process is in the range 50 to 250 psi.

9.5 Water of Dilution

From the operational point of view, it has been reported that the amount of water of dilution (W_D) added in the desalting of crude oils is in the range 5% to 10% by volume, based on the amount of remnant water and its salinity.

The following correlation has been developed for determining W_D as a function of W_R, S_R, S_D, (salinity of the dilution water) and the efficiency of mixing between the two phases, E:

$$W_D = \frac{2.5 \times 10^3 (W_R)^{0.01533}}{(S_D)^{0.2606} (S_R)^{0.0758} E^{0.6305}} \tag{9.3}$$

where S_D and S_R are in parts per million. On the other hand, the following analytical relationship could set the acceptable limits on the salt content in crude oil. A component material balance for the salt gives

$$EW_D S_D + W_R S_R = S_B(W_D + W_R)$$

or

$$S_B = \frac{EW_D S_D + W_R S_R}{W_D + W_R} \tag{9.4}$$

where S_B refers to some average salinity in the bulk of the homogeneous phase as a result of mixing the remnant water with the fresh water.

9.6 Effect of Operating Parameters

Efficiency of desalting of crude oil is normally dependent on the following parameters:

- Water–crude interface level—This level should be kept constant; any changes will change electrical field and perturbs electrical coalescence.
- Desalting temperature—Temperature affects water droplet settling through its effect on oil viscosity; therefore, heavier crude oils require higher desalting temperatures.
- Wash water ratio—Heavy crudes require a high wash water ratio to increase electrical coalescence. A high wash ratio acts similarly to raise temperatures, as illustrated in Table 9.4.
- Pressure drop in the mixing valve—A high-pressure-drop operation results in the formation of a fine stable emulsion and better washing. However, if the pressure drop is excessive, the emulsion might

TABLE 9.4

Average Desalting Conditions

Crude Gravity (°API)	Desalting Temperature (°C)	Minimum Water Ratio (vol%)
>40	110	2–4
30–40	110	4–8
	120	4–7
<30	130	8–10
	140	>10

be difficult to break. The optimum pressure drop is 1.5 bar for light crudes and 0.5 bar for heavy crudes.

- Type of demulsifiers—Demulsifiers are added to aid in complete electrostatic coalescence and desalting. They are quite important when heavy crudes are handled. Levels ranging between 3 and 10 ppm of the crude are used.

9.7 Design Considerations

The following major parameters are considered when designing a desalting system:

- Flow scheme arrangements (conventional one-stage or countercurrent contact desalters)
- Number of desalting stages
- Dehydration levels achieved
- Salinity of the brine in the crude
- Efficiency of valve mixing
- Salinity of dilution water
- Target PTB specification

9.8 Troubleshooting

Table 9.5 lists some tips that are helpful in solving some of the operating problems or troubles that are of significance to the desalting process.

TABLE 9.5

Problems, Causes, and Solutions

Problems	Causes	Solutions
A high salt content in the desalted crude oil	• Feed salt content high • Wash water injection low • Crude oil flow rate exceeds the design flow rate • Insufficient mixing of the crude oil and wash water	• Increase the wash water rate • Reduce the crude oil flow rate • Increase the mix value pressure drop
Oil in the desalter effluent water	• *Interface* level too low • Wide emulsion band at the *interface* • Excessive crude oil wash water mixing • Poor wash water quality • Crude temperature too low	• Increase the interface level • Inject a chemical or dump the emulsion • Reduce the mix valve pressure drop • Check for any waste in the wash water source
High water carry over in desalted crude oil	• Wash water flow rate too high • Excessive formation water in the crude oil	• Reduce the wash water flow rate and commence or increase chemical injection • Reduce the interface level and check the effluent water valve

REVIEW QUESTIONS

1. Crude oil needs to be desalted for the following two main reasons:
   a.
   b.
2. Crude oil should be desalted if its salt content exceeds ________ PBT.
3. Define the following terms:
   a. Remnant water
   b. Dilution water
   c. BS&W
4. What is the main advantage of a two-stage desalting process over a single-stage desalting process?
5. Explain why desalting takes place after and not during the dehydration (emulsion treatment) process.
6. Sketch and briefly describe a two-stage desalting process.
7. An oil field produces 200,000 bbl/day of net oil with a salt content of 10 PTB. The oil out of the emulsion treater contains 0.3% water (salinity = 250,000 ppm and specific gravity = 1.07). In the desalting–dehydration process, the oil is mixed with dilution water (3000 ppm salinity and 1.02 specific gravity) and dehydrated down to 0.1%

BS&W. Assuming that the mixing efficiency is 80%, determine the following:

a. The salt content in PTB of the oil out of the emulsion treater
b. The amount of dilution water required (in bbl/day)
c. The amount of disposed water (in bbl/day) and its salinity (in ppm)

8. A dehydration–desalting system consists of the following:
   a. A first-stage dehydration only (no wash water added), which receives the feed containing 10% by volume water having a salinity of 250,000 ppm and a specific gravity of 1.06. The dehydration is carried out to produce oil with 0.3% water.
   b. A second-stage desalting, which receives the effluent of the first stage and mixed it with 10% by volume of wash water containing 3000 ppm salt. Then, dehydration is carried out to produce oil with 0.2% BS&W.

   Determine the salt content in PTB of the following:
   i. The oil entering the first stage
   ii. The oil leaving the first stage
   iii. The oil leaving the second stage
9. A two-stage desalting unit is used to treat 100,000 BPD of net oil. The oil entering the first stage contains 3% water having 280,000 ppm salt and 1.07 specific gravity. The oil is mixed with recycled water from the second stage (mixing efficiency = 80%) and dehydrated down to 1% water content. In the second stage, the oil is mixed with 7000 bbl of water having 3000 ppm salt and a specific gravity of 1.02 (mixing efficiency = 80%) and is then dehydrated down to 0.1% water content and salt content of 10 PTB.
   a. Draw a flow diagram of the described process.
   b. Determine the salinity (in lb of salt/bbl of water [lb_s/bbl_w]) at all points.
   c. Determine the salt content of the oil (in PTB) at all points.
   d. Determine the amounts or recycled water and disposed water for the two stages.
   e. If only a one-stage desalting unit was used, determine the amount of dilution.

10

Stabilization and Sweetening of Crude Oil

Stabilizer plants are used to reduce the volatility of stored crude oil and condensate. Process units are designed to maximize recovery of hydrocarbon liquids that might otherwise be lost to the natural gas stream. Increasing the volume of production as well as its API gravity against two important constraints imposed by its vapor pressure and the allowable hydrogen sulfide content is the main target behind stabilization.

This chapter deals with the methods of stabilizing the crude oil. By removing dissolved gases and hydrogen sulfide, crude oil stabilization and sweetening processes diminish safety hazards and corrosion problems. Gases are removed using different types of stabilizers that include tray types and stripping types. Sweetening employs stabilization or vaporization processes along with a gas- or steam-based stripping agent.

10.1 Introduction

Once degassed, dehydrated, and desalted, crude oil is pumped to gathering facilities to be stored in storage tanks. However, if there are any dissolved gases that belong to the light or the intermediate hydrocarbon groups (as was explained in Chapter 3), it will be necessary to remove these gases along with hydrogen sulfide (H_2S; if present in the crude) before oil can be stored. This process is described as a *dual process* of both stabilizing and sweetening a crude oil.

In stabilization, adjusting the pentanes and lighter fractions retained in the stock tank liquid can change the crude oil gravity. The economic value of the crude oil is accordingly influenced by stabilization. First, liquids can be stored and transported to the market more profitably than gas. Second, it is advantageous to minimize gas losses from light crude oil when stored.

This chapter deals with methods for stabilizing the crude oil to maximize the volume of production as well as its API gravity against two important constraints imposed by its vapor pressure and the allowable hydrogen sulfide content.

To illustrate the impact of stabilization and sweetening on the quality of crude oil, the properties of oil before and after treatment are compared as follows:

Before treatment

- Water content is up to 3% of crude in the form of emulsions and from 3% to 30% of crude as free water
- Salt content: 50,000–250,000 mg/L formation water
- Gas: Dissolved gases vary in amounts depending on the gas–oil ratio (GOR)
- H_2S: Up to 1000 ppm by weight

After treatment (dual-purpose operation)

Sour wet crude must be treated to make it safe and environmentally acceptable for storage, processing, and export. Therefore, removing water and salt is mandatory to avoid corrosion; separation of gases and H_2S will make crude oil safe and environmentally acceptable to handle.

- Water content (BS&W): 0.3% by volume, maximum
- Salt content: 10–20 lbs salt (NaCl) per 1000 barrels oil (PTB)
- Vapor pressure: 5–20 psia RVP (Reid vapor pressure)
- H_2S: 10–100 ppmw

Crude oil is considered *sweet* if the dangerous acidic gases are removed from it. On the other hand, it is classified as *sour* if it contains as much as 0.05 ft^3 of dissolved H_2S in 100 gal of oil. Hydrogen sulfide gas is a poison hazard because 0.1% in air is toxically fatal in 30 min. Additional processing is mandatory—via this dual operation—in order to release any residual associated gases along with H_2S present in the crude. Prior to stabilization, crude oil is usually directed to a spheroid for storage in order to reduce its pressure to very near atmospheric, as shown in Figure 10.1.

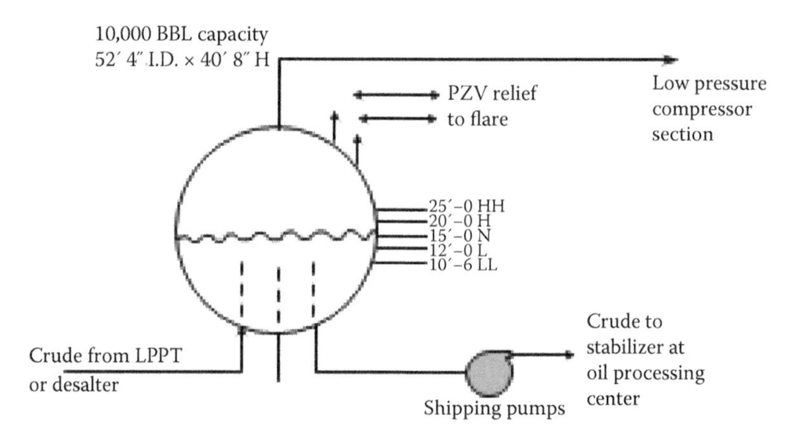

FIGURE 10.1
Typical spheroid for storage prior to stabilization.

10.2 Stabilization Processes

As presented in Chapter 6, the traditional process for separating the crude oil–gas mixture to recover oil consists of a series of flash vessels (gas–oil separation plant [GOSP]) operating over a pressure range from roughly wellhead pressure to nearly atmospheric pressure. The crude oil discharged from the last stage in a GOSP or the desalter has a vapor pressure equal to the total pressure in the last stage. Usually, operation of this system could lead to a crude product with a RVP in the range of 4 to 12 psia. Most of the partial pressure of crude oils comes from the low boiling compounds, which might be present only in small quantities, in particular hydrogen sulfide and low molecular weight hydrocarbons such as methane and ethane.

Now, stabilization is aimed for the removal of these low boiling compounds without losing the more valuable components. This is particularly true for hydrocarbons lost due to vent losses during storage. In addition, high vapor pressure exerted by low boiling point hydrocarbons imposes a safety hazard. Gases evolved from unstable crude are heavier than air and difficult to disperse with a greater risk of explosion.

The stabilization mechanism is based on removing the more volatile components by flashing using stage separation and stripping operations.

As stated in Section 3.1, the two major specifications set for stabilized oil are the Reid vapor pressure (RVP) and hydrogen sulfide content. Based on these specifications, different cases are encountered:

- Case 1—Sweet oil (no hydrogen sulfide); no stabilization is needed. For this case and assuming that there is a gasoline plant existing in the facilities (i.e., a plant designed to recover pentane plus), stabilization could be eliminated, allowing the stock tank vapors to be collected (via the vapor recovery unit [VRU]) and sent directly to the gasoline plant, as shown in Figure 10.2.
- Case 2—Sour crude; stabilization is a must. For this case, it is assumed that the field facilities do not include a gasoline plant. Stabilization of the crude oil could be carried out using one of the approaches outlined in Figure 10.3. Basically, either flashing or stripping stabilization is used.

It can be concluded that the hydrogen sulfide content in the well stream can have a bearing effect on the method of stabilization. Therefore, the recovery of liquid hydrocarbon can be reduced when the stripping requirement to meet the H_2S specifications is more stringent than that to meet the RVP specified. Accordingly, for a given production facility, product specifications must be individually determined for maximum economic return on any investment.

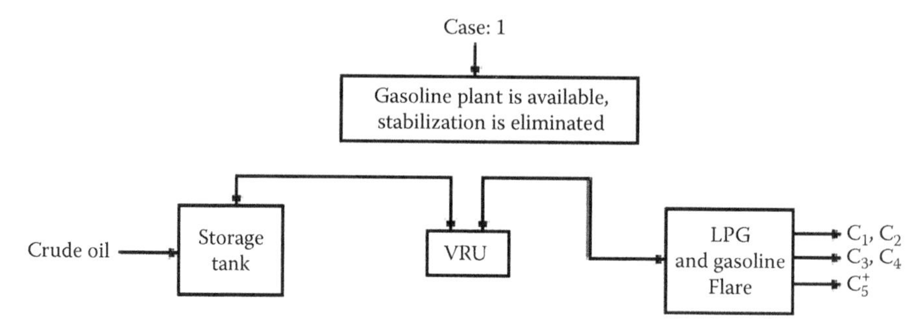

FIGURE 10.2
Field operation with no stabilization.

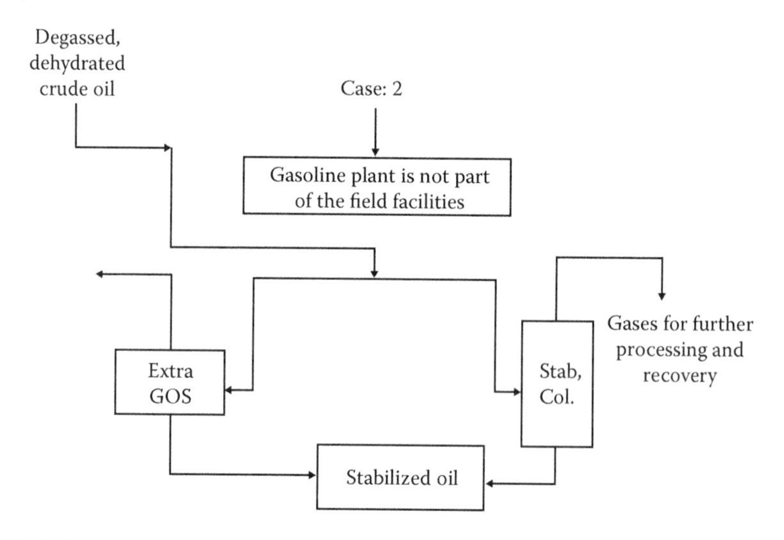

FIGURE 10.3
Alternatives for stabilizing crude oil.

10.2.1 Stabilization by Flashing (Additional Gas–Oil Separator)

Stabilization by flashing utilizes an inexpensive small vessel to be located above the storage tank. The vessel is operated at atmospheric pressure. Vapors separated from the separator are collected using a VRU. This approach is recommended for small-size oil leases handling small volumes of fluids to be processed. The principles underlying the stabilization process are the same as for gas–oil separation covered in Chapter 6.

10.2.2 Stabilization by Stripping

The stripping operation employs a stripping agent, which could be either energy or mass, to drive the undesirable components (low boiling point hydrocarbons and hydrogen sulfide gas) out of the bulk of crude oil. This approach is economically justified when handling large quantities of fluid and in the absence of a VRU. It is also recommended for dual-purpose operations for stabilizing sour crude oil, where stripping gas is used for stabilization. Stabilizer-column installations are used for the stripping operations.

10.3 Types of Trayed Stabilizers

Two basic types of trayed stabilizer are commonly used. Conventional reflux types normally operate from 150 to 300 psia. They are not common in field installations. They are more suitable for large central field processing plants. Nonrefluxed stabilizers generally operate between 55 and 85 psia. These are known as *cold feed* stabilizers. They have some limitations, but they are commonly used in field installations because of their simplicity in design and operation.

10.4 Nonrefluxed Stabilizers

When hydrocarbon liquids are removed from the separators, the liquid is at its vapor pressure or bubble point. With each subsequent pressure reduction, additional vapors are liberated. Therefore, if the liquids were removed directly from a high-pressure separator into a storage tank, vapors generated would cause loss of lighter as well as heavier ones. This explains the need for many stages in a GOSP. Nevertheless, regardless of the number of stages used, some valuable hydrocarbons are lost with the overhead vapor leaving the last stage of separation or the stock tank.

A maximum volume of hydrocarbon liquid could be obtained under stock tank conditions with a minimum loss of solution vapors by fractionating the last-stage separator liquid. This implies using a simple fractionating column, where the vapors liberated by increasing the bottom temperature are counterflowed with the cool feed introduced from the top. Interaction takes place on each tray in the column. The vapors act as a stripping agent and the process is described as stabilization.

10.4.1 Equipment and Operation

In general, a conventional fractionating column would require main auxiliaries such as reflux, pumps, condensers, cooling water, and utilities frequently not available on site in oil fields. Stabilizers or stripping columns, on the other hand, can be operated with a minimum of these auxiliaries. Figure 10.4 depicts a stabilizer in its simplest form.

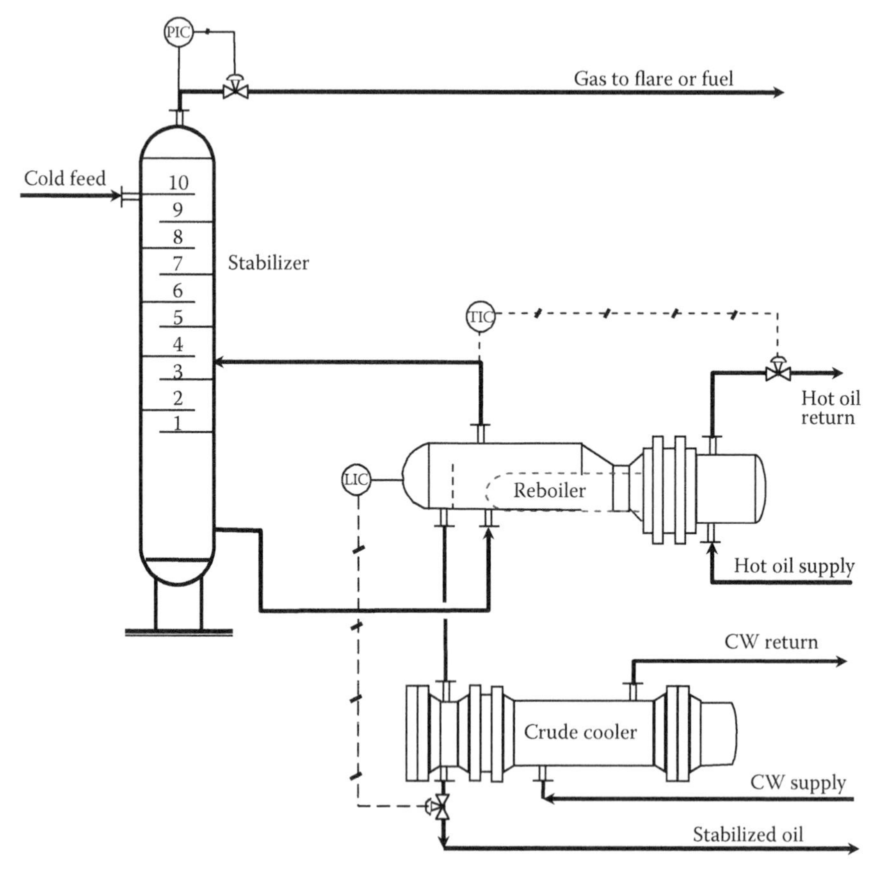

FIGURE 10.4
Typical tray-type stabilizer.

A cold feed stabilizer normally operates with fixed top and bottom temperatures. The former is kept as low as possible to maximize recovery, whereas the latter is controlled to maintain the product bottom pressure. It is of interest to mention that the overhead gas temperature is identical to the liquid feed temperature because the ratio of masses of vapor leaving the column to liquid feed entering is rather small.

Most stabilizers operate above 200 psia and consist of 20 bubble trays. High-pressure stabilizers have more trays because of the higher temperature gradient between the top and the bottom trays. More trays allow the column to operate closer to equilibrium. Columns less than 20 inches in diameter generally use packing rather than trays. A useful rule of thumb is that 1 ft^2 of tower area could handle about 100 bbl/day of stock tank liquid. In some designs, the cold feed is introduced several trays below the top tray, using the upper trays as a scrubber in order to prevent liquid carry over during burping.

Field operation of a stabilizer is described as follows. Relatively cool liquid (oil) exiting the GOSP is fed to the top plate of the column where it contacts the vapor rising from below. The rising vapors strip the lighter ends from the liquid (i.e., acting as a stripping agent). At the same time, the cold liquid—acting as an internal reflux—will condense and dissolve heavier ends from the rising vapor, similar to a rectification process. The net separation is very efficient as compared to stage separation (3%–7% more).

To have a stabilized product of certain specifications, in theory a stabilizer can be operated at multiple combinations of tower pressure and bottom temperature. In general, as the tower pressure is increased, more light ends will condense in the bottom. In normal operation, it is best to operate the tower

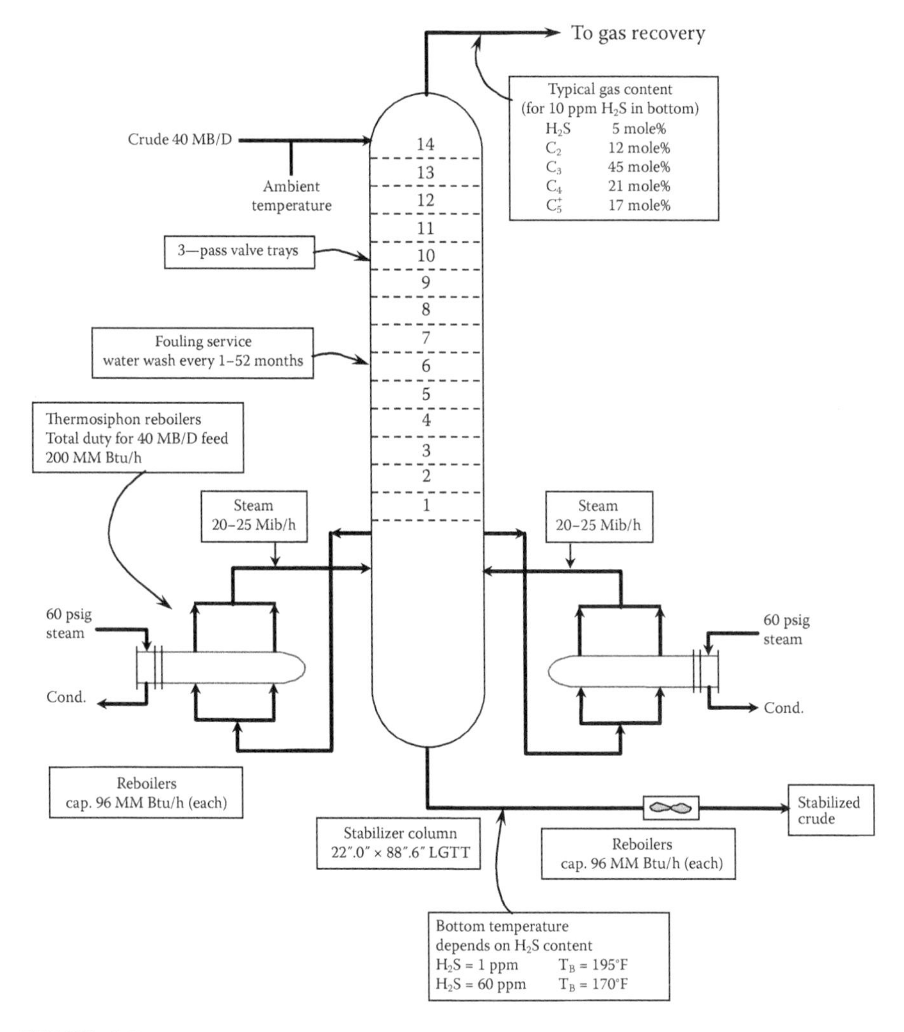

FIGURE 10.5
Typical tray-type stabilizer with operating data.

at the lowest possible pressure without losing too much of the light ends at the initial feed flash. This will minimize burping and cause the column to operate near equilibrium. In addition, lower operating pressures require less reboiler duty with less fuel consumption.

Operating data for a 40,000-bbl/day nonrefluxed stabilizer are given in Figure 10.5.

10.5 Main Features and Applications of Stabilizers

Stabilizers used for oil production field operations should have the following features:

- They must be self-contained and require minimum utilities that are available in the field, such as natural gas for fuel.
- Stabilizers must be capable of unattended operation and withstanding fail-safe operation.
- Stabilizers must be equipped with simple but reliable control system.
- They should be designed in a way to make them accessible for easy dismantling and reassembly in the field.
- Maintenance of stabilizers should be made simple and straightforward.

Applications of stabilizers, on the other hand, are justified over simple stage separation under the following operating conditions:

- The first-stage separation temperature is between 0°F and 40°F.
- The first-stage separation pressure is greater than 1200 psig.
- The liquid gravity of the stock tank oil is greater than 45° API.
- Oil to be stabilized contains significant quantities of pentanes plus, even though the oil gravity is less than 45° API.
- Specifications are set by the market for product compositions—obtained from an oil—that require minimum light ends.

10.6 Crude Oil Sweetening

Apart from stabilization problems of *sweet* crude oil, *sour* crude oils containing hydrogen sulfide, mercaptans, and other sulfur compounds present unusual processing problems in oil field production facilities. The presence of hydrogen sulfide and other sulfur compounds in the well stream impose many constraints. Most important are the following:

- Personnel safety and corrosion considerations require that H_2S concentration be lowered to a safe level.
- Brass and copper materials are particularly reactive with sulfur compounds; their use should be prohibited.
- Sulfide stress cracking problems occur in steel structures.
- Mercaptan compounds have an objectionable odor.

Along with stabilization, crude oil sweetening brings in what is called a *dual operation*, which permits easier and safe downstream handling, and improves and upgrades the crude marketability.

Three general schemes are used to sweeten crude oil at the production facilities:

Process	Stripping Agent
Stage vaporization with stripping gas	Mass (gas)
Trayed stabilization with stripping gas	Mass (gas)
Reboiled tray stabilization	Energy (heat)

10.6.1 Stage Vaporization with Stripping Gas

Stage vaporization with stripping gas, as its name implies, utilizes stage separation along with a stripping agent. Hydrogen sulfide is normally the major sour component having a vapor pressure greater than propane but less than ethane. Normal stage separation will, therefore, liberate ethane and propane from the stock tank liquid along with hydrogen sulfide. Stripping efficiency of the system can be improved by mixing a lean (sweet) stripping gas along with the separator liquid between each separation stage.

Figure 10.6 represents typical stage vaporization with stripping gas for crude oil sweetening/stabilization. The effectiveness of this process depends on the pressure available at the first-stage separator (as a driving force), well stream composition, and the final specifications set for the sweet oil.

10.6.2 Trayed Stabilization with Stripping Gas

In the trayed stabilization with stripping gas process, a tray stabilizer (non-refluxed) with sweet gas as a stripping agent is used, as shown in Figure 10.7.

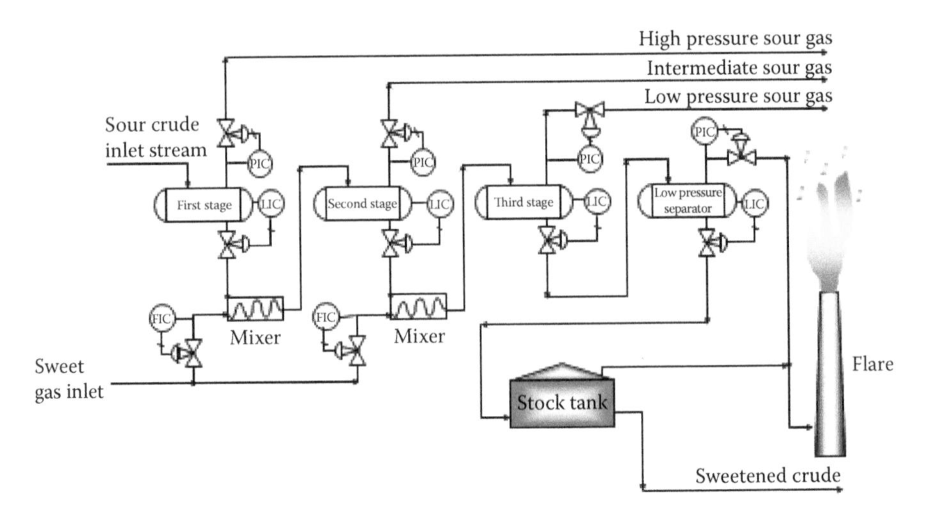

FIGURE 10.6
Sweetening by stage vaporization with stripping gas.

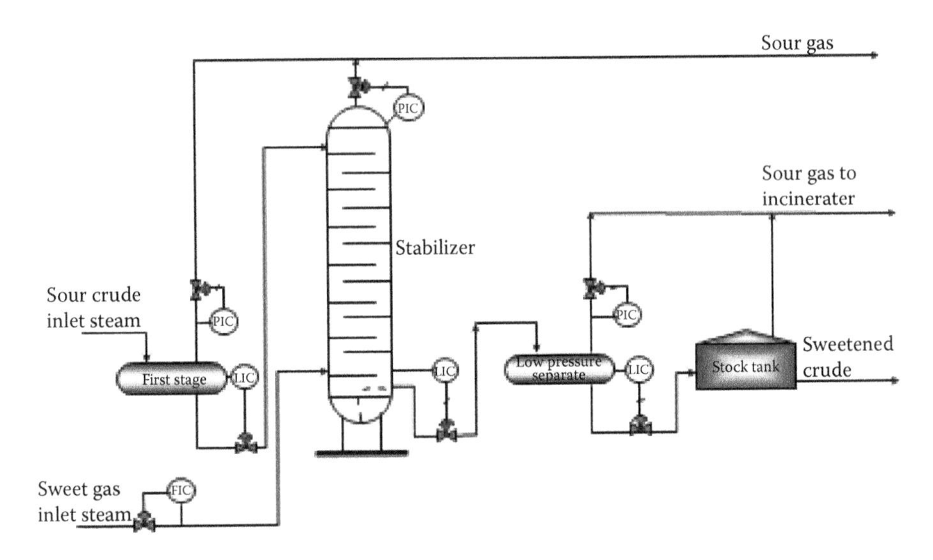

FIGURE 10.7
Crude stabilization by trayed stabilization with stripping gas.

Oil leaving a primary separator is fed to the top tray of the column counter-current to the stripping sweet gas.

The tower bottom is flashed in a low-pressure stripper. Sweetened crude is sent to stock tanks, whereas vapors collected from the top of the gas separator and the tank are normally incinerated. These vapors cannot be vented to the atmosphere because of safety considerations. Hydrogen

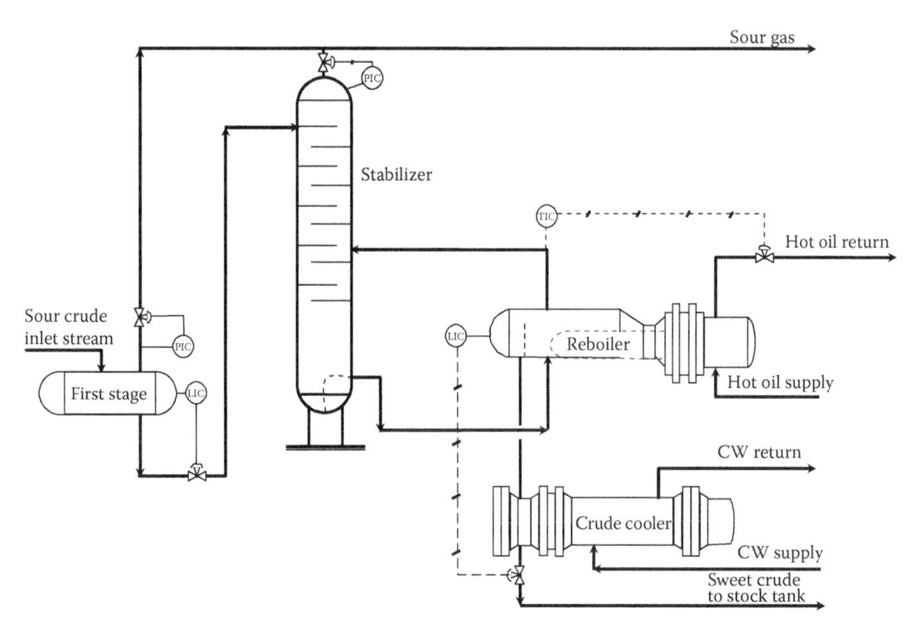

FIGURE 10.8
Crude sweetening by reboiled trayed stabilization.

sulfide is hazardous and slightly heavier than air; it can collect in sumps or terrain depressions.

This process is more efficient than the previous one. However, tray efficiencies cause a serious limitation on the column height. For an efficiency of only 8%, 1 theoretical plate would require 12 actual trays.*

10.6.3 Reboiled Trayed Stabilization

The reboiled trayed stabilizer is the most effective means to sweeten sour crude oils. A typical reboiled trayed stabilizer is shown in Figure 10.8. Its operation is similar to a stabilizer with stripping gas, except that a reboiler generates the stripping vapors flowing up the column rather than using a stripping gas. These vapors are more effective because they possess energy momentum due to elevated temperature.

Due to the fact that hydrogen sulfide has a vapor pressure higher than propane, it is relatively easy to drive hydrogen sulfide from the oil. Conversely, the trayed stabilizer provides enough vapor/liquid contact that little pentanes plus are lost to the overhead.

* Columns are limited to 24–28 ft high, or a maximum of two theoretical trays because trays are spaced about 2 ft apart.

REVIEW QUESTIONS

1. Explain the function of crude oil stabilization operation.
2. Why do lighter crude oils need stabilization but heavier crude oil may not need stabilization?
3. What is the difference between sweet and sour crude oils?
4. What is the function of the crude oil sweetening process?
5. Why it is necessary to remove H_2S from crude oil?
6. Describe how stabilization by flashing is achieved. Under what conditions could this method of stabilization be used?
7. Describe the principle of stripping operation for crude oil stabilization.
8. What are the conditions under which crude oil stabilization should be employed?
9. Describe the principles of operation of the following crude oil sweetening processes:
   a. Stage vaporization with stripping gas
   b. Trayed stabilization with stripping gas
   c. Reboiled trayed stabilization

11

Other Treatment Options

Crude oil once produced and separated undergoes treatment in the chronological order covered in Chapters 8 through 10. These treatment processes are a must in order to have a quality product oil. In this chapter, a different avenue in the treatment process of crude oil is considered. The following criteria are some typical cases that handle crude oil treatment:

- To treat the crude oil to upgrade its quality, shooting for market value added (MVA)
- To treat the crude oil to enhance its quality, aiming for a higher sales value added (SVA)
- Treatment under special conditions

11.1 Economic Treatment for MVA—Case 1

Treatment of crude oil–gas mixture was carried out as we have seen following the traditional procedure usually adopted in field processing. However, crude oil may require treatment as demonstrated in the following cases:

- Case 1: Economic treatment for market value added (MVA)
- Case 2: Economic treatment for sales value added (SVA)
- Case 3: Treatment under special conditions

When we talk about the economic treatment of crude oil in the field, one should be aware of the total costs and expenses incurred in producing the oil and how they are calculated, as was explained in Chapter 5.

In an oil-producing field, the total costs spent in producing one barrel of oil is the sum of two main components:

1. The depletion costs spent in the pre-oil-production phase.
2. Depreciation costs for the physical assets (equipment) plus the operating expenses spent in the post-oil-production phase. Division of the total cost ($) by field production (barrels), should give $/bbl.

In our discussion for case 2, we will be concerned only with the first component, that is, the depletion costs.

11.2 Economic Treatment for SVA—Case 2

In this case, it is envisaged to use a treating process for the crude oil in order to obtain an added market value. MVA is defined as the difference between the current market value of an establishment or an asset, and the original capital investment. If MVA is positive, the firm has added value. If it is negative, the firm has lost value. Mathematically:

$$\mathrm{MVA} = \mathrm{C} - \mathrm{C}_o$$

where MVA is the value added or gained, C is the current value of an asset, and C_o is the original capital investment.

Now, our objective in pursuing an economic treatment for a crude oil is to upgrade its quality by using a predefined process keeping in mind the aforementioned criterion.

The MVA of a crude oil is determined by many factors such as sulfur content, API gravity, and oil viscosity.

The following case study, "Economic Treatment of Whole Crude Oil, for Odor/Corrosion Control & Value-Added Marketing, after COQA" (Merichem Company 2012), is presented in the following condensed form:

Based on customer value proposition, mercapatn treatment of sour cruder oil could lead to numerous benefits. This includes odor control, corrosion control, and toxicity minimization. In return, benefits gained are reported to be a reduction in transportation costs and a drop in odor complaints.

The design capacity of the proposed treating unit is capable of handling 65,000–190,000 BPSD.

PROCESS ANALYSIS

In the study a comparison for the properties of the feed (crude oil), before and after treatment has been reported and is shown next:

Concentration in Crude Oil

	Before Treatment (wppm)	After Treatment (wppm)
CO_2	40	Nil
H_2S	60	1
Mercaptans, as **S**	150–650	30 (max)

The chemical treating process involved the use of caustic soda to extract H_2S, CO_2, and naphethenic acid, while mercaptans are extracted and oxidized by direct contact with caustic soda in the presence of air.

The operational summary for the (treating) sweetening process included the following parameters:

1. Temperature
2. Pressure
3. Chemicals and utilities
4. Plot space requirements and maintenance requirements.

COST ANALYSIS

In this case study, the following information was reported:

1st Initial Investment: No major equipment costs are involved.
2nd Operating Costs: This included costs for caustic soda, oxidation air, electricity, and others.

CONCLUSION

The article as such did not give further details on calculating the market value added as a result of the additional treating (sweetening) process carried out for the crude oil.

11.3 Other Treatments—Case 3

When we talk about the economic treatment of crude oil in the field, one should be aware of the total costs and expenses incurred in producing the oil, and how they are calculated.

Following the presentation given in Chapter 5 on financial measures and profitability analysis in petroleum operations, this method is proposed. The steps involved in calculating the SVA for a proposed crude oil treating process are much simpler than MVA. Once a treatment process is suggested for upgrading the quality of a crude oil, the following procedure is recommended to calculate the SVA to the oil:

- Assume the current selling price of the crude oil before treatment, $\$/bbl = S_1$
- Assume the selling price of the crude oil after treatment, $\$/bbl = S_2$
- Assume the crude oil production, bbl/year = P
- Annual sales gain, $\$/year = [S_2 - S_1]P = S_G$
- Assume the capital cost of equipment used in the treating unit, $\$ = C_1$, and the lifetime of the equipment in years = n

- Depreciation costs of the unit, \$/year = $C_1/n = C_o$
- Assume the operating expenses of the unit, \$/year = C_2
- Total annual expenses of running the treating unit, \$/year = $C_o + C_2 = C_T$

Now, comparing S_G to C_T: If $S_G > C_T$, then the firm has added value to the sales by using the treating process; otherwise do not consider such a process a route for economic treatment.

$$\text{Net value of sales added} = \text{SVA} = S_G - C_T$$

11.4 Case 3: Other Treatments

There are situations where certain crude oils will require specific treatment options in order to market them. Examples are mercaptans and hydrogen sulfide in crude oil, light crude oil, and heavy crude oil.

11.4.1 Sour Crudes: Removal of Hydrogen Sulfide in the Field

According to Wolf Jr. and McLean (1979), a chemical treating process takes place between the emulsion separator equipment and the production tankage. An aqueous ammonium hydroxide solution is used to convert the hydrogen sulfide to a water-soluble ammonium sulfide salt. The aqueous phase separates in the production tank carrying with it the hydrogen sulfide as ammonium sulfide. The primary consideration for treatment is the reduction of the hydrogen sulfide content of the oil phase from 900 ppm to 50 ppm or less.

11.4.2 Cleaning Process for Heavy Crude Oil

Implementing the Crude Oil Quality Association (COQA) technology would lead to the reduction in impurities and improve the quality of heavy crude oil. In this respect Doctor and Mustafa proposed a cleaning process that can be deployed at oil production facilities by adding selective solvent to the crude oil. Solvent is recovered and recycled. They claim that the process will reduce the sulfur content in the crude oil and improve the viscosity of oil, avoiding the price penalty for the transport of high sulfur crude.

11.4.3 Adding Diluents to Heavy Crude Oils

A classification based on the API gravity is made to identify types of crude oils as follows:

- Heavy crude—API gravity of 18 degrees or less
- Intermediate—API greater than 18 degrees and less than 36
- Light—API gravity of 36 degrees or greater

When it comes to viscosity, some crude oils can have viscosities in excess of 15,000 centistokes at 100°F. In order for these crudes to be transferred from the source by pipelines, viscosity must be lower than 150 centistokes at 100°F. It is widely assumed that the asphaltene molecules found in oil agglomerate to form micelle-like clusters. Interaction between these clusters will contribute toward the viscosity of the oil. Breaking these agglomerates is the target of many additives in order to significantly reduce the viscosity.

Treating of heavy crudes by using industrial additives (polymers) or viscosity reducers is practiced in order to bring in drag reduction and reduce the viscosity. This will lead to increased pipeline deliveries of the oil. In other words, more oil to the market.

In some other cases, a diluent could be a natural gas condensate or liquefied natural gas (LNG). This is exemplified by what is known as *dilbit blends,* which are made from heavy crudes and/or bitumen and a diluent. Again, this is a means of transporting highly viscous crude oil.

Section IV

Gas Handling and Treatment

Section IV of this book is devoted to field treatment and processing operations of natural gas and other associated products. These include sour gas treatment (acidic gas removal [hydrogen sulfide, H_2S, and carbon dioxide, CO_2]), dehydration, and the separation and fractionation of liquid hydrocarbons (known as natural gas liquid, NGL), covered in Chapters 12 through 14, respectively. Sweetening of natural gas almost always precedes dehydration and other gas plant processes carried out for the separation of NGL. Dehydration, on the other hand, is usually required before the gas can be sold for pipeline marketing and it is a necessary step in the recovery of NGL from natural gas.

For convenience, a system involving field treatment of a gas plant could be divided into two main stages, as shown next:

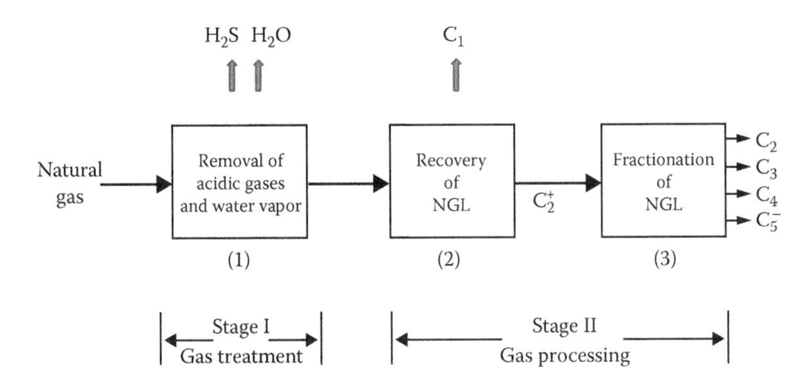

- Stage I, known as gas treatment or gas conditioning
- Stage II, known as gas processing

The gas treatment operations carried out in stage I involve the removal of gas contaminants (acidic gases), followed by the separation of water vapor (dehydration), as will be explained in Chapters 12 and 13, respectively. Gas processing, stage II, on the other hand, comprises two operations: NGL recovery and separation from the bulk of gas and its subsequent fractionation into desired products. The purpose of a fractionator's facility is simply to produce individual finished streams needed for market sales. Fractionation facilities play a significant role in gas plants, as given in Chapter 14.

Gas field processing in general is carried out for two main objectives:

1. The necessity to remove impurities from the gas
2. The desirability to increase liquid recovery above that obtained by conventional gas processing

Natural gas field processing and the removal of various components from it tend to involve the most complex and expensive processes. Natural gas leaving the field can have several components that will require removal before the gas can be sold to a pipeline gas transmission company. All of the H_2S and most of the water vapor, CO_2, and N_2 must be removed from the gas. Gas compression is often required during these various processing steps. To illustrate this point, a sour gas leaving a gas–oil separation plant (GOSP) might require first the use of an amine unit (monoethanolamine, MEA) to remove the acidic gases, a glycol unit (triethylene glycol, TEG) to dehydrate it, and a gas compressor to compress it before it can be sold.

It is also generally desirable to recover NGL present in the gas in appreciable quantities. This normally includes the hydrocarbons known as C_3^+. In many cases, ethane C_2 could be separated and sold as a petrochemical feed stock. NGL recovery is the first operation in stage II, as explained. To recover and separate NGL from a bulk of a gas stream would require a change in phase; that is, a new phase has to be developed for separation to take place by using one of the following:

- An energy-separating agent; examples are refrigeration for partial or total liquefaction and fractionation
- A mass-separating agent; examples are adsorption and absorption (using selective hydrocarbons, 100–180 molecular weight)

The second operation in stage II is concerned with the fractionation of NGL product into specific cuts such as LPG (C_3/C_4) and natural gasoline.

The fact that all field processes do not occur at or in the vicinity of the production operation does not change the plan of the system of gas processing and separation or its basic needs. This requires superimposing the fractionation facilities to the gas field processing scheme.

In the design of a system for gas field processing, the following parameters should be evaluated and considered in the study:

- Estimated gas reserve (both associated and free)
- The gas flow rate and composition of the feed gas
- Market demand, both local and export, for the products
- Geographic location and methods of shipping of finished products
- Environmental factors
- Risks involved in implementing the project and its economics

Of these factors, the gas/oil reserve might be the paramount factor. Several schemes can be recommended for field processing and separation of natural gas, but the specific solution is found to be a function of the composition of the gas stream, the location of this source, and the markets available for the products obtained.

Field processing operations of natural gas, which is classified as a part of gas engineering, generally include the following: removal of water vapor, dehydration; removal of acidic gases (H_2S and CO_2); and separation of heavy hydrocarbons. Before these processes are detailed in Chapters 12 through 14, the effect each of these impurities has on the gas industry, as end user, are briefly outlined:

Water Vapor	H_2S and CO_2	Liquid Hydrocarbons
It is a common impurity. It is not objectionable as such. • Liquid water accelerates corrosion in the presence of H_2S gas. • Solid hydrates, made up of water and hydrocarbons, plug valves, fittings in pipelines, and so forth.	Both gases are harmful, especially H_2S, which is toxic if burned; it gives SO_2 and SO_3, which are a nuisance to consumers. • Both gases are corrosive in the presence of water. • CO_2 contributes a lower heating value to the gas.	Their presence is undesirable in the gas used as a fuel. • The liquid form is objectionable for burners designed for gas fuels. • For pipelines, it is a serious problem to handle two-phase flow: liquid and gas.

Unit operation(s) underlying gas handling and treatment that involves gas sweetening and dehydration are shown in bold in the diagram:

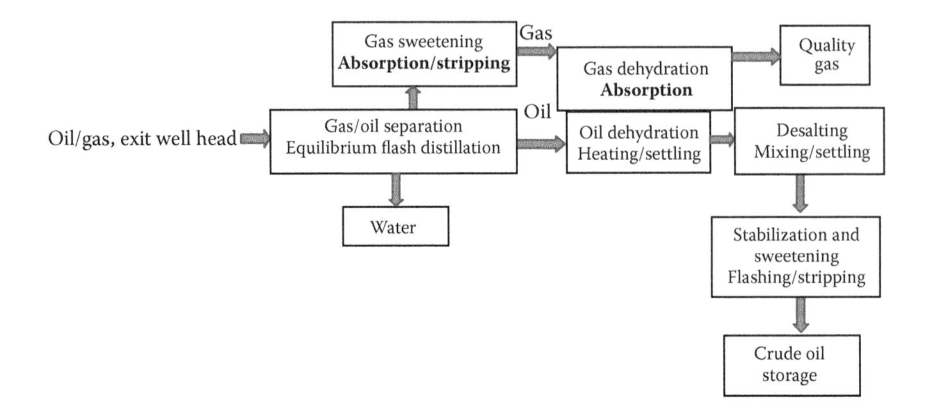

12

Sour Gas Treatment

Natural gas has a wide range of acid concentrations, from parts per million to 50 volume percentage and higher. Gas sales in the United States require to contain no more than 4 ppm (parts per million). The most widely used processes to sweeten sour natural gas are the alkonolamines, and the most common ones are monoethanol amine (MEA) and diethanolamine (DEA).

In this chapter gas sweetening processes are presented and the factors considered in the selection of a given process are identified. Batch sweetening processes including iron sponge, zinc oxide, and molecular sieves are compared versus liquid-phase processes. The latter ones are further subclassified into two categories: chemical adsorption using regenerative alkanolamines solvents (amine processes) and potassium carbonate processes; and physical absorption using physical solvents.

Direct conversion processes for sour gas treating are discussed. These methods utilize an alkaline solution containing an oxidizing agent to convert hydrogen sulfide into sulfur; the Stretford process is a typical example. Membrane processes for the selective permeation of gas species are reviewed as well.

12.1 Introduction

Natural gas usually contains some impurities such as hydrogen sulfide (H_2S), carbon dioxide (CO_2), water vapor (H_2O), and heavy hydrocarbons such as mercaptans. These compounds are known as *acid gases*. Natural gas with H_2S or other sulfur compounds (such as carbonyl sulfide [COS], carbon disulfide [CS_2], and mercaptans) is called *sour gas*, whereas gas with only CO_2 is called *sweet gas*. It is usually desirable to remove both H_2S and CO_2 to prevent corrosion problems and to increase heating value of the gas.

Sweetening of natural gas is one of the most important steps in gas processing for the following reasons:

- Health hazards—At 0.13 ppm, H_2S can be sensed by smell. At 4.6 ppm, the smell is quite noticeable. As the concentration increases beyond 200 ppm, the sense of smell fatigues, and the gas can no longer be detected by odor. At 500 ppm, breathing problems are observed

TABLE 12.1

Natural Gas Pipeline Specifications

Characteristic	Specification	Test Method
Water content	4–7 lb/MMSCF maximum	ASTM (1986), D 1142
Hydrogen sulfide content	0.25 grain/100 SCF maximum	GPA (1968), Std. 2265
		GPA (1986), Std. 2377
Gross heating value	950 Btu/SCF minimum	GPA (1986), Std. 2172
Hydrocarbon dew point	15°F at 800 psig maximum	ASTM (1986), D 1142
Mercaptan content	0.2 grain/100 SCF maximum	GPA (1968), Std. 2265
Total sulfur content	1–5 grain/100 SCF maximum	ASTM (1980), D 1072
Carbon dioxide content	1–3 mol% maximum	GPA (1990), Std. 2261
Oxygen content	0–0.4 mol% maximum	GPA (1990), Std. 2261
Sand, dust, gums, and free liquid	Commercially free	
Delivery temperature (°F)	120°F maximum	
Delivery pressure (psia)	700 psig minimum	

Note: MMSCF ≡ 1,000,000 standard cubic feet.

and death can be expected in minutes. At 1000 ppm, death occurs immediately.

- Sales contracts—Three of the most important natural gas pipeline specifications are related to sulfur content, as shown in Table 12.1. Such contracts depend on negotiations, but they are quite strict about H_2S content.
- Corrosion problems—If the partial pressure of CO_2 exceeds 15 psia, inhibitors usually can only be used to prevent corrosion. The partial pressure of CO_2 depends on the mole fraction of CO_2 in the gas and the natural gas pressure. Corrosion rates will also depend on temperature. Special metallurgy should be used if CO_2 partial pressure exceeds 15 psia. The presence of H_2S will cause metal embrittlement due to the stresses formed around metal sulfides formed.

12.2 Gas-Sweetening Processes

There are more than 30 processes for natural gas sweetening. The most important of these processes can be classified as follows:

- Batch solid-bed absorption—For complete removal of H_2S at low concentrations, the following materials can be used: iron sponge, molecular sieve, and zinc oxide. If the reactants are discarded, then this method is suitable for removing a small amount of sulfur when the gas flow rate is low or the H_2S concentration is also low.

- Reactive solvents—MEA (monoethanolamine), DEA (diethanolamine), DGA (diglycolamine), DIPA (diisopropanolamine), hot potassium carbonate, and mixed solvents. These solutions are used to remove large amounts of H_2S and CO_2, and the solvents are regenerated.
- Physical solvents—Selexol, Recitisol, Purisol, and Fluor solvent. They are mostly used to remove CO_2 and are regenerated.
- Direct oxidation to sulfur—Stretford, Sulferox LOCAT, and Claus. These processes eliminate H_2S emissions.
- Membranes—This is used for very high CO_2 concentrations. AVIR, Air Products, Cynara (Dow), DuPont, Grace, International Permeation, and Monsanto are some of these processes.

12.3 Selection of Sweetening Process

There are many factors to be considered in the selection of a given sweetening process. These include:

- Type of impurities to be removed (H_2S, mercaptans)
- Inlet and outlet acid gas concentrations
- Gas flow rate, temperature, and pressure
- Feasibility of sulfur recovery
- Acid gas selectivity required
- Presence of heavy aromatic in the gas
- Well location
- Environmental consideration
- Relative economics

Figure 12.1 can be utilized for the initial selection of the proper process. It depends on the sulfur content in the feed and the desired product. Several commercial processes are available, as shown in the schematic flow sheet of Figure 12.2. This diagram can help in selecting the suitable process. If the sulfur recovery is not an option, then indirect processes can be considered. These processes can be classified into liquid-phase processes and dry-bed processes. The latter is selected if the sulfur content in the feed is low, as supported by Figure 12.1. In this case, the total sulfur to be removed should not be large. The dry-bed processes can be classified as iron oxide sponge or zinc oxide. In these cases, because the oxides are not expensive, the corresponding sulfide might not be regenerated on site but rather by a contracting

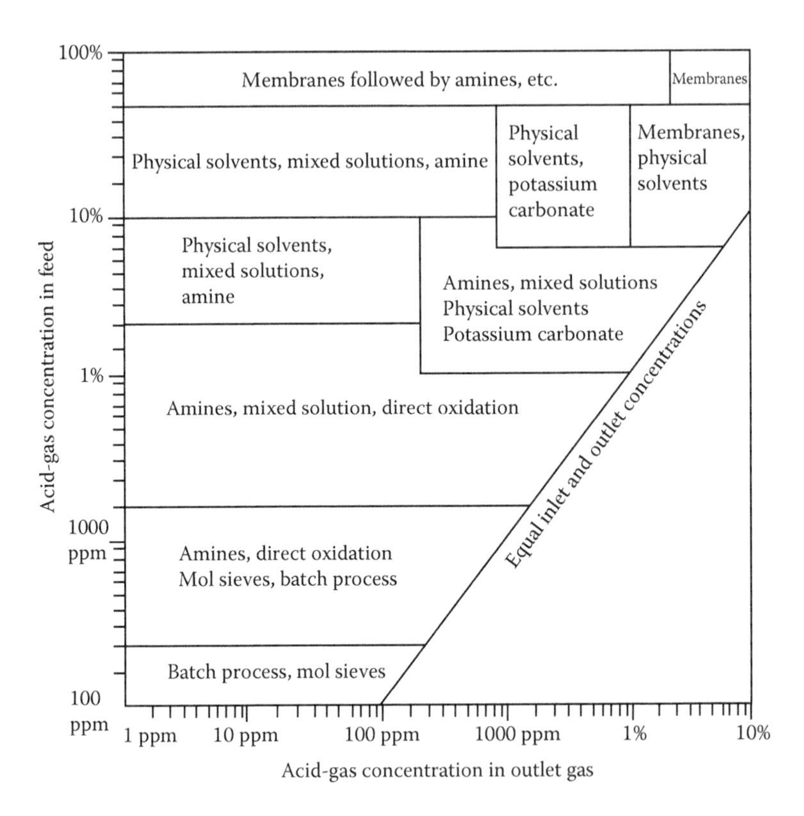

FIGURE 12.1
How to select gas-sweetening process.

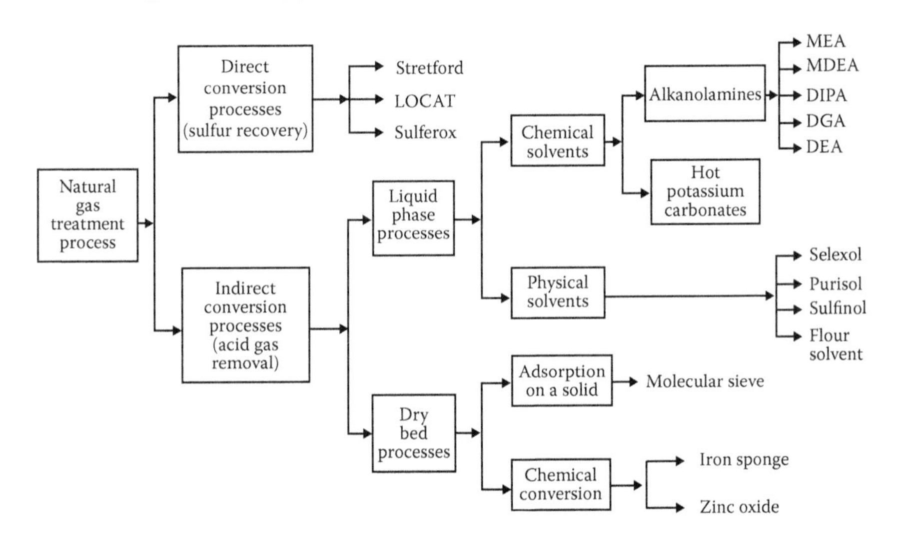

FIGURE 12.2
Alternatives for natural gas sweetening.

company, and this might lower the operational cost and does not require further processing for sulfur recovery. Alternatively, the molecular sieve bed can be used when the bed is regenerated on site.

If the operating conditions require the removal of a large amount of H_2S and CO_2, liquid-phase processes are used. If it is required to selectively remove H_2S, then physical solvents will be suitable; they also remove COS and CS_2. If the feed contains a high amount of heavy hydrocarbons $\left(C_3^+\right)$, then the use of physical solvents may result in a significant loss of these heavy hydrocarbons because they are released from the solvent with the acid gases and cannot be economically recovered. On the other hand, if the feed contains a high amount of H_2S and CO_2 and it is required to remove both of them, then chemical solvents such as amines or carbonates could be used. The amine processes offer good reactivity at low cost and good flexibility in design and operation. However, carbonates can be used if it is required to remove COS and CS_2 from the feed. The carbonate process also can be run at a lower utility cost.

If sulfur recovery is an option, then a direct conversion process can be used. They are only selective for H_2S. Stretford, LOCAT, and Sulferox are some of the processes used for removal of H_2S from natural gas. In the case of the regeneration of solid fixed beds or amine regeneration units, concentrated streams of H_2S are obtained. In these cases, the Claus process will be the most suitable for sulfur recovery.

Figure 12.1 can be used as a guide for the selection of the proper process depending on the acid–gas concentration in the feed and the required degree of acid removal, and its concentration in the outlet stream. However, final selection should consider environmental, economical, and any other concerns.

12.4 Batch Processes

In batch processes, H_2S is basically removed and the presence of CO_2 does not affect the processes. Usually, batch processes are used for low-sulfur-content feeds.

12.4.1 Iron Sponge

Iron sponge fixed-bed chemical absorption is the most widely used batch process. This process is applied to sour gases with low H_2S concentrations (300 ppm) operating at low to moderate pressures (50–500 psig). Carbon dioxide is not removed by this treatment. The inlet gas is fed at the top of the fixed-bed reactor filled with hydrated iron oxide and wood chips. The basic reaction is the formation of ferric sulfide when H_2S reacts with ferric oxide:

$$2Fe_2O_3 + 6H_2S \rightarrow 2Fe_2S_3 + 6H_2O \tag{12.1}$$

The reaction requires an alkalinity pH level 8–10 with controlled injection of water. The bed is regenerated by controlled oxidation as

$$2Fe_2S_3 + 3O_2 \rightarrow 2Fe_2O_3 + 6S \quad (12.2)$$

Some of the sulfur produced might cake in the bed and oxygen should be slowly introduced to oxide this sulfur, as shown next:

$$S_2 + 2O_2 \rightarrow 2SO_2 \quad (12.3)$$

Repeated cycling of the process will deactivate the iron oxide and the bed should be changed after 10 cycles.

The process can be run continuously, and small amounts of air or oxygen are continuously added to the inlet sour gas so that the produced sulfur is oxidized as it forms. The advantage of this process is the large savings in labor cost for loading and unloading of the batch process. In this case, higher sulfur recovery per pound of iron oxide is also obtained.

A typical flow diagram of high-pressure continuous operation of the iron oxide process is shown in Figure 12.3. In this case, one of the towers is on stream removing H_2S from the sour gas, while the second tower is in the regeneration cycle by air blowing. The last regeneration step should be carried out with caution because reaction (12.2) is highly exothermic and the rate of reaction must be controlled. Care must be taken in replacing the exhausted iron sponge beds from the two towers shown in Figure 12.3. On opening the beds, entering air causes a sharp rise in temperature, which can result in spontaneous combustion of the bed. The entire bed should be

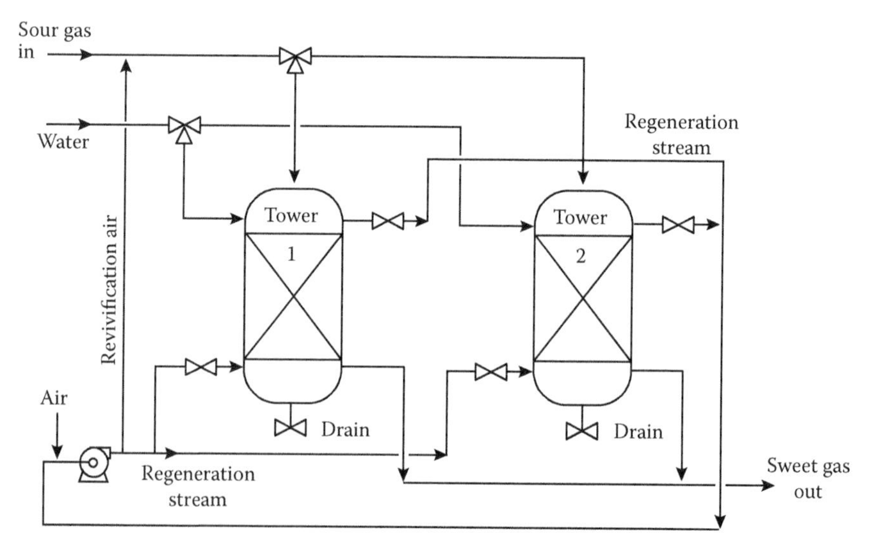

FIGURE 12.3
Typical diagram for iron oxide process.

wetted before recharging. There are only two types of ferric oxide that can be used for gas sweetening: α-Fe_2O_3–H_2O and γ-Fe_2O_3–H_2O. They easily react with H_2S and can be easily regenerated by oxidation to the corresponding ferric oxide forms.

In some cases, it is more economical to operate the process with a single bed and the exhausted bed of iron sulfide is trucked away to a disposal site. The tower is then recharged with a new bed of iron oxide and the tower is put back to service. The spent bed is left to react slowly with oxygen in the air, as shown in reactions (12.2) and (12.3), which can be done by a contracting company.

12.4.1.1 Design of Iron Sponge Bed

It is noted from Figure 12.1 that such a bed (batch process) can handle an acid gas feed of up to 200 ppm and bring it down to 1 ppm in the exit sweet gas stream.

The design of the iron sponge bed requires the following data:

Q_g (MMSCFD) = gas flow rate
P (psig) = operating pressure
T (°F) = operating temperature
X_{AG} (ppm) = H_2S inlet concentration
Z = compressibility factor from charts
SG = specific gravity of the gas (assume 0.7 if not given)

Calculation procedure:

1. Calculate column diameter. First the minimum diameter is calculated as

$$d_m^2(\text{in.}^2) = 360\frac{Q_g TZ}{P}$$

   Then, the maximum diameter to prevent channeling is

$$d_{\max}^2(\text{in.}^2) = 1800\frac{Q_g TZ}{P}$$

   The selected diameter should be between these two values.
2. Calculate daily consumption of iron sponge (R, ft^3/day):

$$R = 0.0133 Q_g (X_{AG})$$

3. Select a bed height between 10 and 20 ft. The bed volume (V) becomes

$$V(\text{ft})^3 = 0.7854\ D^2L$$

where D and L are in feet. The bed is replaced after V/R days.

Example 12.1

Natural gas flowing at 2 MMSCFD has a specific gravity of 0.7 and acid gas concentration of 25 ppm (MMSCFD = 10^6 standard cubic feet per day). The operating pressure and temperature are 1000 psig and 120°F, respectively. Design an iron sponge unit.

Solution

The feed acid concentration is 25 ppm, which is below 200 ppm, so a batch process such as iron sponge unit can be used.

$$d_{\text{min}}^2 = \frac{360(2)(580)(0.86)}{1014.7} = 353.43 \text{ and } d_{\text{min}} = 18.8 \text{ in.}$$

$$d_{\text{max}}^2 = \frac{1800(2)(580)(0.86)}{1014.7} = 1769.7 \text{ and } d_{\text{max}} = 42 \text{ in.}$$

A diameter between 18.8 and 42 in. can be used. Select a diameter of 24 in. (2 ft). Daily iron consumption R is

$$R = 0.0133(2)(25) = 0.665 \text{ ft}^3/\text{day}$$

Select a bed height of 10 ft. The bed volume becomes

$$V = 0.7854(2)^2(10) = 31.416 \text{ ft}^3$$

$$\text{Bed life} = \frac{31.416}{0.665} = 47 \text{ days}$$

Thus, the bed should be replaced in 6 weeks.

12.4.2 Zinc Oxide

Zinc oxide can be used instead of iron oxide for the removal of H_2S, COS, CS_2, and mercaptans. However, this material is a better sorbent and the exit H_2S concentration can be as low as 1 ppm at a temperature of about 300°C. The zinc oxide reacts with H_2S to form water and zinc sulfide:

$$ZnO + H_2S \rightarrow ZnS + H_2O$$

A major drawback of zinc oxide is that it is not possible to regenerate it to zinc oxide on site, because active surface diminishes appreciably by sintering.

Much of the mechanical strength of the solid bed is lost due to fines formation, resulting in a high-pressure-drop operation. The process has been decreasing in use due to these problems and the difficulty of disposing of zinc sulfide; Zn is considered a heavy metal.

12.4.3 Molecular Sieves

Molecular sieves (MSs) are crystalline sodium alumino silicates and have very large surface areas and a very narrow range of pore sizes. They possess highly localized polar charges on their surface that act as adsorption sites for polar materials at even very low concentrations. This is why the treated natural gas could have very low H_2S concentrations (4 ppm). There are many types of sizes available like 3A, 4A, 5A and 13X based on industrial requirement and adsorption capacity of moisture.

In order for a molecule to be adsorbed, it first must be passed through a pore opening and then it is adsorbed on an active site inside the pore. There are four major zones in a sieve bed, as shown in Figure 12.4.

In the presence of water, which is a highly polar compound, H_2O is first adsorbed in the bed, displacing any sulfur compounds. Water equilibrium with the sieves is established in this zone. The second zone is the water–sulfur compound exchange zone, where water is still displacing the sulfur compounds, but some sulfur sites will be left adsorbed. An adsorption front for sulfur compounds is formed and a concentration profile is established. The

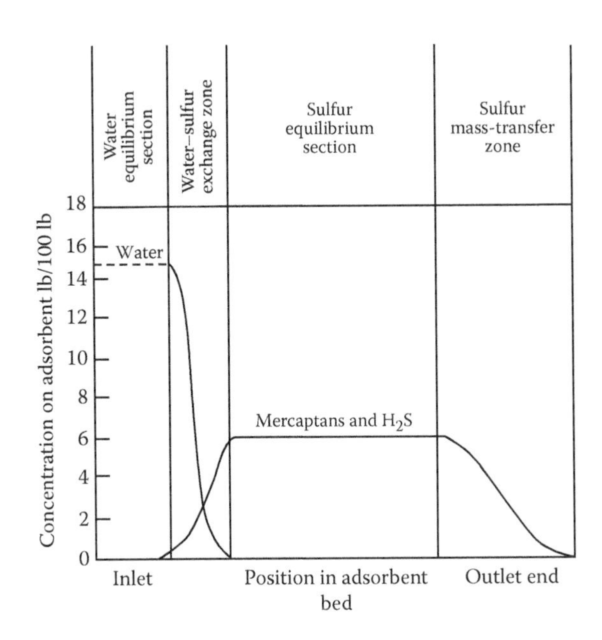

FIGURE 12.4
Adsorption zone in a molecular sieve bed.

concentration profile of sulfur compounds declines along the bed. Zone 3 is the sulfur equilibrium section, showing the highest concentration of sulfur compounds in the bed depending on the MS capacity. Zone 4 is the mass transfer section for sulfur compounds. As this profile reaches the end of the bed, this will mean some sulfur compounds (H_2S) will appear in the gas stream (unabsorbed); this will mean that the bed is exhausted and requires regeneration.

If it is desired to remove H_2S, a MS of 5 A* is selected. However, if it is also desired to remove mercaptans, 13 X* is selected. In either case, a selection is made to minimize the catalytic reaction:

$$H_2S + CO_2 \Leftrightarrow COS + H_2O$$

Olefins, aromatics, and glycols are strongly adsorbed, which may poison the sieves. 5 A* and 13 X* are the basic types of molecular sieves commercially available. Their basic properties and applications are fully described by Campbell.

Commercial applications require at least two beds so that one is always on line while the other is being regenerated. The schematic diagram of the process is shown in Figure 12.5. The sulfur compounds are adsorbed on a cool, regenerated bed in the sweetening. The saturated bed is regenerated by passing a portion of the sweetened gas, preheated to about 400°F–600°F or more, for about 1.5 h to heat the bed. As the temperature of the bed increases, it releases the adsorbed H_2S into the generation gas stream. The sour effluent gas is flared off, with about 1%–2% of the treated gas lost.

An amine unit can be added to this process to recover this loss; in this case H_2S will be flared off from the regenerator of the amine unit. In case

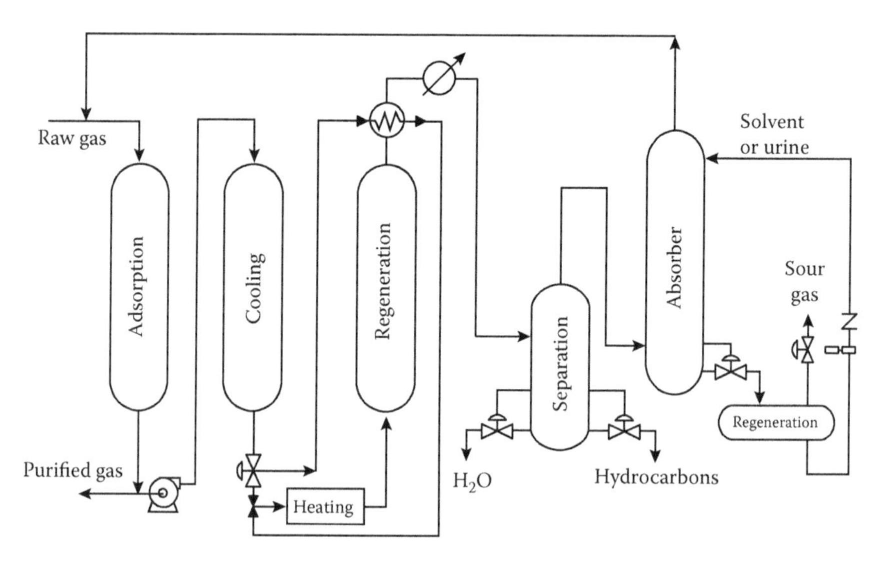

FIGURE 12.5
Sweetening of natural gas by molecular sieves.

this flaring is environmentally prohibited, the H_2S can be sent to a gathering center for the sulfur recovery unit, if it exists on site.

12.5 Liquid-Phase Processes

Liquid-phase processes are one of the most commonly used processes for acid gas treatment. Chemical solvents are used in the form of aqueous solution to react with H_2S and CO_2 reversibly and form products that can be regenerated by a change of temperature or pressure or both. Physical solvents can be utilized to selectively remove sulfur compounds. They are regenerated at ambient temperature by reducing the pressure. A combination of physical and chemical solvents can be used. A comparison of chemical solvents (amines, carbonates) and physical solvents is shown in Table 12.2.

TABLE 12.2

Comparison of Chemical and Physical Processes

	Chemical Absorption		
Feature	**Amine Processes**	**Carbonate Processes**	**Physical Absorption**
Absorbents	MEA, DEA, DGA, MDEA	K_2CO_3, K_2CO_3 + MEA K_2CO_3 + DEA K_2CO_3 + arsenic trioxide	Selexol, Purisol, Rectisol
Operating pressure (psi)	Insensitive to pressure	>200	250–1000
Operating temperature (°F)	100–400	200–250	Ambient temperature
Recovery of absorbents	Reboiled stripping	Stripping	Flashing, reboiled, or stream stripping
Utility cost	High	Medium	Low/medium
Selectivity H_2S, CO_2	Selective for some amines (MDEA)	May be selective	Selective for H_2S
Effect of O_2 in the feed	Formation of degradation products	None	Sulfur precipitation at low temperature
COS and CS_2 removal	MEA: not removed DEA: slightly removed DGA: removed	Converted to CO_2 and H_2S and removed	Removed
Operating problems	Solution degradation; foaming; corrosion	Column instability; erosion; corrosion	Absorption of heavy hydrocarbons

12.5.1 Amine Processes

The most widely used for sweetening of natural gas are aqueous solutions of alkanolamines. They are generally used for bulk removal of CO_2 and H_2S. The properties of several amines are shown in Table 12.2. The low operating cost and flexibility of tailoring solvent composition to suit gas compositions make this process one of most commonly selected. A liquid physical solvent can be added to the amine to improve selectivity.

A typical amine process is shown in Figure 12.6. The acid gas is fed into a scrubber to remove entrained water and liquid hydrocarbons. The gas then enters the bottom of the absorption tower, which is either a tray (for high flow rates) or packed (for lower flow rates). The sweet gas exits at the top of tower.

The regenerated amine (lean amine) enters at the top of this tower and the two streams are countercurrently contacted. In this tower, CO_2 and H_2S are absorbed with the chemical reaction into the amine phase. The exit amine solution, loaded with CO_2 and H_2S, is called rich amine. This stream is flashed, filtered, and then fed to the top of a stripper to recover the amine, and acid gases (CO_2 and H_2S) are stripped and exit at the top of the tower. The refluxed water helps in steam stripping the rich amine solution. The regenerated amine (lean amine) is recycled back to the top of the absorption tower.

The operating conditions of the process depend on the type of the amine used. Some of these conditions are given in Table 12.3. Primary amines are the strongest to react with acid gases; but the stable bonds formed make it difficult to recover by stripping. Secondary amines have a reasonable capacity for acid gas absorption and are easily recovered. Tertiary amines have a lower capacity, but they are more selective for H_2S absorption.

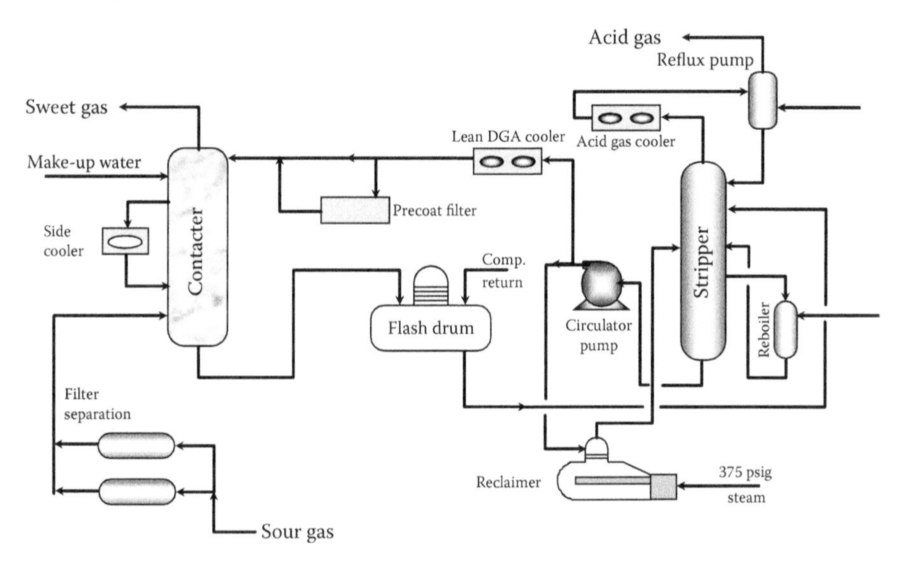

FIGURE 12.6
Flow diagram for the amine process.

TABLE 12.3

Comparison of Amine Solvents

Solvent	MEA	DEA	DIPA	DGA	MDEA
Chemical formula	RNH_2	R_2NH	R'_2NH	$RO(CH_2)_2NH_2$	R_2CH_3N
Molecular weight	61	105	133	105	119
Amine type	Primary	Secondary	Secondary	Primary	Tertiary
Vapor pressure, 100°F (mm Hg)	1.05	0.058	0.01	0.16	0.0061
Freezing point (°F)	15	20	16	–40	–25
Relative capacity (%)	100	58	46	58	51
Solution weight (%)	20	30	35	60	50
K[a]	2.05	1.45	0.95	1.28	1.25
Loading (mol AG/mol amine)	0.35	0.5	0.7	0.3	0.4
H_2S/CO_2 selectivity[b]	1	1	2	1	3
Solvent conc. (wt%)	15–20	20–35	30–40	45–65	40–55
AG (mol/mol)	0.3–0.4	0.5–0.6	0.3–0.4	0.3–0.4	0.3–0.45
Circulation (gal/mol AG)	100–165	60–125	–	50–75	65–110
Steam rate (lb/gal)	1.0–1.2	0.9–1.1	–	1.1–1.3	0.9–1.1
Reboiler temp. (°F)	240	245	255	260	250
Heat of reaction (Btu/lb AG, H_2S)	620	550	–	675	500
CO_2	660	630	0	850	600

[a] Circulation rate (gpm) = KQ_g (MMSCFD) X_{AG} (mol% acid gas) for $P > 400$ psig and $T < 120$°F.
[b] 10 is highly selective; 1 is not selective.

Among the amines discussed here, DEA is the most common. This may be due to the fact it is less expensive to install and operate. Specific details for each amine follows.

12.5.1.1 Monoethanolamine Solvent

Monoethanolamine (MEA) is a primary amine and the strongest amine among others. It can produce pipeline specification gas. It reacts with H_2S and CO_2 as follows:

$$2(RNH_2) + H_2S \Leftrightarrow (RNH_3)_2S \tag{12.4}$$

$$(RNH_3)_2S + H_2S \Leftrightarrow 2(RNH_3)HS \tag{12.5}$$

$$2(RNH_2) + CO_2 \Leftrightarrow RNHCOONH_3R \tag{12.6}$$

These reactions are reversible and are forward in the absorber (at low temperature) and backward in the stripper (at high temperature).

Monoethanolamine will nonselectively react with H_2S and CO_2. Unfortunately, it will irreversibly react with COS and CS_2, which can result in a loss of solution and a buildup of solid products in the circulating solution. These products could also be a source of corrosion.

The MEA process usually uses a solution of 15%–20% MEA (wt%) in water. Loading is about 0.3–0.4 mol of acid removed per mole of MEA. The circulation rate is between 2 and 3 mol of MEA per mole of H_2S, as expected from reactions (12.4) and (12.5). However, commercial plants use a ratio of 3 to avoid excessive corrosion. Solution strength and loading above these limits result in excessive corrosion. MEA easily forms foam due to the presence of contaminants in the liquid phases; this foam results in carryover from the absorber. These contaminants could be condensed hydrocarbons, degradation products, and iron sulfide, as well as corrosion products and excess inhibitors. Solids can be removed by using a filter, hydrocarbons could be flashed, and degradation products are removed using a reclaimer. The number of trays used in absorbers in commercial units is between 20 and 25. However, the theoretical number of trays calculated from published equilibrium data is about 3 to 4. If we assume an efficiency of 35% for each tray, then the actual number of trays is 12. It has been reported that the first 10 trays pick up all of the H_2S and at least another 10 trays are of not much value. Thus, it is suggested to use 15 trays. It is recommended that MEA be used if the feed does not contain COS or CS_2, which form stable products and deplete the amine. If the feed has these compounds, a reclaimer must be used, where a strong base like NaOH is used to regenerate and liberate the amine. This base has to be neutralized later.

12.5.1.2 Diethanolamine

Diethanolamine (DEA) is replacing MEA and becoming the most widely used gas sweetening solvent. It is a secondary amine with lower reactivity and corrosivity than MEA. Moreover, it reacts with COS and CS_2, and the product can be regenerated. It has a lower vapor pressure (lower losses) and lower heat of reaction (easier to generate) than MEA. The basic reactions with CO_2 and H_2S are the same as MEA:

$$2R_2NH + H_2S \Leftrightarrow (R_2NH_2)_2S \tag{12.7}$$

$$(R_2NH_2)_2S + H_2S \Leftrightarrow 2R_2NH_2SH \quad (12.8)$$

$$2R_2NH + CO_2 \Leftrightarrow R_2NCOONH_2R_2 \quad (12.9)$$

Based on these reactions, 1.7 lbs of DEA can be circulated to react with the same amount of acid gas as 1.0 lb of MEA. A higher strength of 35% (wt) can be used because of its lower corrosivity. Loading up to 0.65 mol of acid gas per mole of DEA can cause fewer operational problems than MEA because the elimination of the degradation products and the absence of a reclaimer. Corrosion is less because DEA is weaker than MEA. Foaming is reduced probably due to absence of degradation and corrosion products.

12.5.1.3 Diisopropanolamine

Diisopropanolamine (DIPA) is a secondary amine and is used most frequently in the ADIP process licensed by Shell. DIPA reacts with COS and CS_2, and the products are easily regenerated. At low pressure, DIPA is more selective to H_2S, and at higher pressures, DIPA removes both CO_2 and H_2S. It is noncorrosive and requires less heat for rich amine regeneration.

12.5.1.4 Methyldiethanolamine

Methyldiethanolamine (MDEA) is commonly used in the 20%–50% (wt) range. Lower weight percent solutions are typically used in very low pressure and selectivity applications. Acid gas loading as high as 0.7–0.8 mol acid gas/mol amine is practical in carbon steel equipment. Higher loading may be possible with few problems. Corrosion is much reduced in this case even under these high loadings. In the presence of oxygen, MDEA forms corrosive acids that, if not removed from the system, can result in buildup of iron sulfide. Other advantages include lower vapor pressure, lower heat of reaction, higher resistance to degradation, and high selectivity for H_2S.

The overwhelming advantage of MDEA is its selectivity for H_2S in the presence of CO_2. At high CO_2/H_2S ratios, a major portion of CO_2 can be slipped through the absorber and into the sales gas while removing most of H_2S. The enhanced selectivity of MDEA for H_2S results from its inability to form carbamate with CO_2. Selectivity absorption of H_2S can be enhanced by controlling the residence time per tray to 1.5–3.0 s and increasing the temperature in the absorber. Both of these conditions favor H_2S absorption with CO_2 rejection.

12.5.1.5 Mixed Amines

Mixtures of amines are generally mixtures of MDEA and DEA or MEA and are used to enhance CO_2 removal by MDEA. Such mixtures are referred to as MDEA-based amines. The secondary amine generally comprises less than

20% (mole) of the total amine. At lower MEA and DEA concentrations, the overall amine strength can be as high as 55% (wt), without invoking corrosion problems.

Amine mixtures are particularly useful for low-pressure applications because MDEA becomes less capable of CO_2 pickup sufficient enough to meet pipeline specifications. At higher pressures, amine mixtures appear to have little or no advantage over MDEA. Mixed amines are also useful for cases where the CO_2 content of the feed gas is increasing over time due to field aging.

12.5.1.6 Design of Amine Units

The design of amine units are similar for different solvents; however, a reclaimer should be added in the case of more reactive amines such as MEA to recover these amines. The main equipment in the amine processes is the absorber, and emphasis here will be on the design of the absorber. The main operating conditions such as circulation rate, solvent concentration, acid gas loading and steam rate for regeneration were given in Table 12.3.

12.5.1.7 Absorber Design

For 35 wt% DEA and using 0.5 mol acid gas/mol DEA, the amine circulation rate q_{Am} is

$$q_{\mathrm{Am}}\ (\mathrm{gpm}) = 1.26\, Q_g\, X_{\mathrm{AG}} \tag{12.10}$$

where Q_g is the gas flow rate (MMSCFD) and X_{AG} is the mole percent of acid gas in the feed. Assuming an amine drop size of 100 μm, the diameter d (inches) of the absorber can be calculated as (Arnold and Stewart)

$$d^2 = 504 \frac{TZQ_g}{P} \left[\left(\frac{\rho_g}{\rho_l - \rho_g} \right) C_d \right]^{1/2} \tag{12.11}$$

where T is the temperature (°R), Z is the compressibility factor, P is the pressure (psia), and C_d is the drag coefficient. The density of gas and liquid are ρ_g and ρ_l (lb/ft³), respectively. Usually 20 valve-type trays are used with a spacing of 24 in.

Example 12.2

A natural gas is flowing at 100 MMSCFD at 1000 psia and 100°F. It contains 2 mol% CO_2 and 1.9 mol% H_2S. It is being treated by the amine process using DEA. The outlet gas contains no acid components. Calculate the flow rate of DEA required and the tower diameter and height.

Solution

$$\text{Total acid gas} = 2 + 1.9 = 3.9 \text{ mol\%}$$

$$\text{Amine flow rate } (q_{\text{Am}}) = 126X_{\text{AG}}$$

$$\text{Amine flow rate } (q_{\text{Am}}) = 1.26\, Q_g\, X_{AG} = (1.26)(100)(3.9)$$
$$= 126(3.9) = 491.4 \text{ gpm}$$

The tower diameter is

$$d^2 = 504\left(\frac{560(0.85)(100)}{1000}\right)\left[\left(\frac{\rho_g}{\rho_l - \rho_g}\right)C_d\right]^{1/2}$$

Assuming the drag coefficient (C_d) is 0.7,

$$\rho_g = \frac{PM}{ZRT} = \frac{(1000)(16)}{(0.85)(10.7)(560)} = 3.14\frac{\text{lb}}{\text{ft}^3}$$

where P is the pressure and M is the molecular weight. Assuming an average molecular weight of natural gas the same as methane, then

$$\rho_L = \rho_{\text{water}} \text{ at } 100°\text{F}$$

$$= 65.1\frac{\text{lb}}{\text{ft}^3}$$

$$\frac{\rho_g}{\rho_l - \rho_g} = \frac{3.14}{65.1 - 3.14} = 0.05068 = 0.0507$$

$$d^2 = 504(47.6)[0.0507(0.7)]^{1/2}$$

then

$$d = 67.2 \text{ in.} = 5.6 \text{ ft}$$

Use 24 trays with a spacing of 24 in. with a height of (24)(2) = 48 ft.

12.5.2 Hot Potassium Carbonate Process

In this process, hot potassium carbonate (K_2CO_3) is used to remove both CO_2 and H_2S. It can also remove (reversibly) COS and CS. It works best when the CO_2 partial pressure is in the range 30–90 psi. The following reactions occur in this case:

$$K_2CO_3 + CO_2 + H_2O \Leftrightarrow 2KHCO_3 \tag{12.12}$$

$$K_2CO_3 + H_2S \Leftrightarrow KHS + KHCO_3 \tag{12.13}$$

It can be seen from reaction (12.12) that a high partial pressure of CO_2 is required to keep $KHCO_3$ in solution, and in Equation 12.13 H_2S will not react if the CO_2 pressure is not high. For this reason, this process cannot achieve a low concentration of acid gases in the exit stream and a polishing process is needed (molecular sieve). An elevated temperature is also necessary to ensure that potassium carbonate and reaction products ($KHCO_3$ and KHS) remain in solution. Thus, this process cannot be used for gases containing H_2S only.

The hot carbonate process, which is given in Figure 12.7, is referred to as the *hot* process because both the absorber and the regenerator operate at elevated temperatures, usually in the range 230°F–240°F. In this process, the sour gas enters at the bottom of the absorber and flows countercurrently to the carbonate liquid stream. The sweet gas exits at the top of the absorber. The absorber is operated at 230°F and 90 psia. The rich carbonate solution exits from the bottom of the absorbed and is flashed in the stripper, which operates at 245°F and atmospheric pressure, where acid gases are driven off. The lean carbonate solution is pumped back to the absorber.

The strength of the potassium carbonate solution is limited by the solubility of potassium bicarbonate ($KHCO_3$) in the rich stream. The high temperature of the system increases $KHCO_3$ solubility, but the reaction with CO_2 produces 2 mol of $KHCO_3$ per mole of K_2CO_3 reacted. For this reason, $KHCO_3$ in the rich stream limits the lean solution of K_2CO_3 concentration to 20%–35% (wt).

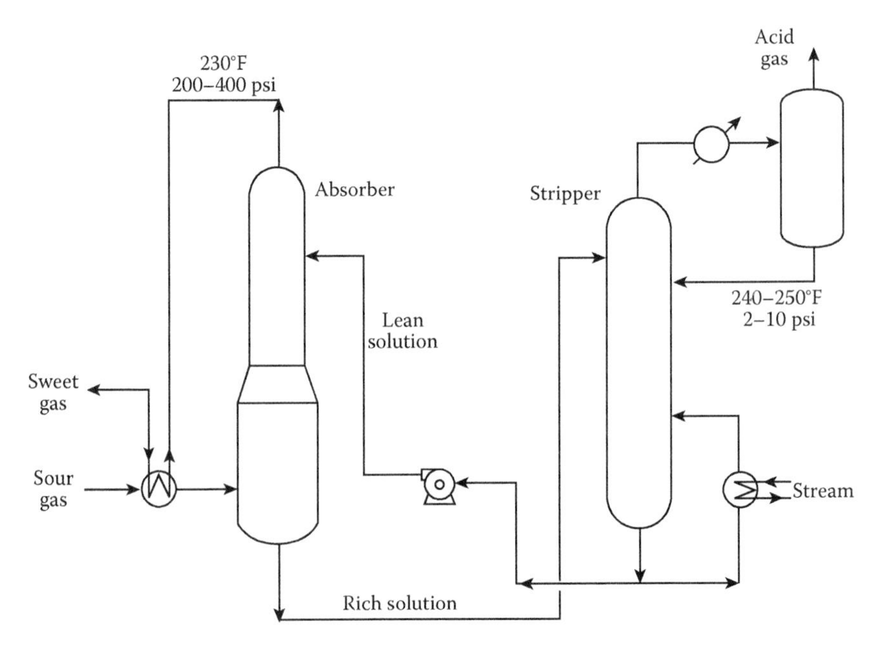

FIGURE 12.7
Hot potassium carbonte process.

12.5.3 Physical Solvent Processes

Organic liquid (solvents) are used in physical solvent processes to absorb H_2S (usually) preferentially over CO_2 at high pressure and low temperatures. Regeneration is carried out by releasing the pressure to the atmosphere and sometimes in vacuum with no heat. Henry's law is

$$P_i = HX_i$$

or

$$X_i = \frac{Y_i}{H}P \tag{12.14}$$

This implies that acid gas absorbed in liquid phase (X_i) is proportional to its gas mole fraction (Y_i) and inversely to Henry's constant (which is constant for a given temperature). Much more important, the solubility is proportional to the total gas pressure (P). This means that at high pressure, acid gases will dissolve in solvents, and as the pressure is released, the solvent can be regenerated.

The properties of four of the important physical solvents used in natural gas processing are given in Table 12.4.

12.5.3.1 Fluor Process

The Fluor process uses propylene carbonate to remove CO_2, H_2S, C_2^+, COS, CS_2, and H_2O from natural gas. Thus, in one step the natural gas can be sweetened and dehydrated.

Figure 12.8 shows a typical process flow sheet with regeneration consisting of three flash drums. The first flash drum gas containing mostly hydrocarbons is compressed and recycled. The second flash drum derives

TABLE 12.4

Properties of Physical Solvents

Process	Fluor	Purisol	Selexol	Sulfinol[a]
Solvent	Propylene carbonate	*N*-Methyl pyrrilidone	Diethylene dimethyl ether	Sulfolane
Structure				
Molecular weight	102.09	99.13	134.17	120.17
Freezing (°F)	–56	–12	–83	77
Boiling (°F)	467	396	324	546
Gas Solubility (cm³ Gas at 1 atm, 75°F/cm³ Solvent)				
H_2S	13.3	43.3	25.5	
CO_2	3.3	3.8	3.6	
COS	6.0	10.6	9.8	
C_3	2.1	3.5	4.6	

[a] Sulfinol is a mixture of Sulfolane + DIPA + H_2O.

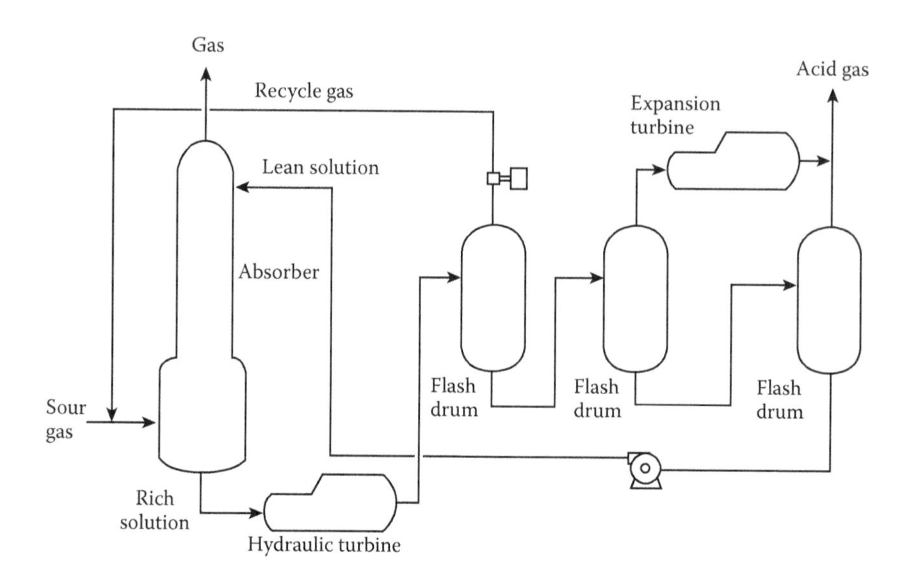

FIGURE 12.8
Flow diagram for the Fluor process.

expansion turbine. The third flash contains mainly acid gases. The process is used to remove the bulk CO_2 down to 3%.

12.5.3.2 Selexol Process

The Selexol process uses a mixture of dimethyl ether of propylene glycols as a solvent. It is nontoxic and its boiling point is not high enough for amine formulation. Figure 12.9 shows the flow sheet of the Selexol process. A cool stream of natural gas is injected in the bottom of the absorption tower operated at 1000 psia. The rich solvent is flashed in a high flash drum at 200 psia, where methane is flashed and recycled back to the absorber and joins the sweet gas stream. The solvent is then flashed at atmospheric pressure and acid gases are flashed off. The solvent is then stripped by steam to completely regenerate the solvent, which is recycled back to the absorber; any hydrocarbons will be condensed and any remaining acid gases will be flashed from the condenser drum.

This process is used when there is a high acid gas partial pressure and no heavy hydrocarbons. DIPA can be added to this solvent to remove CO_2 down to pipeline specifications.

12.5.3.3 Sulfinol Process

The Sulfinol process uses a solvent that is 40% sulfolane (tetrahydrothiophene 1-1 dioxide), 40% DIPA, and 20% water. It is an excellent example of

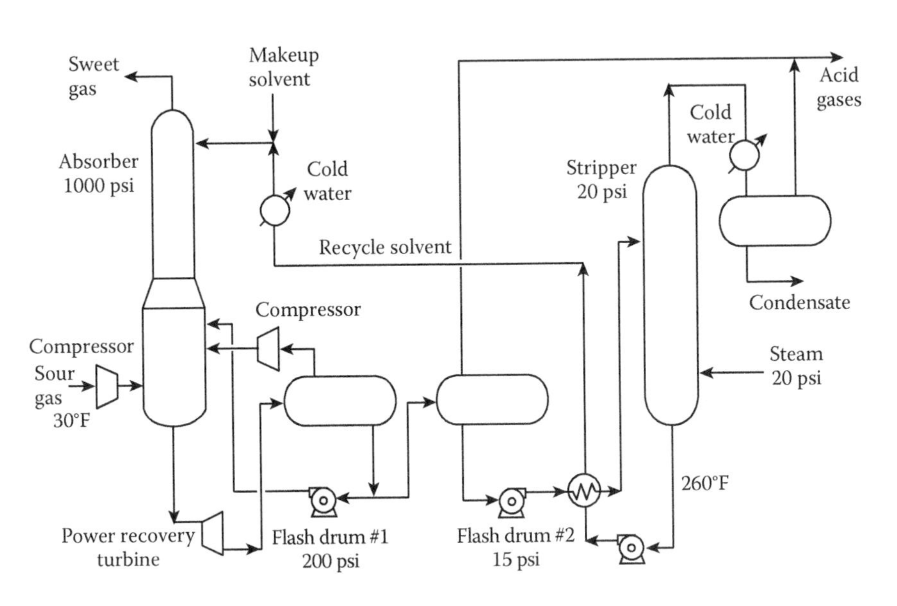

FIGURE 12.9
Flow diagram for the Selexol process.

enhancing amine selectivity by adding a physical solvent such as sulfolane. Sulfolane is an excellent solvent of sulfur compounds such as H_2S, COS, and CS_2. Aromatics, heavy hydrocarbons, and CO_2 are soluble to a lesser extent. Sulfinol is usually used for H_2S/CO_2 ratios greater than 1:1 or where CO_2 removal is not required to the same extent as H_2S. A high loading of 1.5 mol acid gas per mole of Sulfinol can be achieved. The Sulfinol process uses a conventional solvent absorption and regeneration cycle, as shown in Figure 12.10. The sour gas components are removed from the feed gas by countercurrent contact with a lean solvent stream under pressure. The absorbed impurities are then removed from the rich solvent by stripping with steam in a heated regenerator column. The hot lean solvent is then cooled for reuse in the absorber. Part of the cooling may be by heat exchange with the rich solvent for partial recovery of heat energy. In some applications where there is a large amount of hydrogen sulfide in the feed, the overall Sulfinol/Claus plant has a balance that permits the omission of the heat exchanger. The gas flashed from the rich solvent after partial depressuring comes as fuel gas. In some cases, it is desirable to treat this gas stream with Sulfinol solvent to control the acid gas content of the plant fuel supply.

The Sulfinol solvent reclaimer is used in a small ancillary facility for recovering solvent components from higher boiling products of alkanolamine degradation or from other high-boiling or solid impurities. A Sulfinol reclaimer is similar to a conventional MEA reclaimer, but it is much smaller than that in an MEA plant of comparable gas treating capacity. Usually, the Sulfinol reclaimer need not be started up until several months after the treating plant is started up.

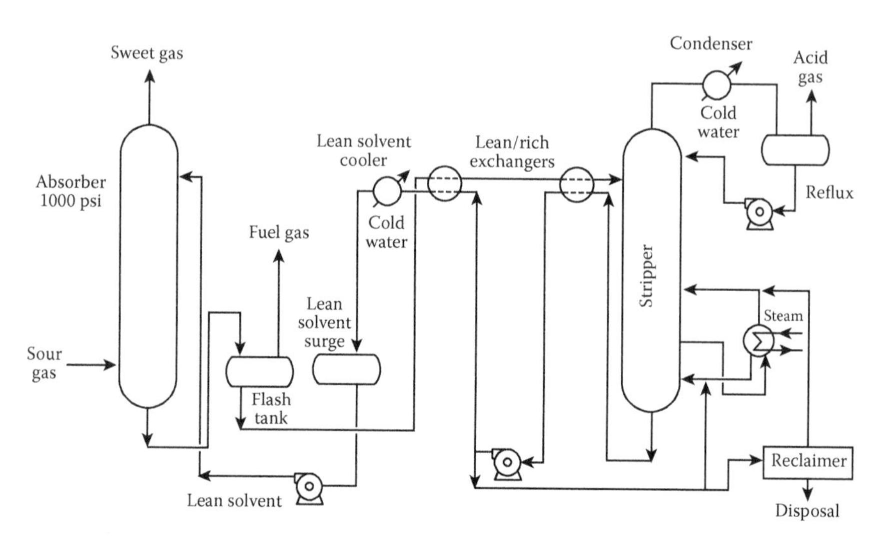

FIGURE 12.10
Flow diagram for the Sulfinol process.

Among this group of physical solvents, Selexol is more selective than Fluor solvent, but it dissolves propane. All solvents exhibit significant affinity for heavy paraffins, aromatics, and water. Water absorption makes them good desiccants. The loading capacity of physical solvent, in general, is much higher than amines, as manifested in Figure 12.11. It

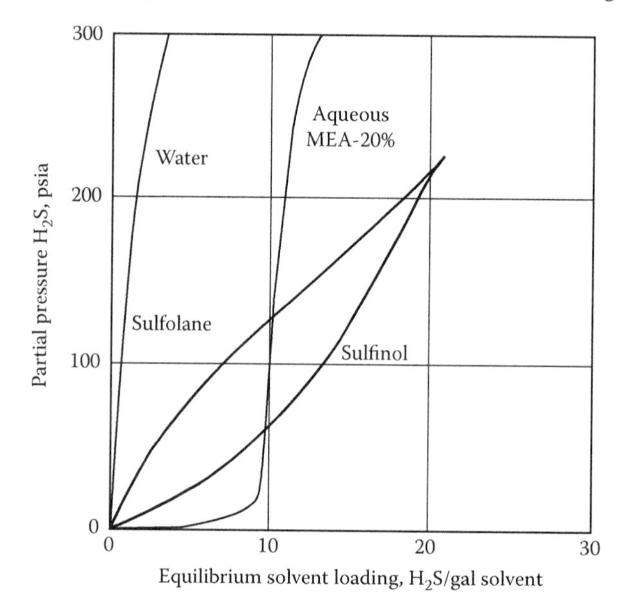

FIGURE 12.11
Equilibrium solvents loading.

is observed that at a partial pressure of H_2S of 200 psia, the loading (mol H_2S/gal solvent) of MEA (20% solution) is about 11.5, and at the same time, sulfolane (physical solvent) is about 18 and Sulfinol (which is a mixed solvent) is about 19.

12.6 Direct Conversion Processes

There are many processes used to convert H_2S to sulfur; however, our discussion here is limited to those processes applied to natural gas. Generally, H_2S is absorbed in an alkine solution containing an oxidizing agent that converts it to sulfur. The solution is regenerated by air in a flotation cell (oxidizer). The following processes are used for this purpose.

12.6.1 Stretford Process

The absorbing solution is dilute Na_2CO_3, $NaVO_3$, and anthraquinone disulfonic acid (ADA). The reaction occurs in four steps:

$$H_2S + Na_2CO_3 = NaHS + NaHCO_3 \tag{12.15}$$

$$4NaVO_3 + 2NaHS + H_2O = Na_2V_4O_9 + 4NaOH + 2S \tag{12.16}$$

$$\begin{aligned} Na_2V_4O_9 + 2NaOH + H_2O + 2ADA(\text{quinone}) = 4NaVO_3 + \\ 2ADA(\text{hydroquinone}) \end{aligned} \tag{12.17}$$

$$AD(\text{hydroquinone}) + \frac{1}{2}O_2 = ADA(\text{quinone}) \tag{12.18}$$

The process uses ADA as the organic oxygen carrier. One of the products is finely divided sulfur and the process is capable of treating natural gas of very low H_2S concentrations. In the oxidizer or regenerator, the reduced anthraquinone disulfonic acid is reoxidized by blowing air, as shown by reaction (12.18). The precipitated sulfur is overflown as froth.

The Stretford process is shown in Figure 12.12. Sour gas enters the bottom of the absorber and sweet gas exits at the top. The Stretford solution enters at the top of the absorber and some time should be allowed for reaction to take place in the bottom part of the absorber, where H_2S is selectively absorbed. The reaction products are fed to the oxidizer, where air is blown to oxidize ADA (hydroquinone) back to ADA (quinone). The sulfur froth is skimmed and sent to either a filtration or centrifugation unit. If heat is used, molten sulfur is produced; otherwise a filter sulfur cake is obtained. The filtrate of these units along with the liquid from the oxidized are sent back to the absorber.

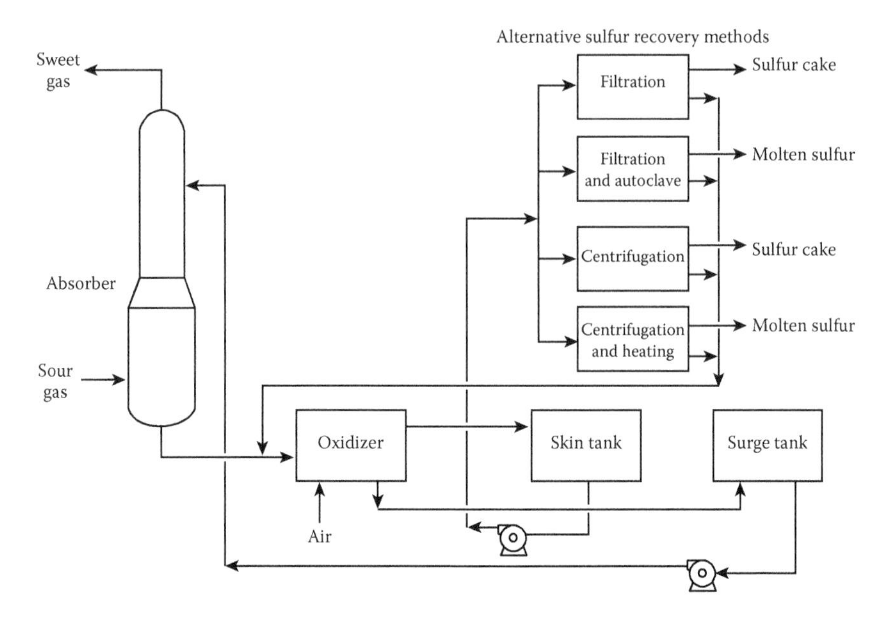

FIGURE 12.12
Flow diagram for the Stretford process.

12.6.2 LOCAT Process

The LOCAT process uses an extremely dilute solution of iron chelates. A small portion of the chelating agent is depleted in some side reactions and is lost with precipitated sulfur. In this process (Figure 12.13), sour gas is contacted with the chelating reagent in the absorber and H_2S reacts with the dissolved iron to form elemental sulfur. The reactions involved are the following:

$$H_2S + 2Fe^{3+} \rightarrow 2H^+ + S + 2Fe^{2+} \tag{12.19}$$

The reduced iron ion is regenerated in the regenerator by blowing air as

$$\frac{1}{2}O_2 + H_2O + 2Fe^{2+} \rightarrow 2(OH)^- + 2Fe^{3+} \tag{12.20}$$

The sulfur is removed from the regenerator to centrifugation and melting. Application of heat is not required because of the exothermic reaction.

12.6.3 Sulferox Process

Chelating iron compounds are also the heart of the Sulferox process. Sulferox is a redox technology, as is the LOCAT; however, in this case, a concentrated iron solution is used to oxidize H_2S to elemental sulfur. Patented organic liquids or chelating agents are used to increase the solubility of iron in the

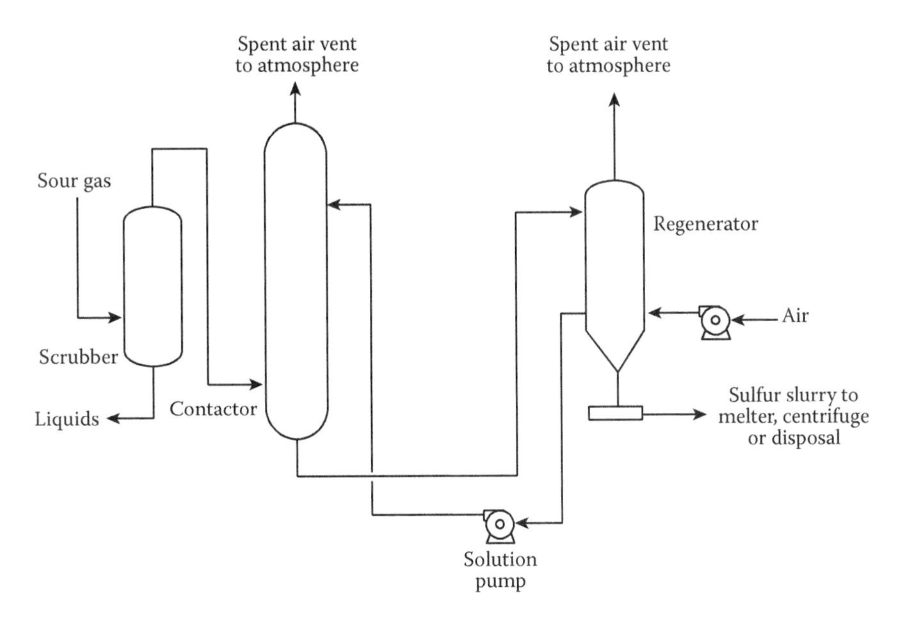

FIGURE 12.13
The LOCAT process.

operating solution. As a result of high iron concentrations in the solution the rate of liquid circulation can be kept low and, consequently, the equipment is small.

As in the LOCAT process, there are two basic reactions; the first takes place in the absorber, as in reaction (12.19), and the second takes place in the regenerator, as in reaction (12.20). The key of the Sulferox technology is the ligand used in the process. The application of this ligand allows the process to use high total iron concentrations (>1% by weight).

In Figure 12.14, the sour gas enters the contactor, where H_2S is oxidized to give elemental sulfur. The treated gas and the Sulferox solution flow to the separator, where sweet gas exits at the top and the solution is sent to the regenerator where Fe^{2+} is oxidized by air to Fe^{3+} and the solution is regenerated and sent back to the contactor. Sulfur settles in the regenerator and is taken from the bottom to filtration, where sulfur cake is produced. At the top of the regenerator, spent air is released. A makeup Sulferox solution is added to replace the degradation of the ligands. Proper control of this degradation rate and purging of the degradation products will ensure smooth operation of the process.

12.6.4 Membrane Processes

Polymeric membranes separate gases by selective permeation of gas species in these membranes. The gas dissolves at the contact surface of the

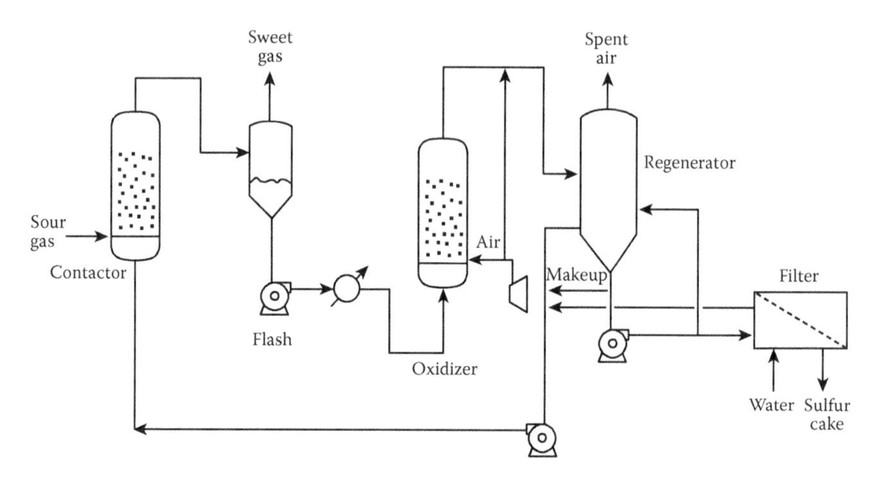

FIGURE 12.14
Sulferox process.

membrane, permeates across the membrane under the partial pressure gradient across the membrane wall. The rate of permeation of gas A(q_A) can be expressed as

$$q_A = \frac{PM}{t} A_m \Delta P_A \tag{12.21}$$

where PM is the gas permeability in the membrane; A_m and t are the surface area and thickness of the membrane, respectively; and ΔP_A is the partial pressure of gas A across the membrane.

The basic idea of the process is to flow sour gas on one side of the membrane where only acid gases diffuse across the membrane to the permeate side and the rest of the gas exits as sweet gas, as shown in Figure 12.15. Two module configurations are usually used: the spiral module and the hollow-fiber module. Spiral-wound membranes consist of a sandwich of four sheets wrapped around a central core of a perforated collecting tube. The whole spiral-wound element is housed inside a metal shell. The feed gas enters at the left end of the shell, enters the feed channel, and flows through this channel in the axial direction of the spiral to the right end of the assembly, as shown in Figure 12.16. The exit sweet gas leaves the shell at this point. The acid gases permeate perpendicularly through the membrane. This permeate then flows through the permeate channel to the perforated collecting tube, where it leaves the apparatus at one end. The direction of flow in the spiral-wound module is shown in Figure 12.17.

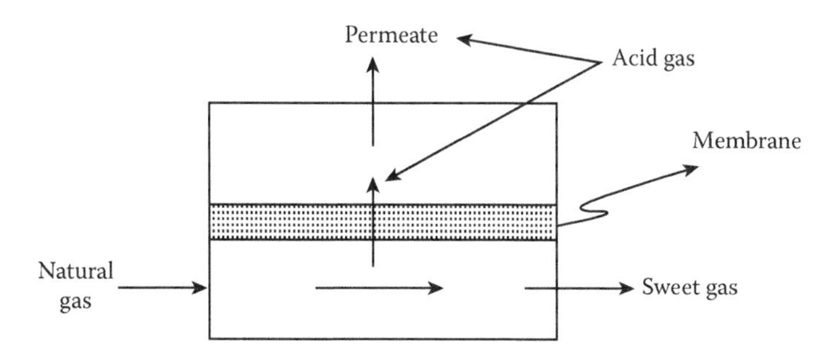

FIGURE 12.15
Basic operation of flow pattern in the membrane.

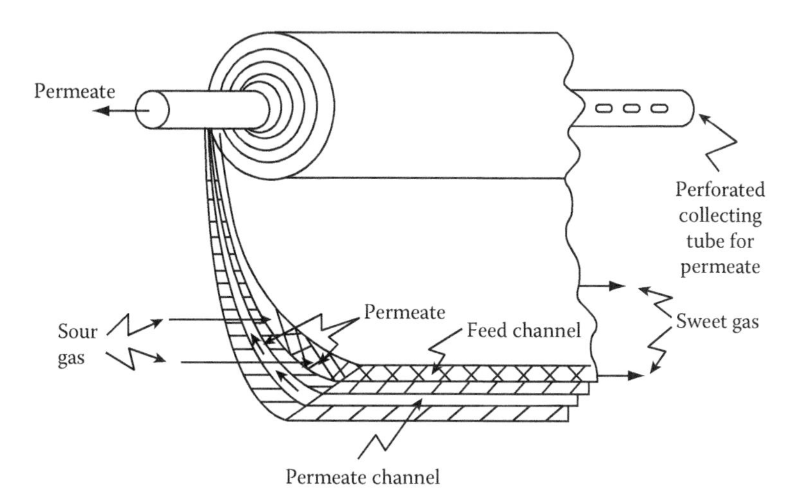

FIGURE 12.16
Spiral-wound elements and assembly.

The hollow-fiber module consists of a bundle of very small-diameter hollow fibers. The module resembles a shell and tube heat exchanger. Thousands of fine tubes are bound together at each end into a tube sheet that is surrounded by metal shell (see Figure 12.18). The membrane area per unit volume is up to 3000 ft^2/ft^3. Acid gases diffuse through the very thin membrane of the tubes and exit at the bottom of the module. Sweet gas exits at the top.

In both modules, high pressure should be maintained to ensure high permeation rates. Table 12.5 lists the permeability (PM) of some gases in different membranes. The permeation rates of different gases in a commercial membrane are given in Table 12.6. It is noted from Table 12.6 that H_2S and CO_2 will permeate much faster than other hydrocarbons. However, the permeate gas

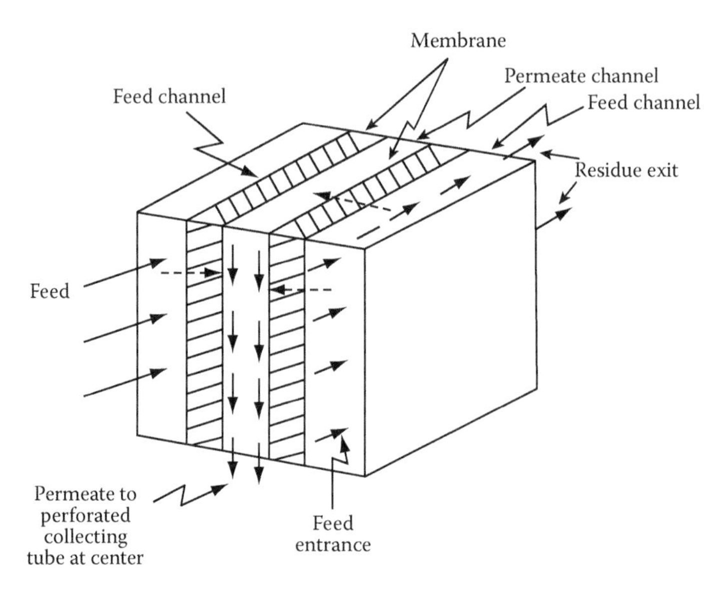

FIGURE 12.17
Gas flow paths for the spiral-wound module.

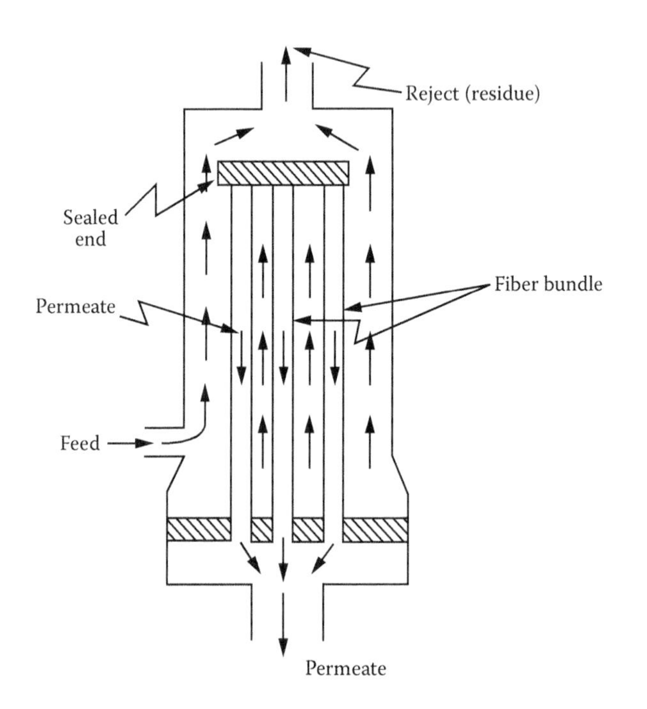

FIGURE 12.18
Hollow-fiber separator assembly.

TABLE 12.5

Permeabilities of Various Gases in Membranes

		Permeability (PM) $\frac{cm^3(STP) \cdot cm}{s \cdot cm^2/cmHg} \times 10^{10}$					
Material	**Temp. (°C)**	**He**	**H_2**	**CH_4**	**CO_2**	**O_2**	**N_2**
Silicone rubber	25	300	550	800	2700	500	250
Natural rubber	25	31	49	30	131	24	8.1
Polycarbonate (Lexane)	25–30	15	12	–	5.6, 10	1.4	–
Nylon 66	25	1.0	–	–	0.17	0.034	0.008
Polyester (Permasep)	–	–	1.65	0.035	0.31	–	0.031
Silicone–polycarbonate copolymer (57% silicone)	25	–	210	–	970	160	70
Teflon FEP	30	62	–	1.4	–	–	2.5
Ethyl cellulose	30	35.7	49.2	7.47	47.5	11.2	3.29
Polystyrene	30	40.8	56.0	2.72	23.3	7.47	2.55

TABLE 12.6

Gas Permeation Rates

Gas	Spiral-Wound Membrane
H_2	100.0
He	15.0
H_2O	12.0
H_2S	10.0
CO_2	6.0
O_2	1.0
Ar	–
CO	0.3
CH_4	0.2
N_2	0.18
C_2H_6	0.1

will also contain hydrogen and water vapor if they were present in the natural gas feed. It is possible to enhance the mass transfer rates by either blowing an inert gas in the permeate side or using amines on the permeate side to chemically react with the acid gas permeated and removed from the surface of the membrane. Several membrane modules may be connected in series and parallel arrangements to meet specific requirements. In Figure 12.19, a two-stage system is shown. If a gas stream contains a high acid content (50%),

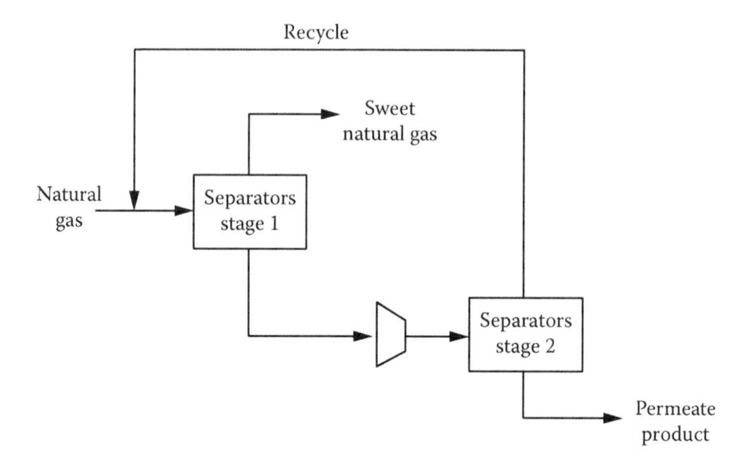

FIGURE 12.19
Two-stage membrane process.

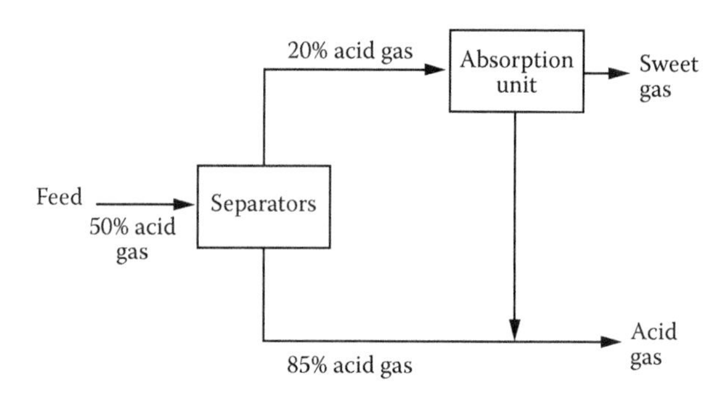

FIGURE 12.20
Bulk acid gas removal by the membrane process.

it is possible to reduce it in a single module to about 20% and to use amine absorption unit to reduce it to natural gas acid concentration, as shown in Figure 12.20.

12.6.5 Environmental Aspects

Hydrogen sulfide is one of the major pollutants in petroleum field processing, and acid gas treatment is one of the most important steps to eliminate H_2S. However, in the course of this treatment, some side products can cause pollution problems and must be eliminated.

In amine treatment, the following pollutants are produced and the corresponding countermeasures are given in Table 12.7.

TABLE 12.7

Pollution Prevention Alternatives

Waste	Pollution Prevention Alternatives
Amines	Use an amine reclaimer in the system to allow reuse of amine and minimize the volume of waste generated.
	Use an amine filter to extend the life of solution and maintain efficiency.
Filters	Change filters only when necessary; use differential pressure as indicator of needed change.
	Use reusable filters.
	When handling filters, care should be taken to prevent oil spillage.
	Isolate all drained fluids in tight container for recycling.
Iron sponge and iron sulfide scale	Consider alternative methods of removing H_2S from gas stream.
	Treat production streams with biocide or scale inhibitor to reduce iron sulfide formation.

REVIEW QUESTIONS

1. Name the main categories of the sweetening processes.
2. What constituents from natural gas are removed by the sweetening process?
3. What is the difference between *acid gases* and *sour gases*?
4. The Sulfinol process is an example of:
   a. Physical
   b. Dry bed
   c. Chemical
   d. Chemical/physical (hybrid)
5. Some sweetening processes have a *dual function*; they bring in gas dehydration as well. Which of the following sweetening processes fall into this category?
   a. MEA
   b. DGA
   c. DEA
   d. Sulfinol
   e. Molecular sieves
6. A sour crude oil contains 12 SCF of dissolved H_2S per 1000 gallons of crude oil (API = 35). If the buyer's specs for crude oil is 50 ppm maximum, calculate:
   a. The H_2S content of the sour crude oil in ppm
   b. The amount of H_2S gas (SCF) to be removed per 1000 bbl of oil in order to meet the specs

7. Why is the selective removal of H_2S with respect to CO_2 in some special cases desirable?
8. Most amine solvents are regenerated by:
   a. Lowering both T and P
   b. Increasing both T and P
   c. Lowering T and increasing P
   d. Lowering P and increasing T

13

Gas Dehydration

Natural gas dehydration is the process of removing water vapor from the gas stream to lower the dew point of that gas. Water is the most common contaminant of hydrocarbons. It is always present in the gas–oil mixtures produced from wells. The dew point is defined as the temperature at which water vapor condenses from the gas stream. The sale contracts of natural gas specify either its dew point or the maximum amount of water vapor present.

Dehydration is usually required before the gas can be sold for pipeline marketing and it is a necessary step in the recovery of NGL from natural gas.

Chapter 13 contains an elaborate discussion and presentation on the following topics as related to gas dehydration:

- The operating conditions leading to hydrate formation
- Procedures used to prevent hydrate formation
- Dehydration methods by absorption using liquid desiccants, by adsorption using solid desiccants, and by cooling/condensation

Solved examples as well as field case studies are presented including a simulation problem using HYSYS process simulator for the triethylene glycol (TEG) dehydrator.

13.1 Introduction

There are three basic reasons for the dehydration of natural gas streams:

- To prevent hydrate formation—Hydrates are solids formed by the physical combination of water and other small molecules of hydrocarbons. They are icy hydrocarbon compounds of about 10% hydrocarbons and 90% water. Hydrates grow as crystals and can build up in orifice plates, valves, and other areas not subjected to full flow. Thus, hydrates can plug lines and retard the flow of gaseous hydrocarbon streams. The primary conditions promoting hydration formation are the following: gas must be at or below its water (dew) point with *free* water present, low temperature, and high pressure.

- To avoid corrosion problems—Corrosion often occurs when liquid water is present along with acidic gases, which tend to dissolve and disassociate in the water phase, forming acidic solutions. The acidic solutions can be extremely corrosive, especially for carbon steel, which is typically used in the construction of most hydrocarbon processing facilities.
- Downstream processing requirements—In most commercial hydrocarbon processes, the presence of water may cause side reactions, foaming, or catalyst deactivation. Consequently, purchasers typically require that gas and liquid petroleum gas (LPG) feedstocks meet certain specifications for maximum water content. This ensures that water-based problems will not hamper downstream operations.

13.2 Prediction of Hydrate Formation

In this section, methods for determining the operating conditions leading to hydrate formation are presented. In particular, methods are presented to determine the following:

- Hydrate formation temperature for a given pressure
- Hydrate formation pressure for a given temperature
- Amount of water vapor that saturates the gas at a given pressure and temperature (i.e., at the dew point)

At any specified pressure, the temperature at which the gas is saturated with water vapor is being defined as the *dew point*. Cooling of the gas in a flow line due to heat loss can cause the gas temperature to drop below the hydrate formation temperature.

Two methods are discussed next for predicting the conditions leading to hydrate formation: approximate methods and analytical methods.

13.2.1 Approximate Methods

The first approximate method determines the hydrate formation pressure or temperature. This method utilizes the chart shown in Figure 13.1. It involves the following steps:

1. Calculate the average molecular weight of the gas mixture M_G.
2. Calculate the specific gravity of the gas, γ_G, where $\gamma_G = M_G/M_{air}$.

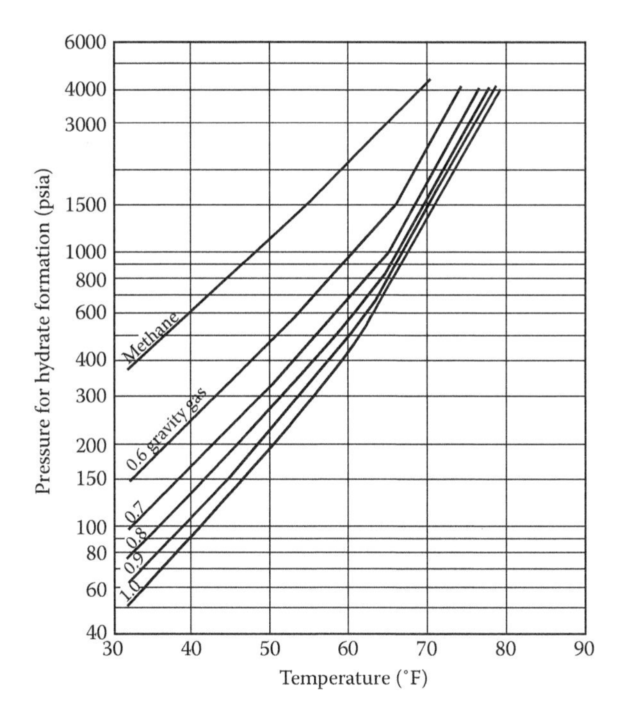

FIGURE 13.1
Pressure–temperature curves to predict hydrate formation.

3. Given the gas temperature (T) and its specific gravity (γ_G), find the corresponding gas pressure for hydrate formation using Figure 13.1.
4. Similarly, at a given gas pressure, the hydrate formation temperature can be determined from Figure 13.1.

Example 13.1

If the specific gravity of a natural gas stream is calculated and found to be 0.69, find the hydrate formation pressure (in psia) at a temperature of 50°F.

Solution

From Figure 13.1, a pressure of about 330 psia is determined for the corresponding conditions of temperature and specific gravity stated.

The second approximate method determines the amount of water vapor at dew point condition. This method utilizes the McKetta–Wehe chart, which relates the gas temperature (T) and its pressure (P) to the water vapor content of the natural gas at saturation conditions, as presented in Figure 13.2.

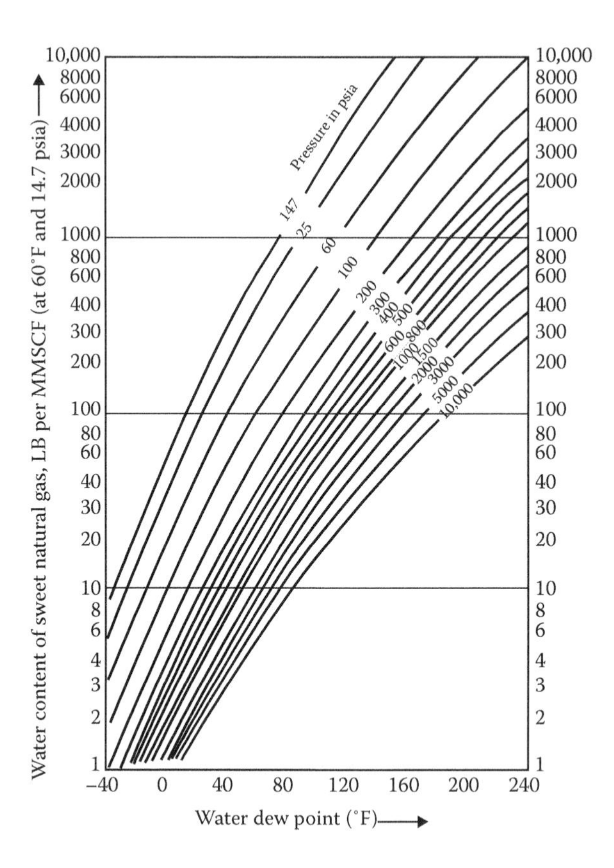

FIGURE 13.2
Water content of sweet natural gas function of pressure and temperature.

Example 13.2

Natural gas, saturated with water vapor at conditions of 1000 psia and 90°F, is exposed to cooling in a flow line due to heat losses, where the temperature reaches 35°F and the pressure remains the same.

a. Calculate how much liquid water will drop out of the gas.
b. Assuming that the gas flowing through the pipeline is to reach a delivery point at 300 psia pressure, find the corresponding dew point of the gas.

Solution

a. From Figure 13.2, we obtain the following: water content of the gas stream at the initial conditions (1000 psia and 90°F) is 46 lb/MMCF (MMCF: 10^6 cubic feet); water content of the gas stream at conditions of 1000 psia and 35°F is 7.6 lb/MMCF; and water to be separated is 46 – 7.6 = 38.4 lb/MMCF (MMCF: 10^6 cubic feet).

b. The natural gas, once it reaches the delivery point at 300 psia, carries with it a water content of 7.6 lb/MMCF. Applying these two parameters to Figure 13.2, one can read the dew point temperature of 12°F.

13.2.2 Analytical Methods

The calculations presented in this section are concerned with finding the hydrate formation temperature (T) at a given pressure (P), or the pressure (P) at which hydrate formation takes place for a given operating temperature (T). Knowledge of the temperature and pressure of a gas stream at the wellhead is important for determining whether hydrate formation can be expected when the gas is expanded into the flow lines. In general, the temperature at the wellhead can change as the reservoir conditions or production rate change over the production life of the well. Wells, therefore, that initially flowed at conditions where no hydrate formation occurred in downstream equipment may require hydrate inhibition, or vice versa. The computational approach is analogous to the one used in the dew point calculation for a multicomponent mixture of hydrocarbon gases. The basic equation is given by

$$\sum_{i=1}^{n} \frac{y_i}{K_i} = 1 \quad (13.1)$$

where y_i is the mole fraction of component i in the gas phase, on a water–free basis; K_i is the vapor–solid equilibrium constant for component i, defined by $K_i = y_i/x_i$; and x_i is the mole fraction of component i in the solid phase, on a water-free basis.

Given the gas pressure (P) and the mole fraction of its components y_i, the hydrate formation temperature (T) is obtained through a trial-and-error procedure according to the following steps:

1. Assume a value for T.
2. Find the values of K for the components of the gas at P and T using the K charts.
3. Calculate the values of x_i, where $x_i = y_i/K_i$.
4. Find the sum of y_i/K_i for all components.
5. Convergence to the desired value of T is obtained by trial-and-error until

$$\sum_{i=1}^{n} \frac{y_i}{K_i} = 1$$

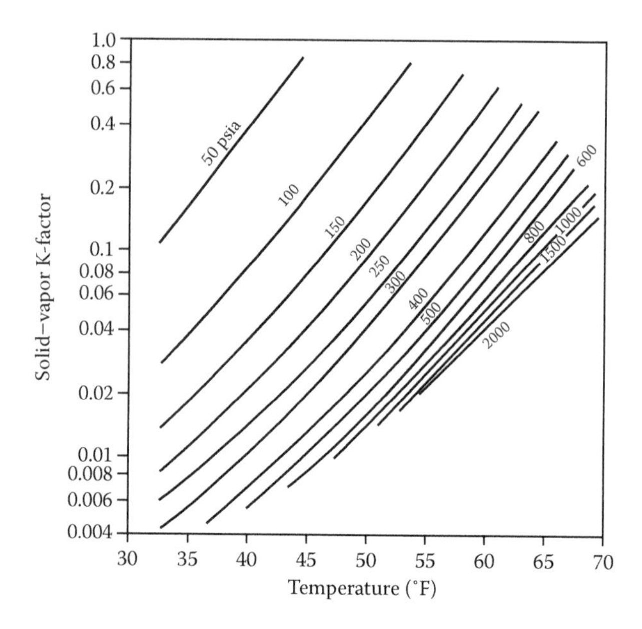

FIGURE 13.3
Vapor–liquid equilibrium constants for isobutane (K_i).

A sample chart for the vapor–solid equilibrium constant for isobutane is given in Figure 13.3.

13.3 Methods Used to Inhibit Hydrate Formation

Hydrate formation in natural gas is promoted by high-pressure, low-temperature conditions and the presence of liquid water. Therefore, hydrates can be prevented by the following: raising the system temperature and/or lowering the system pressure (temperature/pressure control); injecting a chemical such as methanol or glycol to depress the freezing point of liquid water (chemical injection); or removing water vapor from the gas liquid–water drop out that is depressing the dew point (dehydration).

13.3.1 Temperature/Pressure Control

Methods recommended for temperature control of natural gas streams include downhole regulators or chokes, and indirect heaters. In this first method, a pressure regulator (choke) is installed downhole (in the well). This causes the largest portion of the desired pressure drop from the bottom-hole flowing pressure to the surface flow line pressure to occur where the gas temperature is still high. The bottom-hole temperature will be sufficiently high to prevent hydrate formation as the pressure is reduced. At the surface, little or no pressure reduction may be required, thus hydrate formation is also avoided at the surface.

Both wellhead and flow line indirect heaters are commonly used to heat natural gas to maintain the flowing temperature above the hydrate formation temperature. The primary purpose of the wellhead heater is to heat the flowing gas stream at or near the wellhead, where chocking or pressure reduction frequently occurs. Flow line heaters, on the other hand, provide additional heating if required. They are particularly used for cases where the conditions necessitate a substantial reduction in pressure between the wellhead stream and the next field processing facility. A typical example is an offshore production field with its treating facilities on land. Heat is utilized to compensate for the loss of temperature as the gas expands. Another significant factor that contributes to the reduction of the gas temperature in a gas line is the ground temperature. A temperature drop of about 80°F is experienced by a gas flow line travelling a distance of about 5000 ft.

13.3.2 Chemical Injection

Methanol and glycols are the most commonly used chemicals, although others (such as ammonia) have been applied to lower the freezing point of water, thus reducing (or preventing) hydrate formation.

The application of hydrate inhibitors should be considered for such cases:

- A system of gas pipelines, where the problem of hydrate formation is of short duration
- A system of gas pipelines that operate at a few degrees below the hydrate formation temperature
- Gas-gathering systems found in pressure-declining fields
- Gasolines characterized by hydrate formation in localized points

Inhibitors function in the same manner as *antifreeze* when added to liquid water. Thus, the principle underlying the use of hydrate inhibitors is to lower the formation of the hydrate by causing a depression of the hydrate formation temperature.

13.3.2.1 Methanol Injection

Methanol is the most commonly used nonrecoverable hydrate inhibitor. It has the following properties:

- It is noncorrosive.
- It is chemically inert; no reaction with the hydrocarbons.

- It is soluble in all proportions with water.
- It is volatile under pipeline conditions, and its vapor pressure is greater than that of water.
- It is not expensive.

Methanol is soluble in liquid hydrocarbons (about 0.5% by weight). Therefore, if the gas stream has high condensate contents, a significant additional volume of methanol will be required. This makes this method of hydrate inhibition unattractive economically because methanol is nonrecoverable. In such a situation, it will be necessary to first separate the condensate from the gas. Some methanol would also vaporize and goes into the gas. The amount of methanol that goes into the gas phase depends on the operating pressure and temperature.

In many applications, it is recommended to inject methanol some distance upstream of the point to be protected by inhibition, in order to allow time for the methanol to vaporize before reaching that point.

13.3.2.2 Glycol Injection

Glycol functions in the same way as methanol; however, glycol has a lower vapor pressure and does not evaporate into the vapor phase as readily as methanol. It is also less soluble in liquid hydrocarbons than methanol. This, together with the fact that glycol could be recovered and reused for the treatment, reduces the operating costs as compared to the methanol injection.

Three types of glycols can be used: ethylene glycol (EG), diethylene glycol (DEG), and triethylene glycol (TEG). The following specific applications are recommended:

- For natural gas transmission lines, where hydrate protection is of importance, EG is the best choice. It provides the highest hydrate depression, although this will be at the expense of its recovery because of its high vapor pressure.
- Again, EG is used to protect vessels or equipment handling hydrocarbon compounds, because of its low solubility in multicomponent hydrocarbons.
- For situations where vaporization losses are appreciable, DEG or TEG should be used, because of their lower vapor pressure.

It is of importance to mention that when hydrate inhibitors in general are injected in gas flow lines or gas gathering networks, installation of a high-pressure free-water knockout at the wellhead is of value in the operation. Removing of the free water from the gas stream ahead of the injection point will cause a significant savings in the amount of the inhibitor used.

TABLE 13.1

Properties of Chemical Inhibitors

Inhibitor	M	K
Methanol	32.04	2335
Ethylene glycol	62.07	2200
Propylene glycol	76.10	3590
Diethylene glycol	106.10	4370

The amount of chemical inhibitor required to treat the water in order to lower the hydrate formation temperature may be calculated from the Hammerschmidt equation:

$$\Delta T = \frac{KW}{M(100 - W)} \quad (13.2)$$

where ΔT is the depression in hydrate formation temperature (°F), W is the weight percent of the inhibitor for water treatment, K is a constant that depends on the type of inhibitor, and M is the molecular weight of the inhibitor. Values of M and K for various inhibitors are given in Table 13.1.

Example 13.3

A gas well produces 10 MMSCF/day along with 2000 lbs of water and 700 barrels per day (BPD) of condensate having a density of 300 lbs/bbl. The hydrate formation temperature at the flowing pressure is 75°F. If the average flow line temperature is 65°F, determine the amount of methanol needed to inhibit hydrate formation in the flow line given that the methanol solubility in condensate is 0.5% by weight and that the ratio of the pounds of methanol in vapor/MMSCF of gas to the weight percent of methanol in water is 0.95.

Solution

To prevent hydrate formation in the flow line, we need to lower the hydrate formation temperature to 65°F or less. Therefore, the depression in hydrate formation temperature, ΔT, is

$$\Delta T = 75 - 65 = 10°\text{F}$$

Using Equation 13.2,

$$10 = \frac{2335W}{32.04(100 - W)}$$

Therefore, $W = 12.07\%$.

$$\text{Required methanol in water} = (0.1207)(2000) = 241.4 \text{ lb/day}$$

$$\text{Pounds of methanol in vapor} = (0.95)(12.07) = 11.47 \text{ lbs/MMSCF}$$

$$= (11.47)(10) \text{ MMSCF/day}$$

$$= 114.7 \text{ lbs/day}$$

$$\text{Methanol dissolved in condensate} = (0.005)(300) \text{ lbs/bbl} \times 700 \text{ bbl/day}$$

$$= 1050 \text{ lbs/day}$$

$$\text{Total amount of methanol} = 241.4 + 114.7 + 1050 = 1406 \text{ lbs/day}$$

From the results, we see that 1050 lbs of methanol (75% of the total) are dissolved in the condensate and thus do not contribute to the treatment. Such treatment is, therefore, economically unacceptable. It is evident that the condensate must be separated first before the treatment.

13.3.3 Dehydration Methods

The most common dehydration methods used for natural gas processing are as follows: absorption, using the liquid desiccants (e.g., glycols and methanol); adsorption, using solid desiccants (e.g., alumina and silica gel); and cooling/condensation below the dew point, by expansion and/or refrigeration. This is in addition to the hydrate inhibition procedures described in Section 13.3.1. Classification of dehydration methods is given in Figure 13.4.

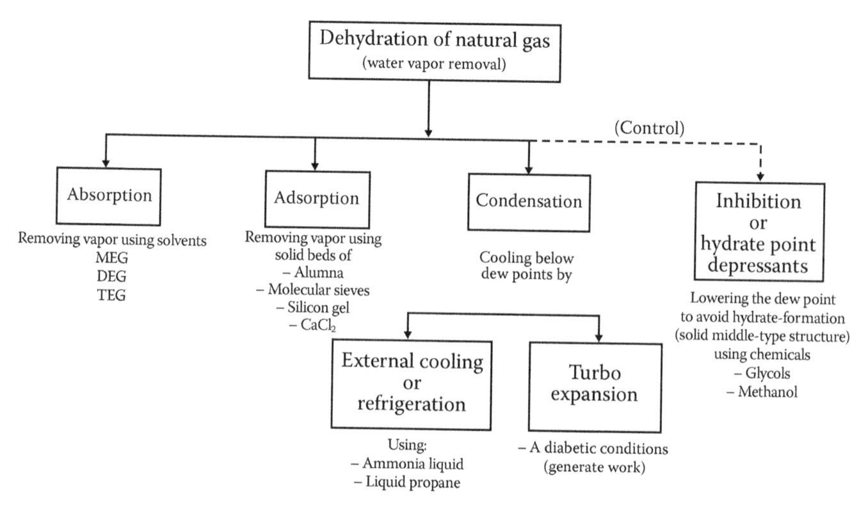

FIGURE 13.4
Classification of gas dehydration methods.

13.4 Absorption (Glycol Dehydration Process)

13.4.1 Basic Principles

The basic principles of relevance to the absorption process are as follows:

1. In this process, a hygroscopic liquid is used to contact the *wet* gas to remove water vapor from it. Triethylene glycol (TEG) is the most common solvent used.
2. Absorption, which is defined as the transfer of a component from the gas phase to the liquid phase, is more favorable at a *lower temperature and higher pressure*. This result is concluded by considering the following relationship (which is a combination of Raoult's law and Dalton's law):

$$\frac{P_i}{P} = \frac{Y_i}{X_i} = K_i$$

   where P_i is the pressure of pure component *i*, *P* is the total pressure of the gas mixture (system), X_i is the mole fraction of component *i* in the liquid phase, Y_i is the mole fraction of component *I* in the vapor phase, and K_i is the equilibrium constant, increasing with temperature and decreasing with pressure.

   Now, if temperature decreases (where *P* is constant), P_i decreases, which means that the water vapor concentration in the gas, Y_i, decreases, allowing more absorption of water in the liquid phase. The same conclusion is reached if the total pressure *P* is increased.
3. The actual absorption process of water vapor from the gas phase using glycol is dynamic and continuous. Therefore, the gas flow cannot be stopped to let a vapor and the liquid reach an equilibrium condition. Accordingly, the system under consideration must be designed to allow for a close approach to equilibrium while the flow continues. Two means are provided to accomplish this task for a countercurrent flow of the feed natural gas and liquid (solvent or glycol):

   a. Tray column, or stage wise operation (equilibrium concept)

   b. Packed column or continuous-contact operation (rate concept)

   This countercurrent system allows for the wet gas to enter the bottom of the column and contact the *rich* glycol (high water content) at its exit point. On the other hand, as the gas works its way up the column, it encounters the *leanest* glycol (lowest water content) before the gas leaves the column.

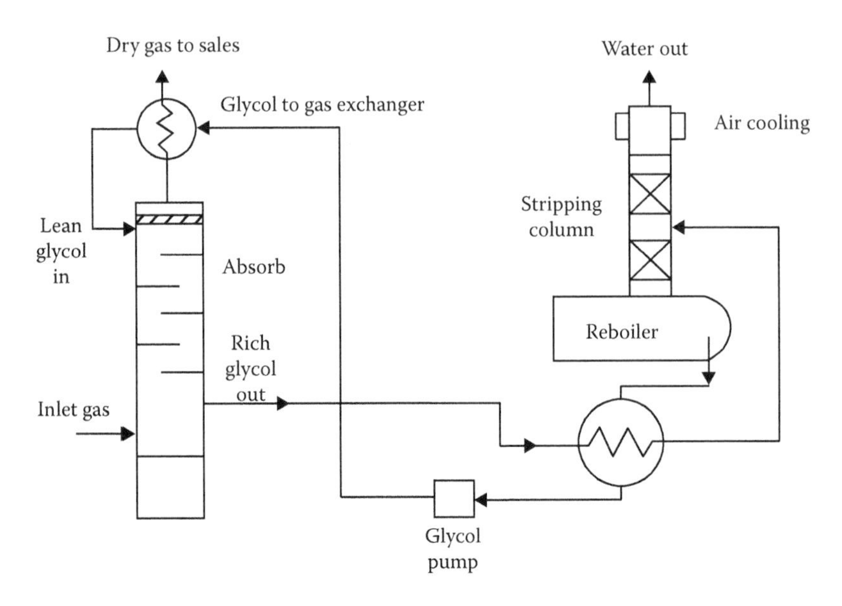

FIGURE 13.5
Flow diagram of TEG dehydration.

13.4.2 Absorption System

The absorption process is shown schematically in Figure 13.5. The wet natural gas enters the absorption column (glycol contactor) near its bottom and flows upward through the bottom tray to the top tray and out at the top of the column. Usually six to eight trays are used. Lean (dry) glycol is fed at the top of the column and it flows down from tray to tray, absorbing water vapor from the natural gas. The rich (wet) glycol leaves from the bottom of the column to the glycol regeneration unit. The dry natural gas passes through mist mesh to the sales line.

The glycol regeneration unit is composed of a reboiler where steam is generated from the water in the glycol. The steam is circulated through the packed section to strip the water from glycol. Stripped water and any lost hydrocarbons are vented at the top of the stripping column. The hydrocarbon losses are usually benzene, toluene, xylene, and ethyl benzene (BTXE), and it is important to minimize these emissions. The rich glycol is preheated in heat exchangers, using the hot lean glycol, before it enters the still column of the glycol reboiler. This cools down the lean glycol to the desired temperature and saves the energy required for heating the rich glycol in the reboiler.

13.4.3 Pressure and Temperature Considerations

The absorption process improves at higher pressures because the higher-pressure gas will contain less water vapor as compared to a lower-pressure gas at the same temperature. The effect of pressure on the process, however, is not significant for pressures below about 3000 psi. However, the gas pressure should not be too high, as this increases the pressure rating of the column and, consequently, increases its cost. A high operating pressure would also require high glycol pumping power. On the other hand, if the gas pressure is too low, the column size would be too large. Normally, most operations are designed at pressures between 1000 and 1200 psi.

Glycol regeneration is better achieved at lower pressures. Usually, the glycol regeneration takes place at atmospheric pressure. In some cases, the process takes place under a vacuum to achieve higher lean glycol concentrations; this, however, makes the system too complicated and very expensive.

The inlet gas temperature should not be too low in order to avoid condensation of water vapor and hydrocarbons. Also a low gas temperature means a low glycol temperature. At low temperatures (below 50°F), glycol becomes too viscous and more difficult to pump. Also, at low temperatures (below 60°F–70°F), glycol can form a stable emulsion with the hydrocarbon in the gas and may also cause foaming. On the other hand, high gas temperatures increase the gas volume, thus requiring a large-size column, and increase the water vapor content of the gas. Also, a high gas temperature results in high glycol losses. The inlet glycol temperature should not be lower than the gas temperature in order to avoid condensation of water and hydrocarbons. Normally, the gas temperature is maintained between 80°F and 110°F. The inlet glycol temperature is normally kept at about 10°F above the exiting gas temperature.

In the glycol regenerator, the glycol temperature is normally raised up to between 370°F and 390°F. This results in a lean glycol concentration of about 98.5%–98.9%. A higher temperature will cause degradation of the glycol. To achieve higher glycol concentrations, stripping gas may be used.

13.4.4 Process Design

The procedure for determining the absorber column diameter and height along with solved example for a case study are presented next.

- Column diameter—The column diameter can be calculated using the Souder–Brown correlation:

$$U_{max} = C_{SB}\left(\frac{\rho_l - \rho_v}{\rho_v}\right)^{1/2} \quad (13.3)$$

where U_{max} is the maximum gas superficial velocity (ft/h), C_{SB} is the Souder–Brown coefficient (ft/h) (assume a value 660), ρ_l is the glycol density (lb/ft^3), and ρ_v is the gas density (lb/ft^3).

The cross-section area of the column can be calculated as

$$A = \frac{V}{U_{max}} \tag{13.4}$$

where V is the gas flow rate at operating conditions.

$$V = \frac{nZRT}{P} \tag{13.5}$$

The compressibility factor Z can be calculated from the corresponding state method as

$$Z = 1 + \frac{B_r P_r}{T_r} \tag{13.6}$$

$$B_r = B^\circ + wB'$$

$$B^\circ = 0.083 - \frac{0.422}{T_r^{1.6}}$$

$$B' = 0.139 - \frac{0.172}{T_r^{4.2}}$$

$$P_r = \frac{P}{P_c} \quad \text{(reduced pressure of natural gas)}$$

$$T_r = \frac{T}{T_c} \quad \text{(reduced temperature of natural gas)}$$

Thus the diameter is

$$D = \sqrt{\frac{4A}{\pi}} \tag{13.7}$$

Column height—The number of theoretical plates can be calculated using the McCabe–Thiele method. The distribution of water between TEG and natural gas (equilibrium line) is almost linear, as shown in Figure 13.6:

$$Y = 0.1042X \tag{13.8}$$

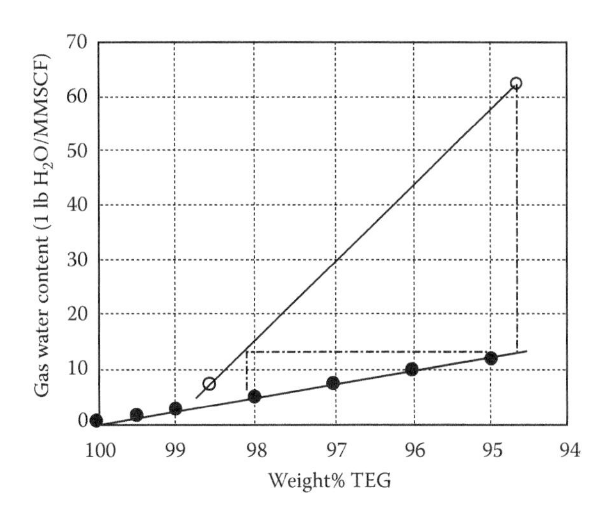

FIGURE 13.6
McCabe–Thiele diagram.

where Y is the water content in the gas and X is the rich glycol concentration. If the operating line can also be assumed to be linear and the tray efficiency (η) can be assumed to be 25%, then the actual number of trays can be calculated. If the tray spacing is assumed to be 2 ft, the height (H) is calculated as

$$H = \frac{N(\text{spacing})}{\eta}$$

Example 13.4: Case Study

Calculate the (a) diameter and (b) height for an absorption column for the dehydration of wet natural gas under the following operating conditions:

- Natural gas flow rate = 98 MMSCFD
- Saturated with water at 1000 psig, 100°F gas
- Target gas water content = 7 lb/MMSCF
- Use TEG for dehydration (98.5% purity, the balance is water)
- No stripping gas used

Solution

A. ABSORBER DIAMETER

Calculate Z, the compressibility factor, where $T_r = 1.6317$, $P_r = 1.5217$, $B^\circ = -0.1100$, $B' = 0.1170$, and $B_r = -0.1092$.

$$Z = 1 + \frac{(-0.1092)(1.5217)}{1.6317} = 0.898$$

Calculate the volumetric flow rate, V:

$$PV = nZRT$$

$$V = \frac{(98 \times 10^6/379.5)(0.9)(10.7)(560)}{1015}$$

$$V = 953 \frac{\text{ft}^3}{\text{min}}$$

Applying the Souder–Brown equation:

$$U_{max} = 660\left(\frac{(70 - 3.79)}{3.79}\right)^{1/2} = 2759 \frac{\text{ft}}{\text{h}} = 46 \frac{\text{ft}}{\text{min}}$$

The cross-section area of the column is now found to be

$$A = \frac{953}{46} = 20.7 \text{ ft}^2$$

$$D = \sqrt{\frac{4(20.7)}{\pi}} = 5.1 \text{ ft}$$

B. ABSORBER HEIGHT

To calculate the number of theoretical plates, first plot the operating line: At tower bottom, calculate Y_1 from Figure 13.6:

$$Y_1 = \text{water content of inlet gas} = 63 \text{ lbs } H_2O/\text{MMSCF}$$

Calculate X_1 from column material balance, where X_1 is the rich glycol weight percent.

At tower top,

$$Y_2 = \text{water content in exit gas} = 7 \text{ lbs } H_2O/\text{MMSCF (given)}$$

$$X_2 = \text{lean glycol weight percent} = 98.5 \text{ (given)}$$

$$V(Y_1 - Y_2) = L(X_2 - X_1)$$

where L is the circulation (ft^3/h).

If we assume that 2 gal TEG/lb H_2O are needed, then the TEG circulation rate is found to be

$$L = G(Y_1 - Y_2)(2)$$
$$= \frac{98}{\text{day}}\left(\frac{1\,\text{day}}{24\,\text{h}}\right)\left(\frac{1\,L}{60\,\text{min}}\right)(63-7)(2\,\text{gal})$$
$$= 7.6\frac{\text{gal}}{\text{min}}(\text{gpm})$$

We can find X_1 from the material balance equation:

$$98\,\text{MMSCD}(Y_1 - Y_2) = (X_2 - X_1)\frac{\text{ft}^3}{\text{day}}$$

$$98(63-7) = \frac{\text{ft}^3}{9.5}\left(\frac{7.6\,\text{gal}}{\text{min}}\right)\left(\frac{60\,\text{min}}{1\,\text{h}}\right)\frac{24\,\text{h}}{1\,\text{day}}(98.5 - X_1)$$
$$5488 = 1459.2(98.5 - X_1)$$
$$3.761 = 98.5 - X_1$$
$$X_1 = 98.5 - 3.76$$
$$= 94.7$$

From Figure 13.6, we can find the theoretical number of stages: $N = 1.5$ stages. Assuming 25% efficiency,

$$N_{act} = \frac{1.5}{0.25} = 6\text{ stages}$$

Assuming a spacing of 2 ft, the height H is found to be 12 ft.

The dew point of the exit sales gas can be estimated from Figure 13.2 and found to be equal to –10°F.

13.5 Adsorption: Solid-Bed Dehydration

When very low dew points are required, solid-bed dehydration becomes the logical choice. It is based on fixed-bed adsorption of water vapor by a selected desiccant. A number of solid desiccants could be used such as silica gel, activated alumina, or molecular sieves.

The properties of these materials are shown in Table 13.2. The selection of these solids depends on economics. The most important property is the capacity of the desiccant, which determines the loading design expressed as the percentage of water to be adsorbed by the bed. The capacity decreases as temperature increases.

TABLE 13.2
Properties of Solid Desiccants

Desiccant Reference	Silica Gel	Activated Alumina	Molecular Sieves
Pore diameter (Å)	10–90	15	3, 4, 5, 10
Bulk density (lb/ft^3)	45	44–48	43–47
Heat capacity (Btu/lb °F)	0.22	0.24	0.23
Minimum dew point (°F)	–60 to –90	–60 to –90	–100 to –300
Design capacity (wt%)	4–20	11–15	8–16
Regeneration stream temperature (°F)	300–500	350–500	425–550
Heat of adsorption (Btu/lb)	–	–	1800

13.5.1 Operation of Solid-Bed Dehydrator

The system may consist of two-bed (as shown in Figure 13.7), three-bed, or multibed operation. In the three-bed operation, if two beds are loading at different stages, the third one would be regenerated. The feed gas in entering the bed from the top and the upper zone becomes saturated first. The second zone is the mass transfer zone (MTZ) and is being loaded. The third zone is still not used and active. The different saturation progress and representation of different zones is shown in Figure 13.8. While the bed is in operation, the outlet concentration has very low water concentration (C_s) and the MTZ moves downward. At a certain point, the outlet water content rises to the point that is equivalent to the initial wet gas content (C_0) as if the bed is not present. Thus, the beginning of this period is called the breakthrough period (θ_B). This situation is shown in Figure 13.9.

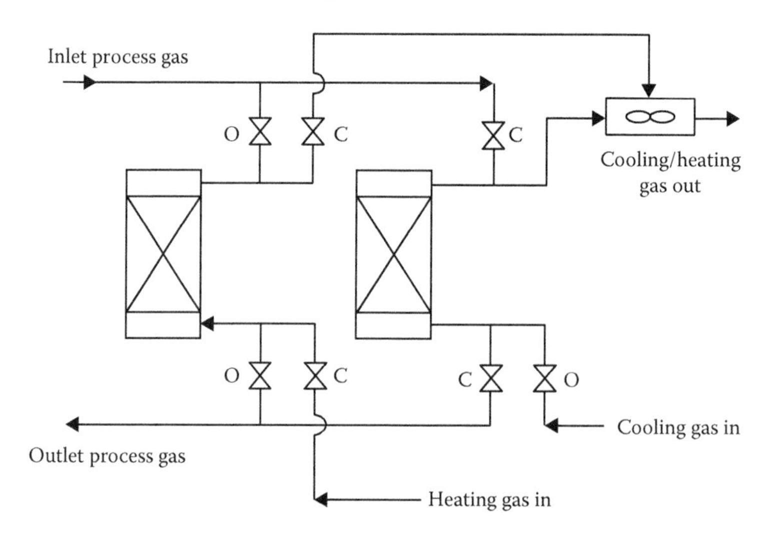

FIGURE 13.7
Solid-bed dehydration process.

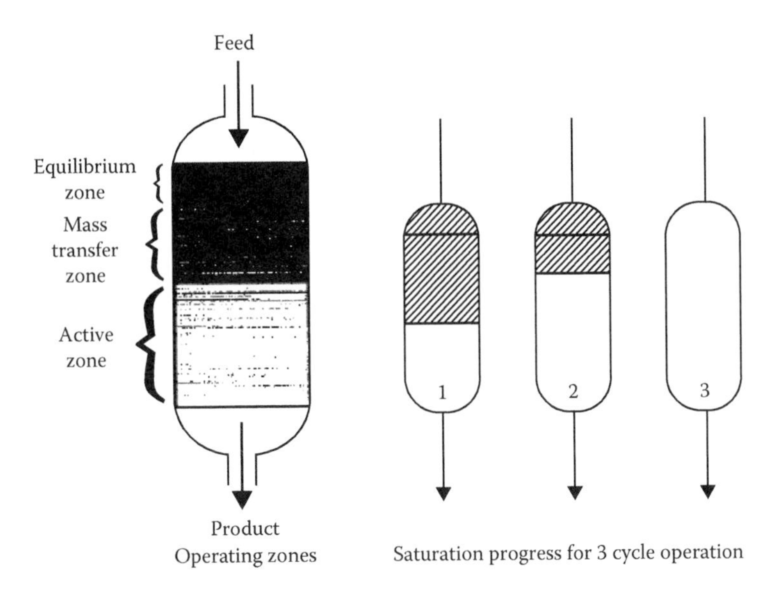

FIGURE 13.8
Mode operation.

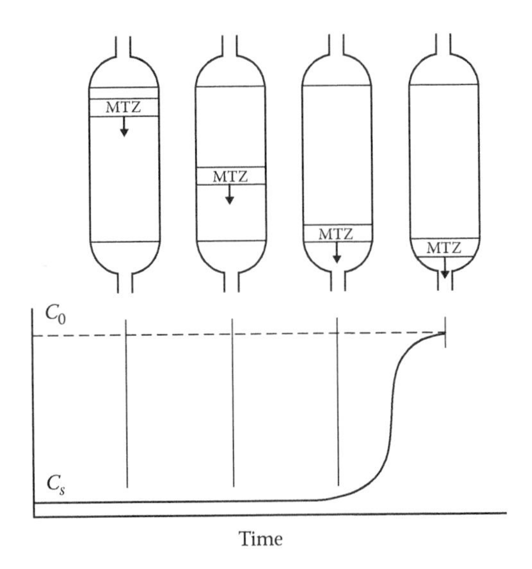

FIGURE 13.9
Breakthrough diagram in a fixed-bed dehydrator.

The operation of the process is controlled by the opening valve (O) and closing valve (C), which are shown in Figure 13.9. After the bed has been used and loaded with water, then it is regenerated by hot gas (say 6 h, as heating time θ_H) and then cooled by switching to cold gas (say for 2 h, θ_C). The approximate temperature profile for inlet gas temperature during heating

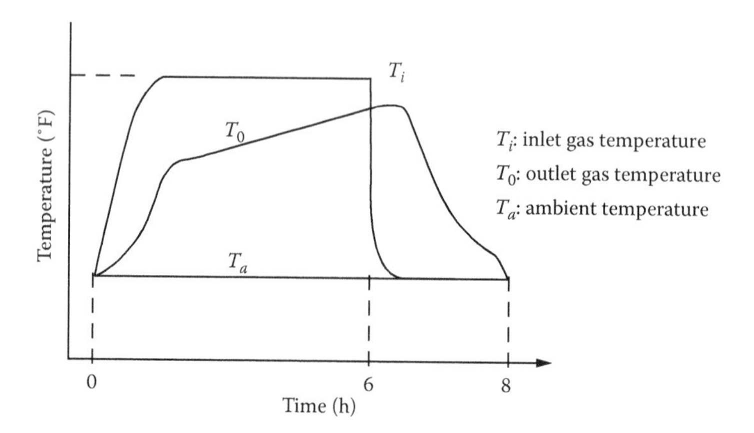

FIGURE 13.10
Temperature profile for two-tower operation.

and cooling of the bed (T_i) to desorb water, and then to cool the bed to prepare it for the next cycle and the outlet gas temperature (T_0) in the same period are shown in Figure 13.10.

13.5.2 Design of an Adsorber for Gas Dehydration

Usually, the given process conditions are gas flow rate V (MMSCF/d), temperature T (°R), gas water content Y_1 (lbs H_2O/MMSCF), desiccant particle size D_P (ft), gas density ρ_g (lbs/ft^3), and operating pressure P (psia). The maximum useful desiccant capacity X_C (lbs H_2O/100 lbs desiccant) is given by the vendor. The capacity will change if the storage conditions are not good. It should be assumed and then verified in the calculations. The bed operation cycle time is usually assumed as 8 h; thus, one day has three cycles.

1. For given gas flow rate V or G (MMSCF/d) and water content Y_1 or W (1bH_2O/MMSCF), calculate the water to be adsorbed from the gas as

$$\frac{\text{Water adsorbed}}{\text{Cycle}} = \frac{VY_1}{3} \tag{13.9}$$

   Assuming three cycles, each cycle is 8 h in one day.

2. If the particle size of the bed is known D_P (ft), calculate the superfacial velocity

$$V_{ug}\left(\frac{\text{ft}}{\text{min}}\right) \text{ as } V_{ug} = 540\left(\frac{D_P}{\rho_g}\right)^{1/2} \tag{13.10}$$

   where ρ_g is the gas density (lb/ft^3).

3. Calculate the bed diameter d (ft) from

$$d = \left(\frac{25\, V Z_g T_g}{P V_g} \right)^{1/2} \tag{13.11}$$

4. Calculate the water loading q (lbs/ft^{2}h) using

$$q = 0.053 \left(\frac{V Y_1}{d^2} \right) \tag{13.12}$$

5. Calculate the mass transfer zone height h_z (ft) using

$$h_z = 31.25 \left(\frac{q^{0.7895}}{V_g^{0.5506} \mathrm{RS}^{0.2646}} \right) \tag{13.13}$$

where RS is relative saturation.

6. Calculate the bed length h_B (ft). The bed weight W_B (lb) is calculated using

$$W_B = \frac{\text{Amount of water adsorbed(lbs } H_2O)}{\text{Capacity of desiccant(lbs } H_2O/\text{lb desiccant)}}$$

$$W_B = \frac{V Y_1}{3} X_C^{-1} \tag{13.14}$$

The desiccant density ρ_B (lbs/ft^3) is used to calculate bed volume V_B (ft^3) using

$$V_B = \frac{W_B}{\rho_B} \tag{13.15}$$

The bed length h_B (ft) is calculated using

$$h_B = \frac{V_B}{\pi d^2/4} \tag{13.16}$$

7. The maximum useful desiccant capacity (lbs H_2O/100 lbs desiccant) is calculated from

$$X_C = \frac{X'_S (h_B - 0.45 h_Z)}{h_B} \tag{13.17}$$

where X'_S is the dynamic desiccant capacity at saturation and is a function of temperature as

$$X'_S = X_S (1.518 - 0.00673\, T_B) \tag{13.18}$$

X_S is the dynamic capacity at 77°F and T_B (°F) is the bed temperature and can be calculated using gas relative saturation (RS) at 77°F:

$$X_S = 7407 + 0\ 09357\ \text{RS} \tag{13.19}$$

The calculated value of X_C from Equation 13.17 is compared with the assumed value and is readjusted so that assumed and calculated values become close.

8. Calculate the breakthrough time θ_b from

$$\theta_B = \frac{0.01\ X_C \rho_B h_B}{q} \tag{13.20}$$

The cycle time θ_C should be greater than the breakthrough time:

$$\theta_C > \theta_b$$

9. Check pressure drop across the bed for a 1/8 in particle size of packed material. The pressure drop can be calculated as

$$\frac{\Delta P}{h_B} = 0.056_{\mu}^{*} V_g + 0.0000889 + V_g^2 \tag{13.21}$$

It is recommended that the pressure drop should be less than 8 psia. The design procedure is illustrated in Figure 13.11. (*μ is the gas viscosity in centipoise, Cp.)

Example 13.5

It is given that 35 MMSCF of wet gas is fed into a packed bed of desiccant. The gas molecular weight is 18.0 and its density is 1.5 lbs/ft³. The operating temperature is 110°F and pressure is 700 psia. The inlet dew point is 100°F and the desired outlet dew point is 1 ppm H_2O.

Solution

Assuming

$$X_C = 0.12$$

$$W_B = \frac{35(90)}{3(0.12)} = 8750$$

$$V_B = \frac{8750}{45} = 195\ \text{ft}^3$$

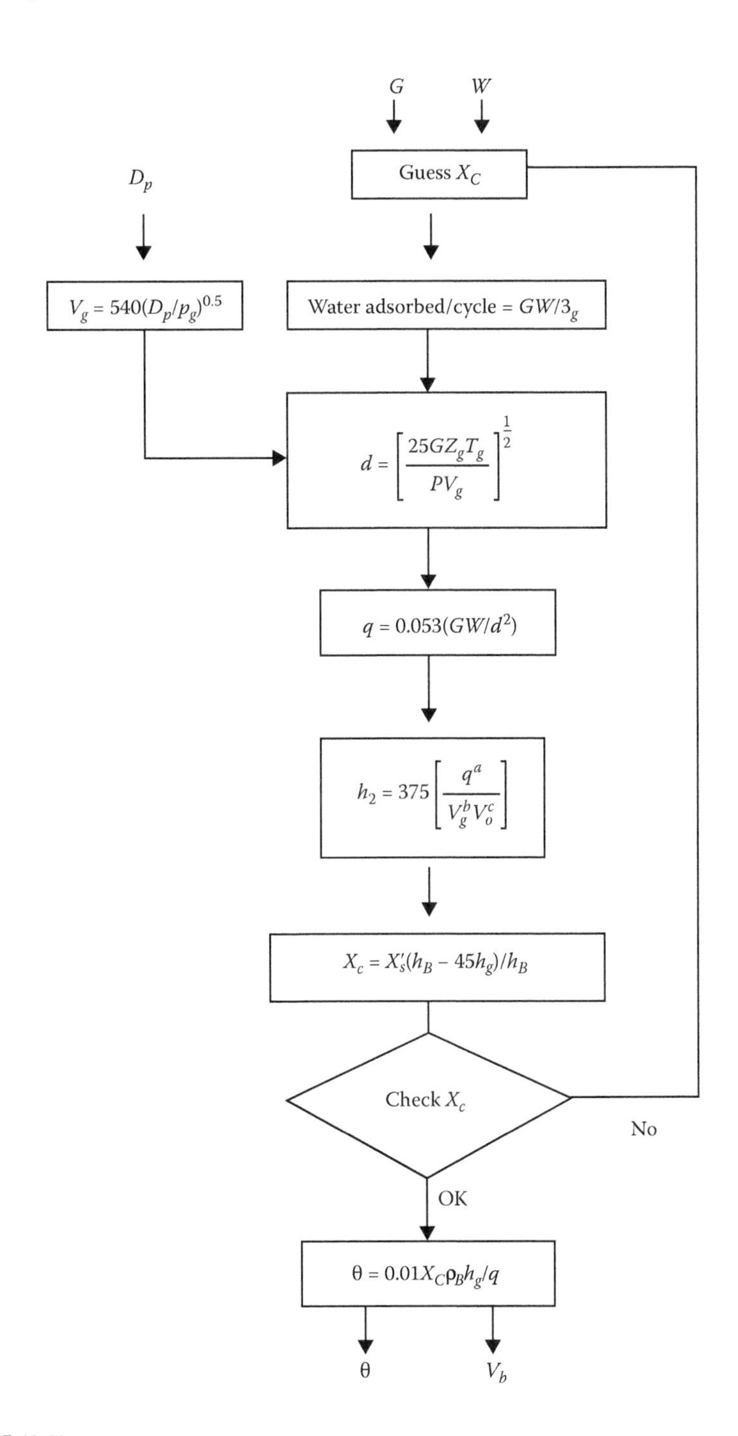

FIGURE 13.11
Design procedure for solid dehydrator.

Assume 1/8 in particle size.

$$D_P = \frac{1}{8(12)}$$

$$V_{ug} = 540\left(\frac{1}{8(12)(1.5)}\right)^{1/2} = 45\,\frac{\text{ft}}{\text{min}}$$

$$d = \left(\frac{25(35)(1)(570)}{700(45)}\right)^{1/2} = 3.9\text{ ft}$$

$$h_B = \frac{V_B}{\pi d^2/4} = \frac{195(4)}{\pi(3.9)^2} = 16.3\text{ ft}$$

$$q = 0.053\left(\frac{GW}{d^2}\right) = \frac{35(90)(0.053)}{(3.9)^2} = 10.98$$

$$h_Z = 31.25\left(\frac{11^{0.7895}}{(45)^{0.5505}(100)^{0.2646}}\right) = 7.5\text{ ft}$$

$$X_C = \frac{X_S'(h_B - 0.45\,h_Z)}{h_B}$$

$$X_S = 7.407 + 0.09357(100) = 16.764$$

$$X_S' = 16.764[1.518 - 0.00673(100)] = 14.17$$

$$X_C = \frac{14.17[16.3 - 0.45(7.5)]}{16.3}$$

$$= 11.236\text{ lbs } H_2O/100\text{ lbs solid}$$

or

$$X_C = 0.1124\text{ lbs } H_2O/\text{lb solid (per lb solid)}$$

Initially, we assumed $X_C = 0.12$, which is close enough.

$$\theta_B = \frac{0.01 X_C \rho_B h_B}{q}$$

$$\theta_B = \frac{0.01(11.2)(45)(16.3)}{11} = 7.468$$

So, assuming three cycles each 8 h will be fine.
The pressure drop ΔP (psia) is calculated as

$$\frac{\Delta P}{h_B} = 0.0560\,\mu V_g + 8.89\times 10^{-5}\rho V_g^2$$

$$\Delta P = 0.0560(0.01)(45) + 8.89\times 10^{-5}(1.5)(45)^2(16.3) = 4.8\text{ psi}$$

which is less than 8 psia (recommended).

13.5.3 Regeneration Heat Requirements

The heat required during the heating part of the cycle is

$$Q_H = m_d(T_m - T_a)C_d + m_t(T_m - T_a)C_t + w_1 \Delta H_d + w_1(T_{av} - T_a)$$
$$T_{av} = \frac{T_m + T_a}{2} \quad (13.22)$$

where w_1 is the water adsorbed/cycle (lbs/cycle); Q_H is the heat required/cycle (Btu/cycle); m_d and C_d is the mass and specific heat of desiccant; m_t and C_t is the mass and specific heat of tower, ΔH_d is the heat of water desorbed (given) (Btu/lb), T_m is the maximum outlet temperature (°F), and T_a is ambient temperature (°F).

To calculate gas flow rate V_H (lbs/cycle) required to desorb water, you need to raise heating gas to T_H from ambient temperature T_a to supply Q_H.

Thus,

$$Q_H = V_H(T_H - T_{av})C_g$$

where C_g is specific heat of gas; or

$$V_H = \frac{Q_H}{(T_H - T_{av})C_g} \frac{\text{lbs}}{\text{cycle}} \quad (13.23)$$

The gas heater load, Q_G (Btu/cycle), is

$$Q_G = V_H(T_H - T_a)C_g \frac{\text{Btu}}{\text{cycle}}$$

If the heating period is θ_H, h for each cycle is then

$$Q_G = \frac{V_H(T_H - T_a)C_g}{\theta_H} \frac{\text{Btu}}{\text{h}} \quad (13.24)$$

where Q_G is the gas heater load and θ_H is the heating time per cycle. Usually, 25% excess load is assumed for losses.

For cooling part cooling requirements, H_C (Btu/cycle),

$$H_C = m_d(T_m - T_a)C_d + m_t(T_m - T_a)C_t \quad (13.25)$$

and the gas flow rate for cooling, V_C (lbs/cycle),

$$V_C = \frac{H_C}{(T_{av} - T_a)C_g} \frac{\text{lbs}}{\text{cycle}}$$

or

$$V_C = \frac{H_C}{(T_{av} - T_a)C_g\theta_C} \frac{\text{lbs}}{\text{h}} \tag{13.26}$$

where θ_C is the cooling time, (h/cycle).

Example 13.6

Find the flow rate of a heating gas at 400°F to desorb 2000 lbs of water from a solid desiccant bed weighing 30,000 lbs and has C_d = 0.24 Btu/lb °F. The tower weight is 50,000 lbs and its C_t = 0.1 Btu/lb °F. The ambient temperature is 110°F and the maximum outlet temperature is 360°F. The 8 h cycle is operated as 6 h heating and 2 h cooling. Assume C_g = 0.6 Btu/lb °F.

Solution

Using Equation 13.22, the following results are obtained:

$$T_m = 360, T_a = 110, T_{av} = \frac{360 + 110}{2} = 235$$

$$\begin{aligned} Q_H &= 30{,}000(360 - 110)(0.24) + 50{,}000(360 - 110)(0.1) \\ &\quad + 2000(1100) + 2000(235 - 110) \\ &= 5{,}500{,}000 \text{ Btu/cycle} \end{aligned}$$

The flow rate of heating gas V_H is obtained from Equation 13.23:

$$V_H = \frac{5{,}500{,}000}{(400 - 235)(0.6)} = 55{,}555 \frac{\text{lbs}}{\text{cycle}}$$

or

$$V_H = 9259 \frac{\text{lbs}}{\text{h}}$$

The gas heater load Q_G is obtained from Equation 13.24:

$$Q_G = \frac{55{,}555(400 - 110)(0.6)}{6} = 1.6 \times 10^6 \frac{\text{Btu}}{\text{h}}$$

For cooling, use Equation 13.25 to find H_C:

$$H_C = 30{,}000(360-110)(0.24)+50{,}000(360-110)(0.1)$$

$$= 3.05\times 10^6 \frac{\text{Btu}}{\text{cycle}}$$

The gas flow rate for cooling is calculated from Equation 13.26; for 2 h cooling, the value of V_C is found to be

$$V_C = \frac{3.05\times 10^6}{(235-110)(0.6)(2)} = 20{,}233\frac{\text{lbs}}{\text{h}}$$

Additional heat requirements of 25% can be used for each cycle.

13.6 Simulation Problem for Triethylene Glycol (TEG) Dehydrator

In a gas field, natural gas is available at a flow rate of 10 MMSCFD and at operating conditions of 900 psia and 80°F. The gas composition is given as follows (mole fractions):

N_2	0.001	C_3H_8	0.0148
CO_2	0.0284	iC_4H_{10}	0.0059
H_2S	0.0155	nC_4H_{10}	0.0030
CH_4	0.898	iC_5H_{12}	0.001
C_2H_6	0.031	nC_5H_{12}	0.005

The operating and design conditions of the TEG absorber are

- The flow rate of the TEG fed to the column is 2 gpm (U.S. gallon) at 900 psia and 120°F.
- Number of trays are 14 with a tray efficiency of 50% except the top and bottom trays are 100% each.

Solution

Using the HYSYS process simulator licensed by Hypotech, the process flow sheet for the proposed dehydration plant along with material and energy balance calculations for all streams are produced as presented in Figure 13.12 and Table 13.3.

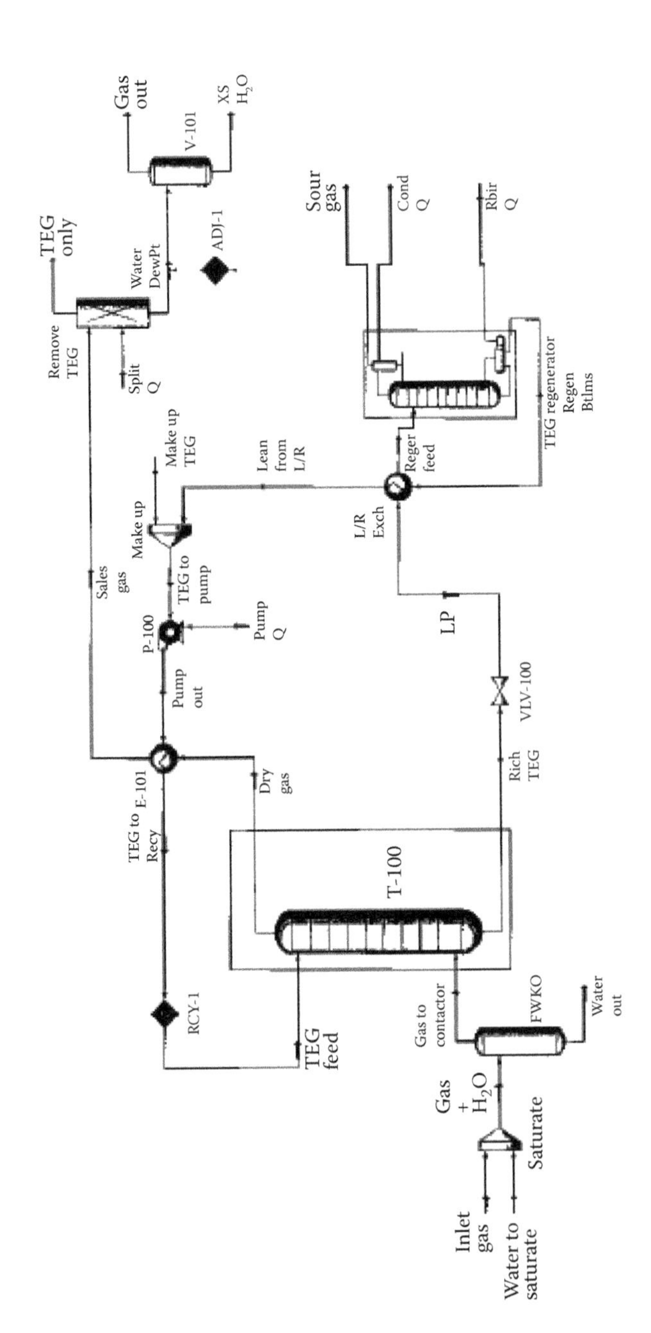

FIGURE 13.12
Computer similation of TEG dehydration.

TABLE 13.3

Stream Material and Energy Balance for Flow Sheet in Figure 13.11

Name	Inlet Gas	TEG Feed	Water to Saturate	Gas + H_2O	Gas to Contactor
Vapor fraction	1.0000	0.0000	0.5420	0.9999	1.0000
Temperature (°C)	29.44[a]	48.89[a]	277.1	29.44[a]	29.44
Pressure (bar)	62.05[a]	62.05[a]	62.05	62.05	62.05
Molar flow (kgmole/h)	498.1[a]	3.649[a]	0.4990[a]	498.6	498.5
Mass flow (kg/h)	9183	512.0	8.989	9192	9191
Liquid column flow (m^3/h)	27.58	0.4543	0.009007	27.59	27.59
Heat flow (kcal/h)	–1.018e+07	–6.562e+05	–2.961e+04	–1.021e+07	–1.021e+07
Name	**Water Out**	**Dry Gas**	**Rich TEG**	**LP TEG**	**Regen Bttms**
Vapor fraction	0.0000	1.0000	0.0000	0.0362	0.0000
Temperature (°C)	29.44	31.14	30.17	35.51	204.4
Pressure (bar)	62.05	62.05	62.05	1.793	1.034
Molar flow (kgmole/h)	0.06651	497.9	4.270	4.270	3.648
Mass flow (kg/h)	1.200	9177	525.6	525.6	511.9
Liquid volume flow (m^3/h)	0.001203	27.57	0.4727	0.4727	0.4542
Heat flow (kcal/h)	–4527	–1.017e+07	–6.949e+05	–6.949e+05	–6.071e+05
Name	**Lean from L/R**	**Regen Feed**	**Sour Gas**	**Make Up TEG**	**TEG to Pump**
Vapor fraction	0.0000	0.0531	1.0000	0.0000	0.0000
Temperature (°C)	145.1	104.4[a]	101.7	15.56[a]	145.1
Pressure (bar)	1.027	1.103[a]	1.014	1.027	1.027
Molar flow (kgmole/h)	3.648	4.270	0.6216	0.0004591	3.649
Mass flow (kg/h)	511.9	525.6	13.71	0.06834	512.0
Liquid volume flow (m^3/h)	0.4542	0.4727	0.01854	6.057e–05	0.4543[a]
Heat flow (kcal/h)	–6.283e+05	–6.737e+05	–2.861e+04	–87.51	–6.264e+05

(Continued)

TABLE 13.3 (CONTINUED)

Stream Material and Energy Balance for Flow Sheet in Figure 13.11

Name	Pump Out	TEG To Recy	Sales Gas	TEG Only	Water Dew Pt
Vapor fraction	0.0000	0.0000	1.0000	–	1.0000
Temperature (°C)	144.0	48.89[a]	36.19	–	–13.56[a]
Pressure (bar)	62.74[a]	62.05	61.71	–	62.05[a]
Molar flow (kgmole/h)	3.649	3.649	497.9	0.0001055	497.9
Mass flow (kg/h)	512.0	512.0	9177	0.01584	9177
Liquid volume flow (m^3/h)	0.4543	0.4543	27.57	1.403e–05	27.57
Heat flow (kcal/h)	–6.274e+05	–6.562e+05	–1.014e+07	–	–1.043e+07
Name	**Gas Out**	**XS H_2O**			
Vapor fraction	1.0000	0.0000			
Temperature (°C)	–13.56	–13.56			
Pressure (bar)	62.05	62.05			
Molar flow (kgmole/h)	497.9	0.0005366			
Mass flow (kg/h)	9177	0.009708			
Liquid volume flow (m^3/h)	27.57	9.746e–06			
Heat flow (kcal/h)	–1.043e+07	–36.93			

[a] Specified by user.

REVIEW QUESTIONS

1. When it is necessary to reduce gas pressure in a gas production system, the largest pressure drop should be taken at __________ to minimize hydrate formation.
2. To prevent hydrate formation, the following methods may be used:
   a. ____________________
   b. ____________________
   c. ____________________
3. Given that the hydrate formation pressure of a gas at 60°F is 1000 psia, state whether hydrate will form if the gas is under the following conditions:
   a. $T = 60°$ F and $P = 1200$ psia
   b. $T = 70°$ F and $P = 1000$ psia
   c. $T = 50°$ F and $P = 1000$ psia
4. A gas well produces with a high condensate–gas ratio:
   a. Is it practical to use methanol injection to prevent hydrate formation?
   b. What is the reason for your answer?
5. What are the conditions that must exist for hydrate formation?
   a. ____________________
   b. ____________________
   c. ____________________
6. In a central separation and treatment plant, 100 MMSCFD of gas containing 10% H_2S is collected at 500 psia and 80°F. The gas is first treated using a DEA process to reduce its H_2S content to 3 ppm. The sweet gas is then compressed to 1500 psia and cooled to 110°F. The gas is then dehydrated using a glycol dehydration process that reduces its water content to 7 lbs/MMSCF. Finally, a solid desiccant process is used to further dehydrate (polish) the gas to reduce its water content to less than 1 ppm. (See Figure 13.10.)
   a. Assuming that the gas is fed to the compressor at 500 psia and 80°F and that it is saturated with water vapor, determine the amount of free water in the gas after its compression and cooling (1500 psia and 110°F).
   b. Why is it necessary to compress and cool the gas before the dehydration process?
   c. Determine the required diameter of the glycol contactor using $C_D = 0.852$ and $d_m = 120$ μm.

d. Determine the required glycol circulation rate in gallons per minute and the reboiler duty assuming the lean glycol concentration to be 98.5%.

e. Discuss the limitations on the temperatures of the glycol and gas.

f. Draw the process flow diagram of the solid desiccant dehydration process used for polishing the gas.

g. Determine the volume of the desiccant needed and the diameter and height of the absorber given that the maximum superficial velocity allowed in the tower is 40 ft/min and using a cycle time of 8 h.

14

Separation and Production of Natural Gas Liquids (NGLs)

Natural gas liquids (NGLs) are hydrocarbons—in the same family of molecules as natural gas and crude oil—composed exclusively of carbon and hydrogen. Ethane, propane, butane, isobutane, and pentane are all NGLs.

The material presented in this chapter includes two parts: (1) the recovery and separation of NGL constituents, and (2) methods of fractionation into finished product streams suitable for sale. In the first part, several alternatives for the separation and recovery of NGL are detailed. They are essentially based on phase change either by using an energy separating agent or mass separating agent. Thus, partial liquefaction or condensation of some specific NGL constituents will lead to their separation from the bulk of the gas stream. Total condensation is also a possibility. The role of the operating parameters that influence phase change, hence NGL separation from the bulk of gas, is discussed. The second part of this chapter covers materials on fractionation facilities that are recommended to produce specification quality products from NGLs. Types of fractionators with recommended feed streams as well as produced products are highlighted in this part.

14.1 Introduction

Production of NGLs topped 2 million b/d in 2010 in the United States and is set to grow faster, with ethane and propane contributing much of the increase. Even OPEC production of NGLs, unfettered by quotas, was expected to rise by more than 7% to 6 million b/d in 2012.

An interesting new trend is that increased drilling for gas to access NGLs, such as ethane, propane, and butane, is set to add to the pressure on industry infrastructure, at least in the United States.

The chemical composition of these NGL hydrocarbons is similar, yet their applications vary widely as shown next.

NGL attribute summary				eia
Natural gas liquid	Chemical formula	Applications	End use products	Primary sectors
Ethane	C_2H_6	Ethylene for plastics production; petrochemical feedstock	Plastic bags; plastics; anti-freeze; detergent	Industrial
Propane	C_3H_8	Residential and commercial heating; cooking fuel; petrochemical feedstock	Home heating; small stoves and barbeques; LPG	Industrial, residential, commercial
Butane	C_4H_{10}	Petrochemical feedstock; blending with propane or gasoline	Synthetic rubber for tires; LPG; lighter fuel	Industrial, transportation
Isobutane	C_4H_{10}	Refinery feedstock; petrochemical feedstock	Alkylate for gasoline; aerosols; refrigerant	Industrial
Pentane	C_5H_{12}	Natural gasoline; blowing agent for polystyrene foam	Gasoline; polystyrene; solvent	Transportation
Pentanes plus*	Mix of C_5H_{12} and heavier	Blending with vehicle fuel; exported for bitumen production in oil sands	Gasoline; ethanol blends; oil sands production	Transportation

Source: U.S. Energy Information Administration, Bentek Energy LLC.
C indicates carbon, H indicates hydrogen; Ethane contains two carbon atoms and six hydrogen atoms.
*Pentanes plus is also known as *natural gasoline*. Contains pentane and heavier hydrocarbons.

Components of NGL are extracted from the natural gas production stream in natural gas processing plants. Statistics on the total annual U.S. field production of natural gas, crude oil, and NGLs for the period 2000 to 2011 demonstrates a trend in increase of NGL production.

14.2 Recovery and Separation of NGLs

14.2.1 Options of Phase Change

To recover and separate NGL from a bulk of gas stream, a change in phase has to take place. In other words, a new phase has to be developed for separation to occur. Two distinctive options are in practice depending on the use of an energy separating agent (ESA) or a mass separating agent (MSA).

14.2.1.1 Energy Separating Agent

The distillation process best illustrates a change in phase using ESA. To separate, for example, a mixture of alcohol and water heat is applied. A vapor phase is formed in which alcohol is more concentrated and then separated by condensation. This case of separation is expressed as follows:

$$\text{A mixture of liquids} + \text{Heat} \rightarrow \text{Liquid} + \text{Vapor}$$

For the case of NGL separation and recovery in a gas plant, removing heat (by refrigeration), on the other hand, will allow heavier components to condense; hence, a liquid phase is formed. This case is represented as follows:

$$\text{A mixture of hydrocarbon vapor} - \text{Heat} \rightarrow \text{Liquid} + \text{Vapor}$$

Partial liquefaction is carried out for a specific cut, whereas total liquefaction is done for the whole gas stream.

14.2.1.2 Mass Separating Agent

To separate NGLs, a new phase is developed by using either a solid material in contact with the gas stream (adsorption) or a liquid in contact with the gas (absorption). These two cases are represented in the following sections.

14.2.2 Parameters Controlling NGL Separation

A change in phase for NGL recovery and separation always involves control of one or more of the following three parameters:

- Operating pressure, P
- Operating temperature, T
- System composition or concentration, x and y

To obtain the right quantities of specific NGL constituents, a control of the relevant parameters, shown next, has to be carried out:

- For the case of separation using ESA, pressure is maintained by direct control. Temperature, on the other hand, is reduced by refrigeration using one of the following techniques:
  - Compression refrigeration
  - Cryogenic separation; expansion across a turbine
  - Cryogenic separation; expansion across a valve

- For separation using MSA, a control in the system composition or concentration x, y of the hydrocarbons to be recovered as NGL, is obtained by using adsorption or absorption methods.

Adsorption provides a new surface area through the solid material that entrains or *adsorbs* the components to be recovered and separated as NGLs. Thus, the components desired as liquids are deposited on the surface of the selected solid, regenerated off in a high concentration; hence, their condensation efficiency is enhanced. About 10%–15% of the feed is recovered as liquid.

Adsorption is defined as a concentration (or composition) control process that precedes condensation. Therefore, refrigeration methods may be coupled with adsorption to bring in condensation and liquid recovery. *Absorption*, on the other hand, presents a similar function of providing a surface or *contact* area of liquid–gas interface.

The efficiency of condensation, hence NGL recovery, is a function of P, T, gas and oil flow rates, and contact time. Again, absorption could be coupled with refrigeration to enhance condensation.

To summarize, a proper design of a system implies the use of the optimum levels of all operating factors plus the availability of sufficient area of contact for mass and heat transfer between phases.

14.2.3 Selected Separation Processes

In this section a brief description is given for the absorption, refrigeration, and cryogenic (Joule–Thomson turbo expansion) processes recommended to separate NGL constituents from a gas stream. Details are illustrated in the corresponding flow diagrams.

14.2.3.1 Absorption Process

The absorption unit consists of two sections: the absorption and regeneration, as illustrated in Figure 14.1. An up-flow natural gas stream is brought in direct contact, countercurrently with the solvent (light oil in the kerosene boiling range) in the absorber. The column—a tray or packed one—operates at about 400–1000 psia and ambient or moderately subambient temperatures. The rich oil (absorbed NGL plus solvent) is directed to a distillation unit to separate and recover the NGL, whereas the lean oil is recycled back to the absorber. In addition to natural gasoline, C_3/C_4 could be recovered as well in this process. Provision is made to separate ethane from rich oil using a deethanizer column. Ethane recovery, however, is quite small. This process is being phased out.

14.2.3.2 Refrigeration Process

The production of NGLs at low temperature is practiced in many gas processing plants in order to condense NGLs from gas streams. As indicated

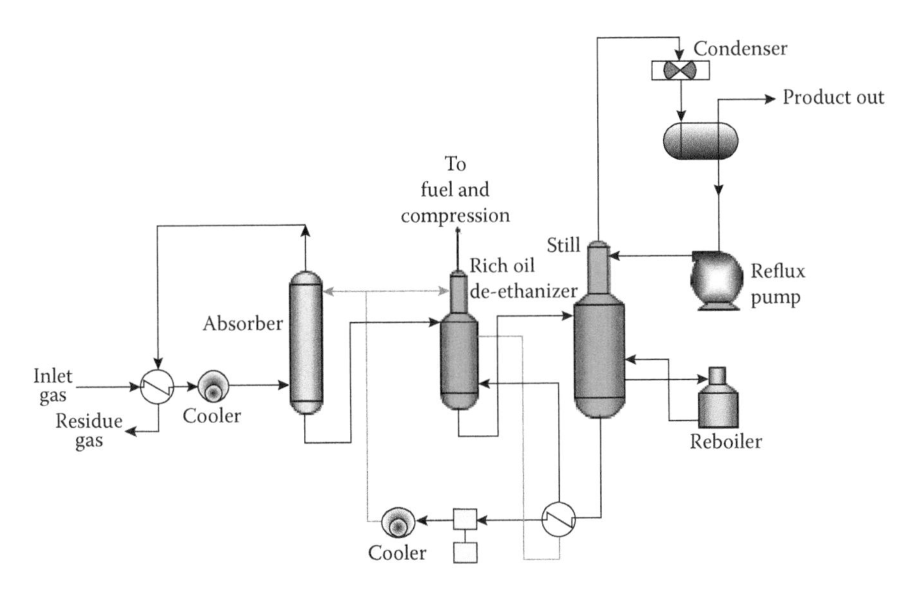

FIGURE 14.1
Separation of NGLs by absorption. (From U.S. Energy Information Administration Form EIA-816, Form EIA-914, Petroleum Supply Monthly.)

in Figure 14.2, using nontoxic and noncorrosive refrigerants to chill the feed natural gas to a temperature between 0°F and –40°F using a low-level one-component refrigerant system provides external refrigeration. When using a high-level cascade refrigerant system, a much lower temperature in the range of –100°F to –150°F is reached. Liquids are separated from the residue gas at multiple temperatures and then fractionated into final products. Ethane recovery is a strong function of the operating temperatures as is explained next.

The following operating conditions are important in the separation of NGL constituents. Two main objectives are usually targeted when the desired temperature of the gas–liquid stream leaving the chiller is specified: (1) dew point control of the hydrocarbons and (2) liquid recovery. Now, if the dew point control of the hydrocarbon is the primary target, then the temperature of the gas–liquid stream is set at about 6°C–10°C below the desired dew point temperature. On the other hand, if the liquid recovery is the main objective, then condensing the least amount of nonsalable components should be achieved. In other words, condensing methane is not cost effective.

As far as the operating pressure for the refrigeration system, the recommended operating pressure for maximum liquid recovery is set between 400 and 600 psia. Condensation of methane increases with higher pressure; therefore, optimum pressure must minimize the total cost of the system. In general, separation is carried out at a pressure corresponding to the sales gas pressure simply to eliminate the cost of gas recompression. To summarize, for a given selected separation pressure the corresponding operating temperature is chosen based on the type of product:

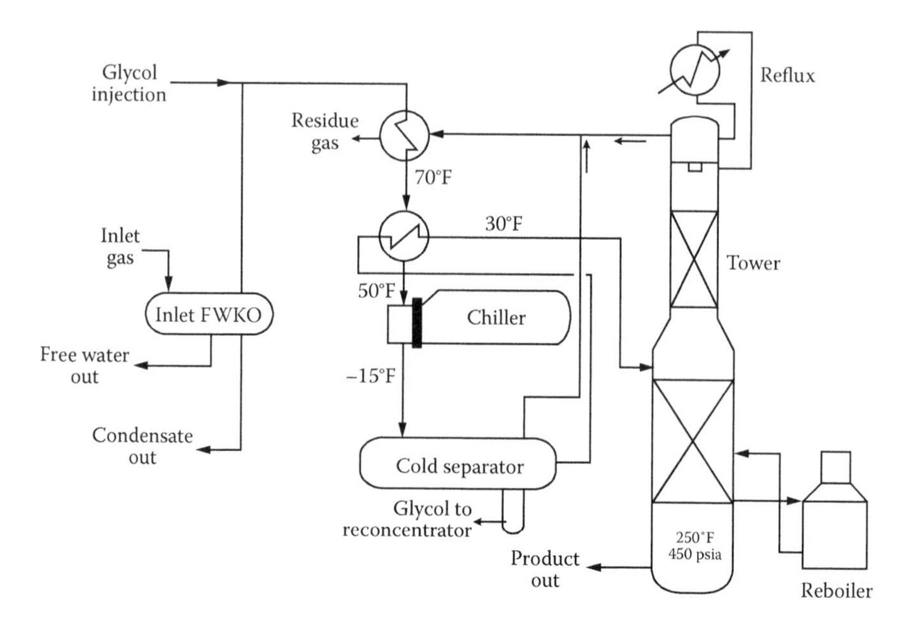

FIGURE 14.2
Separation of NGLs using refrigeration plant.

- If the liquid product is to be sold as *crude oil*, then the separation temperature is between 0°C and 5°C.
- If the liquid product contains propane as the lightest component, then the temperature is about −30°C to −18°C. In this case, the temperature depends on whether absorption or adsorption is combined with refrigeration.
- If the operating temperature is set below −30°C, a cryogenic range of ethane recovery is encountered.

14.2.3.3 Cryogenic Processes

Natural gas liquid could be separated from natural gas using two approaches based on cryogenic expansion (autorefrigeration):

1. An expander plant produces refrigeration to condense and recover the liquid hydrocarbons contained in the natural gas by using a turboexpander. In this process, the enthalpy of the natural gas is converted into useful work, behaving thermodynamically as an approximate isentropic process.
2. Expansion across the valve will lead to a similar result. However, the expansion is described in this case as *isenthalpic*.

Temperatures produced by turboexpansion are much lower than those of valve expansion.

A schematic presentation for the turboexpansion process is presented in Figure 14.3. The process operates at –100°F to –160°F and 1000 psia. The process represents a new development in the gas processing industry. Increased liquid recovery (especially ethane) is an advantage of this process. Figure 14.4 illustrates the condensation process using ethane/propane, followed by demethanization to produce NGL as a final product. A concise comparison among the three options (absorption, refrigeration, and cryogenic separation processes) recommended for NGL recovery is given by Abdel-Aal (1998).

14.3 Fractionation of NGLs

14.3.1 Goals and Tasks

In general, and in gas plants in particular, fractionating plants have common operating goals: the production of on-specification products, the control of impurities in valuable products (either top or bottom), and the control in fuel consumption.

As far as the tasks for system design of a fractionating facility, the goals are as follows: fundamental knowledge on the process or processes selected to carry out the separation, in particular, distillation; and guidelines on the order of sequence of separation (i.e., synthesis of separation sequences).

14.3.2 Fundamentals of Distillation

For separation to take place, say by distillation, the selection of an exploitable chemical or physical property difference is very important. Factors influencing this are as follows:

- The physical property itself
- The magnitude of the property difference
- The amount of material to be distilled
- The relative properties of different species or components; purity required
- The chemical behavior of the material during distillation and its corrosiveness

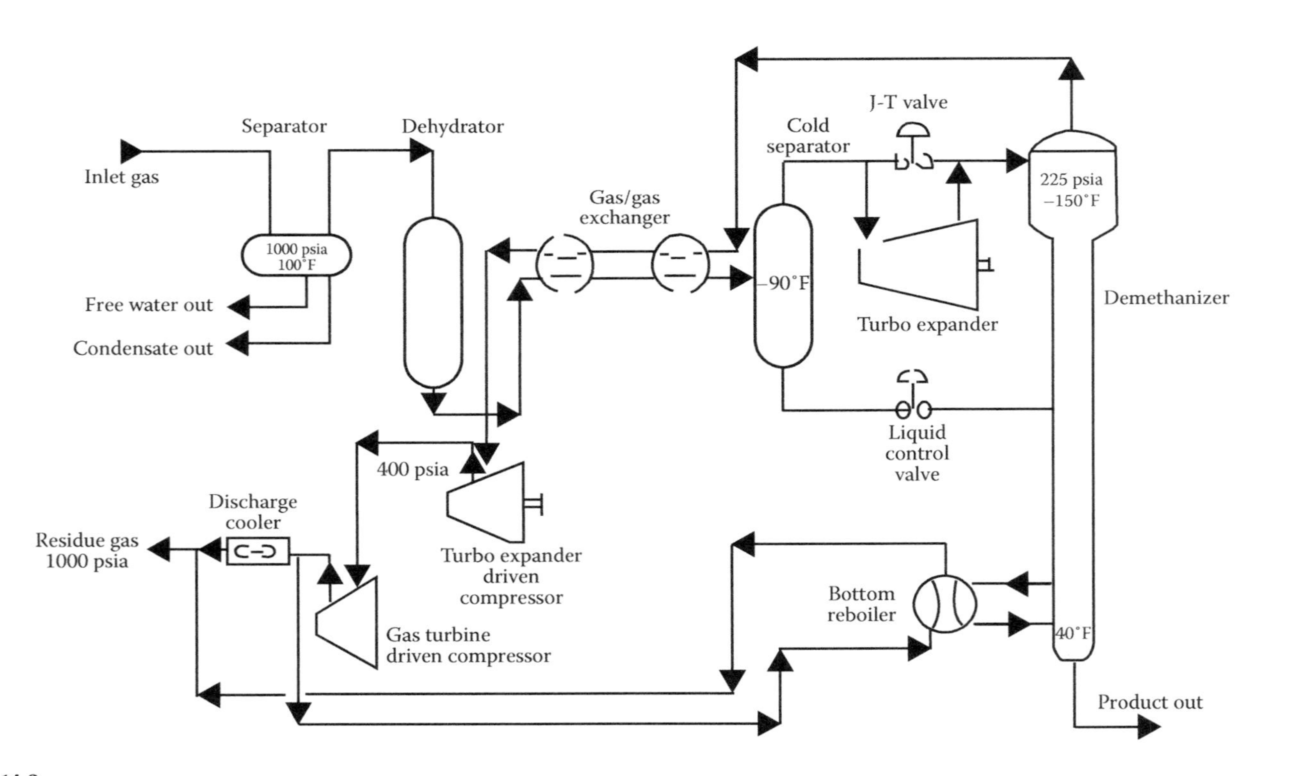

FIGURE 14.3
Cryogenic separation of NGLs.

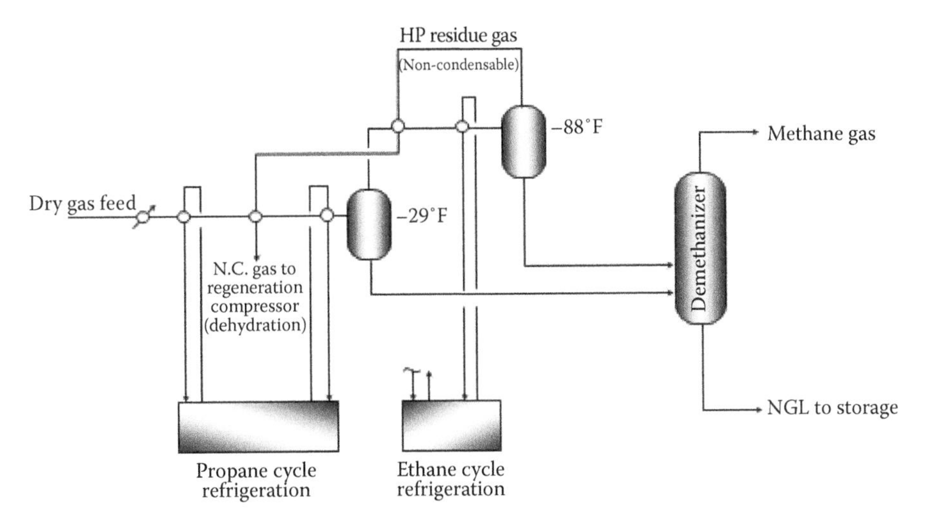

FIGURE 14.4
NGL condensation/demethanization.

A measure of the ease of separation of one component *A* from another *B* is known as the separation factor, SF, defined as

$$SF = \frac{(C_A/C_B)_{\text{Top product}}}{(C_A/C_B)_{\text{Bottom product}}}$$

where C is the concentration. A high value of SF means an easy separation. A good example is the separation of salt from seawater by *evaporation*. Here, the value of SF is found, by intuition, to be infinity, because we are separating water (volatile component) from salt (nonvolatile).

Consider a raw hydrocarbon stream to be fractionated, which may contain the following hydrocarbons:

Selected Hydrocarbons with Corresponding Boiling Point

Component	**Boiling Point (°F) at Atmospheric Pressure**
Ethane	–128
Propane	–44
Isobutane	11
N-Butane	31
Pentanes plus	82–250

From the data, it is evident that in order to separate a mixture of NGL, say propane plus, as a bottom product, then there must be a difference in boiling point between the top product (ethane) and the bottom product. This difference is an indication of the degree of difficulty of separation or the value of SF. As an equilibration separation process, SF should be much greater than unity for ethane to concentrate in the top and propane plus to concentrate in the bottom.

A difficult separation implies the following:

- Higher number of distillation trays, which affects column size
- Higher reflux ratio, which influences pump size and power consumption
- Additional reboiler heat duty, which influences reboiler size and energy consumption

14.3.3 Distillation Processes and Types of Fractionators

The separation of NGLs may require different modes of distillation as well as other methods of separation techniques.

Fractionators of different types are commonly used in gas plants. The following types are typical ones:

Type of Fractionator	Feed	Top Product	Bottom Product
Demethanizer	C_1/C_2	Methane	Ethane
Deethanizer	LPG	Ethane	Propane plus
Depropanizer	Deethanizer bottoms	Propane	Butanes plus
Debutanizer	Depropanizer bottoms	Butanes (iso + n)	Natural gasoline (pentanes plus)
Deisobutanizer	Debutanizer top	Isobutane	Normal butane

Control of the following key operating parameters will ensure efficient performance of fractionation operations:

- Top tower temperature, which sets the amount of the heavy hydrocarbons in the top product. This is controlled by the reflux ratio. Increasing the reflux rate will decrease this amount. The reader should observe that reflux liquid is produced as a result of overhead condensation of vapors. For columns using total condensers, such as depropanizers and debutanizers, all vapors are condensed to produce reflux and liquid product. On the other hand, for columns employing partial condensers, such as deethanizers, product is produced as vapor (ethane).
- Bottom reboiler temperature, which sets the amount of light hydrocarbons in the bottom product. Adjusting the heat input to the reboiler controls this.

- Tower operating pressure, which is fixed by the type of condensing medium (i.e., its temperature). Product quality is not affected, to a great extent, by changing the operating pressure.

14.4 Synthesis of Separation (Distillation) Sequences

In the separation and the recovery of NGL constituents from a gas stream, it is required to find the optimal arrangement of the separation steps that will be both economically and technically feasible. In this regard, solutions are reached through some rules or heuristics, after Rudd, as explained next.

14.4.1 Cost of Separation (Fractionation)

The cost of fractionation is influenced by the following:

- The quantity or feed input to be processed, load (L)
- The boiling point difference between two components (Z), defined also as the property difference between the two species on each side of the separation breaking point

One can simply relate the cost of separation to these two variables by the following relationship: cost of separation is a function of L/Z, or

$$\text{Cost of separation} = k\left(\frac{L}{Z}\right)$$

Since the value of Z is fixed for a given mixture, our main objective is to select an optimal arrangement of separation sequence that will minimize L, thus, the cost of separation.

14.4.2 Rules (Heuristics) and Examples

In this subsection, some examples are presented along with the relevant rules that apply. Some important formulas are presented first:

$$\text{Number of separating columns} = \text{Number of components} - 1 \tag{14.1}$$

The number of sequences (S) is related to the number of components (c) by the relationship

$$S = \frac{[2(c-1)]!}{c!(c-1)!} \tag{14.2}$$

Example 14.1 (Introductory Example)

Find the possible technical sequences for separating a mixture of BTX (benzene, toluene, xylene).

Solution

First, applying Equation 14.1, the number of columns to be used is two. Similarly, solving Equation 14.2,

$$S = \frac{[2(2)]!}{(3\times2\times1)(2\times1)}$$

$$= \frac{4\times3\times2\times1}{3\times2\times2} = 2$$

The number of feasible sequences is found to be 2. However, in Figure 14.5, four sequences are shown; but not all of them are feasible. Both arrangements (b) and (c) are excluded because the former does not produce pure products and the latter employs three columns. Thus, arrangements (a) and (b) are both feasible and equivalent.

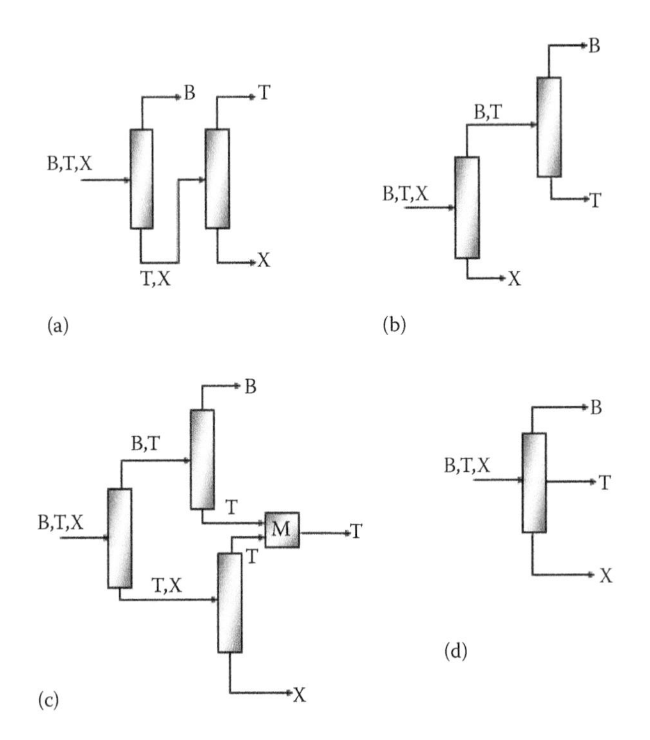

FIGURE 14.5
(a) Two columns. (b) Two columns, modified. (c) Three columns. (d) One single column.

14.4.2.1 Statement of Rules

- **Rule 1:** All other things being equal, aim to separate the more plentiful components early.
- **Rule 2:** Difficult separations are best saved last.
- **Rule 3:** When using distillation or similar schemes, choose a sequence that will finally remove the most valuable species or desired product as a distillate, all other things being equal.
- **Rule 4:** During distillation, sequences that remove the components one by one in column overheads should be favored, all other things being equal.
- **Rule 5:** All other things being equal, avoid excursions in temperature and pressure, but aim high rather than low.

Example 14.2

Suppose we want to separate a mixture of four components: components 1–4 into pure products. Assume that they are present in the feed in equal amounts: D_1, D_2, D_3, and D_4.

Solution

Let us calculate the value of S by applying Equation 14.2. S is found to be 5; the five schemes are shown in Figure 14.6. Now, in order to select the best scheme the total load of all sequences have to be calculated:

Total load of a sequence = Sum of individual loads of all components undergoing separation
= sum of L_i

Because $L_i = D_i n$, where n is defined as the number of columns or separators component i, goes through before final separation,

$$\text{Total load of a sequence} = \text{Sum of } D_i n \tag{14.3}$$

Calculations are carried out as shown in Figure 14.6. It can be concluded that sequence 3 has less load ($8D$).

Example 14.3

This example is a direct industrial application for some of the rules or heuristics presented in the course of separation of NGL constituents. It deals with ethylene and propylene manufacture. Figure 14.7 describes a sequence of distillation separation processes for a gas mixture produced

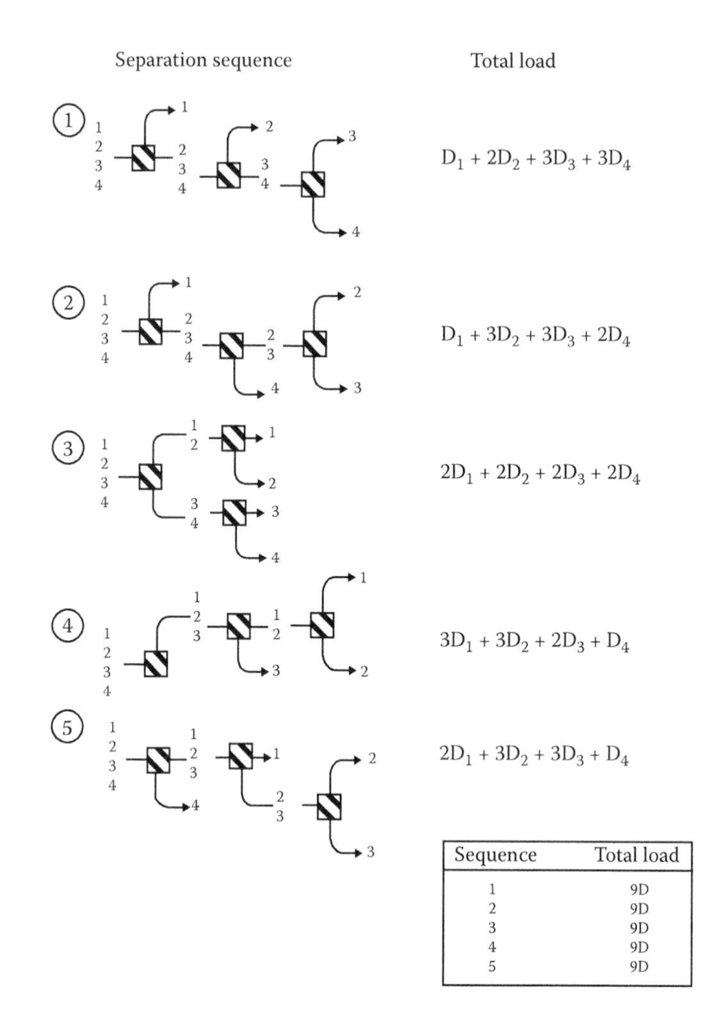

Sequence	Total load
1	9D
2	9D
3	9D
4	9D
5	9D

FIGURE 14.6
Solution of Example 14.2.

by a catalytic cracking plant of natural gas. The plant built by Sinclair produces 500 million lbs/year of ethylene. Ethylene and propylene are the valuable products formed by catalytic cracking of the hydrocarbons in the natural gas.

The following observations are cited in accordance with the heuristics stated:

- Difficult separation is last—The close boiling points of propane and propylene makes the separation between them, the very last (splitter). The next most difficult separation is between ethane and ethylene. Again, it is kept to a last position (splitter).

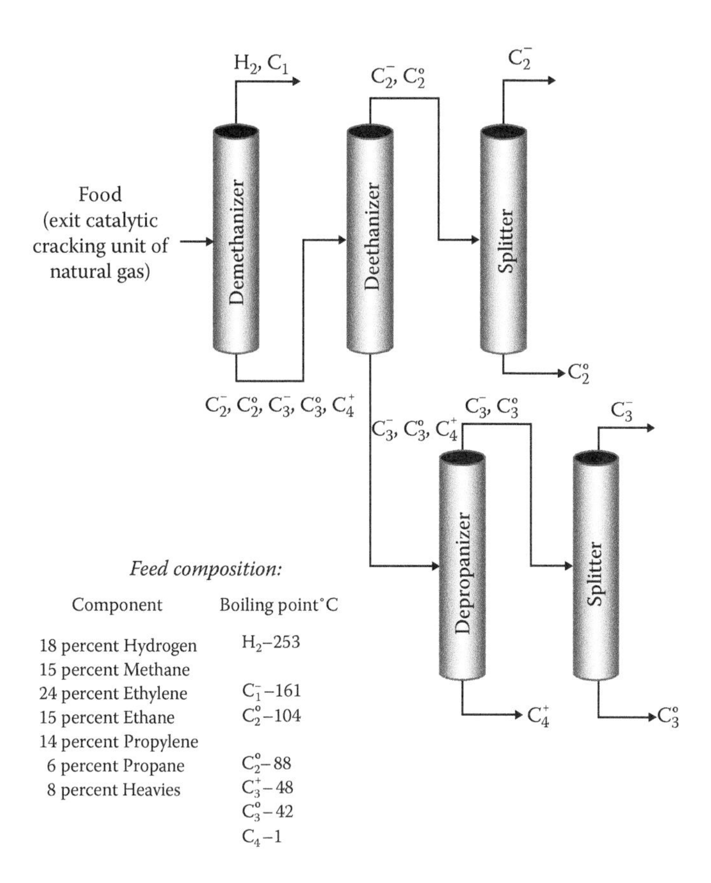

FIGURE 14.7
Separation of products in the manufacture of ethylene and propylene.

- Favor overhead removal in plentiful quantities—In the first distillation column, the demethanizer separates the volatile components hydrogen and methane (18% and 15%, respectively).
- Remove valuable products as distillates—Both ethylene and propylene are separated as top products. This ensures that the materials do not experience discoloration and separate in pure form.

14.5 NGL Products

Following are some commercial finished products that are obtained by fractionation of NGL stream and sold in a gas plant: ethane product, propane product, butanes product, and pentanes and heavier (natural gasoline) product. Field stabilization is required for liquid products recovered as explained before, in order to remove the large quantities of methane and other volatile hydrocarbons that remain locked in the condensed liquid at high pressure. (Chapter 10 should be consulted.)

REVIEW QUESTIONS

1. To recover and separate NGLs from a bulk of gas stream, a change in phase must occur. What are the two main options that bring in a change in phase that lead to separating NGLs?
2. A change in phase for NGL recovery, thus its separation from the gas stream always involves the control of one or more of the following parameters:
   a. Operating pressure
   b. Acid gas content
   c. Operating temperature
   d. Water vapor content
   e. System composition of hydrocarbons
3. For separation using an energy separating agent (ESA), _____ is maintained by direct control. _____, on the other hand, is reduced by refrigeration using one of the three techniques: _____, _____, or _____.
4. For separation using a mass separating agent (MSA), a control in the _____ of the hydrocarbons to be recovered as NGLs is obtained by using _____ or _____.
5. In the absorption process for the recovery of NGL, the up-flow gas stream is brought in direct contact countercurrent with the solvent which is _____. The rich oil is made up of _____ plus _____.
6. Refrigeration methods for NGL recovery imply using _____ to chill the feed (natural gas) to a temperature between 0°F and –40°F.
7. Name the two well-known approaches of cryogenic processes that lead to the separation of NGL, that is *autorefrigeration*.

Section V

Surface Production Facilities

Section V includes topics that would serve as facilities for surface production operations. For convenience they are grouped in this section under the heading of surface production facilities. This section is a new addition to the second edition of this book. It includes four chapters that cover the following topics:

- Produced water management and disposal
- Field storage tanks, vapor recovery unit (VRU) and tank blanketing
- Oil field chemicals (OFCs): Evaluation and selection
- Piping and pumps

These facilities are best recognized as the four pillars in surface field facilities, as shown next:

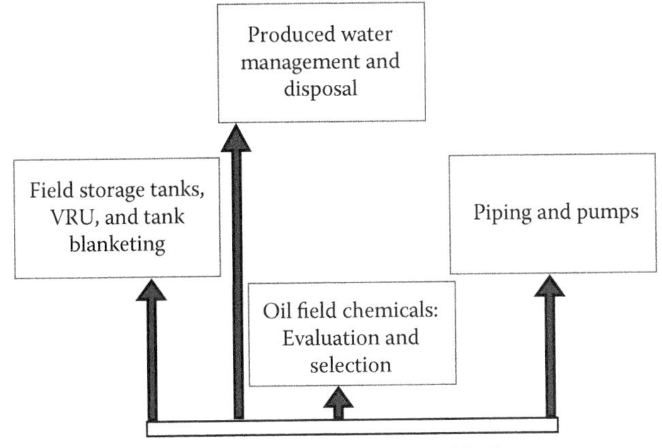

The four pillars in surface field facilities.

15

Produced Water Management and Disposal

Produced water is a large and growing problem. It is defined as the water that exists in the subsurface formation, and is brought to the surface during oil and gas production. Eventually, the production of crude oil and natural gas is usually associated with the production of water. During the early life of the petroleum fields, water-free production of oil and gas is normally experienced. However, water will eventually be produced later. The produced water may be water that exists within the petroleum reservoir as connate water or bottom water. Alternatively, water may be produced as a result of water-flooding operations, where water is injected into the reservoir to enhance the recovery.

In this chapter, a produced water treatment system is described first along with water treatment equipment. Parameters involved in the design and the sizing of equipment used in water treatment and disposal devices are presented. Concepts and procedures recommended for the selection of equipment are discussed.

15.1 Background

Water production presents serious operating, economic, and environmental problems. Production of water with the crude oil or natural gas reduces the productivity of the well due to the increased pressure losses throughout the production system. This may either result in reduced production or necessitate the installation of costly artificial lifting systems to maintain the desired production levels. Production of water also results in serious corrosion problems, which add to the cost of the operation. As discussed in previous chapters in the book, production of water with the crude oil or natural gas requires the use of three-phase separators, emulsion treatment, and desalting systems, which further add to the cost of the operation.

In most situations, the produced water has no value and should be disposed of. In other situations, the produced water may be used for water flooding or reservoir pressure maintenance. The produced water, collected from the separation, emulsion treatment, and desalting systems, contains hydrocarbon concentrations that are too high for environmentally safe disposal. The presence of the hydrocarbon droplets in the water makes it difficult to inject the water into disposal wells or into water-injection wells for enhanced recovery operations. This is because the hydrocarbon droplets cause severs plugging of the formation. In all cases, the produced water must be treated to lower its hydrocarbon content to acceptable limits. For the heavy oil field, produced water may be used to generate the steam needed for oil recovery. In this case, additional chemical treatment will be needed to reduce the concentration of the salt and other minerals to make the water quality adequate for steam generation.

The purpose of this chapter is to present the concepts and procedures used for selecting and sizing the equipment used for removal of oil from the produced water.

15.2 Produced Water Quality Characterization

The quality of treated water (i.e., the maximum allowable oil concentration and maximum allowable oil droplet size) is determined to meet water injection or disposal requirements. From an environmental point of view, it should be desirable to remove all the oil from the produced water or at least allow the technically minimum possible. This, however, can impose substantial additional operating costs. Therefore, operators would usually provide the necessary water treatment to achieve the maximum allowable oil content. To properly design an efficient and economical treatment system that achieves this objective, knowledge of the produced water quality (oil concentration and droplet size distribution) is necessary. This is best determined from laboratory analysis of actual field samples. Such samples, however, are not normally available, especially when designing a treatment system for new field development.

Theoretically speaking, it is possible to determine the droplet size distribution throughout the various components of the production system, and the separation and oil treatment equipment. However, most of the parameters needed to solve the governing equations, especially those involving dispersion and coalescence, are normally unknown. As discussed in previous chapters in the book, the design of separation and oil treatment equipment determines the maximum oil droplet size remaining in the water. Several attempts have been made to determine the oil concentration in water for properly designed separation and treatment equipment. The results showed that the dispersed oil content ranges from 1000 to 2000 mg of oil per liter of water. Unfortunately, as the water leaves the separation and treatment equipment, it flows through various restrictions (such as valves and bends) in the piping system before it reaches the water treatment facility. In its journey, the oil droplets are subjected to a series of dispersion and coalescence that makes it difficult to exactly determine the oil droplet size distribution in the water to be treated. Experience showed that a conservative assumption for design purposes would be to represent the droplet size distribution by a straight-line relationship between the droplet size and cumulative oil concentration, with the maximum oil droplet diameter being between 250 and 500 μm.

15.3 Produced Water Treatment System

In general, produced water always has to be treated before it is disposed of or injected into the reservoir. The purpose of the treatment is to remove enough oil from the water such that the remaining amount of oil in the water and the oil droplet size are appropriate for the disposal or injection of the water. For example, for water disposal into underground formations and water injection into the producing reservoir, the pore size of the formation determines the allowable oil droplet size in the treated water. The maximum droplet size of the remaining oil in the water should be less than the minimum pore size of the formation to avoid plugging of the formation by the oil droplets. For water disposal into the sea, as is normally practiced in offshore operations, the amount and droplet size of the oil in the water is governed by environmental constraints.

Depending on the amount and droplet size of the oil in the produced water, the required quality of the treated water, and the operating conditions, water treatment may be achieved through a single or two stages of treatment. The single, or first, stage of treatment is normally known as the *primary treatment* stage; the second stage of treatment is known as the *secondary treatment* stage.

The equipment used for water treatment serves the function of allowing the oil droplets to float to the surface of the water, where they are skimmed and removed. For primary treatment, this may be achieved by using skim tanks for atmospheric treatment or skim vessels for treatment under pressure. Plate coalescers such as the *parallel plate interceptor* and *corrugated plate interceptor* are used to promote coalescence of the oil droplets to increase their size and thus speeds their floatation to the surface. Another device, known as the *serpentine-pipe (SP) pack*, is also used to promote coalescence of the oil droplets. For secondary treatment, plate interceptors, SP packs, and flotation units are normally used.

For offshore operations, water disposal must be through a disposal pile, skim pile, or SP pile. The deck drains normally contains free oil and must be treated before disposal. This can be done either in similar primary treatment equipment or directly through the various disposal piles.

15.4 Water Treatment Equipment

The various produced water treatment equipment are described in this section. The main function of the treating equipment is to separate the free oil droplets from the water. The fluid may contain some dissolved gas, which will be liberated in the treating equipment and must be removed. Therefore, the produced water treatment equipment is, in essence, similar to the three-phase oil–water–gas separators. The main difference is that for water treatment equipment, water is the main and continuous phase, and oil represents a small volume of the fluid mixture.

15.4.1 Filters

One of the very efficient ways of removing oil droplets from water is the use of filters. In this method of water treatment, produced water is made to flow through a bed of porous medium, normally sand, where the oil droplets are trapped in the filtering medium. At least two filters arranged in parallel are used. As the filter in use gets clogged, the flow is directed to the other filter and the clogged filter is backwashed using water or solvent. The backwash fluid must be treated or disposed of properly, which adds more complications and cost to the water treatment process. Several onshore successful operations have been reported in which sand filters were used to yield treated produced water with oil content as little as 25 mg/L of water.

15.4.2 Precipitators

In this method of treatment, the produced water is directed through a bed of porous material, such as excelsior, placed inside a horizontal vessel that is similar in design to the three-phase separator to promote the coalescence of oil droplets. The coalesced large oil droplets flow upward, countercurrent to the downward flow of the water where it can be skimmed out of the vessel. Although this method has been effective in treating produced water to desired quality, clogging of the coalescing medium represented a serious problem, which limited the use of such precipitators.

15.4.3 Skim Tanks and Vessels

Skim tanks and vessels are the simplest equipment used for primary treatment of produced water. Skim tanks and skim vessels are generally similar in shape, components, and function. However, the designation of skim tanks is associated with atmospheric treatment, whereas skim vessels are used when water treatment is performed under pressures above the atmospheric pressure. The equipment is normally large in volume to provide residence time that is sufficiently long (10–30 min) for the coalescence and gravity separation of the oil droplets.

Pressure vessels are more expensive than atmospheric tanks. However, the choice is controlled by the overall requirements of the water treatment system. Pressure vessels are normally preferred over atmospheric tanks for the following reasons:

- To avoid the potential gas venting problems associated with atmospheric tanks
- To eliminate the potential danger of overpressure that may occur in an atmospheric tank
- To eliminate the need for pumps that may be required to deliver the treated water to other secondary treating equipment or to other locations for disposal

The technical aspects, benefits, and cost should all be considered in deciding on the pressure rating of the skimmers.

Skimmers can be either horizontal or vertical in configuration. The shape and internal components of the skimmers are generally similar to those of the three-phase separators. Figure 15.1 shows a schematic of a horizontal skimmer. As shown, the produced water enters the skimmer below the water–oil interface and flows horizontally along the length of the vessel. The oil droplets coalesce and rise to the oil pad perpendicular to the direction of the water flow. The oil flows over the weir into the oil collection section and out of the skimmer. The height of the oil pad is controlled by the weir as shown. Alternatively, the height of the oil pad may be controlled by an interface level controller or by an external water leg. The treated water is withdrawn from the skimmer at the bottom of the vessel. The liberated gas leaves the vessel at the top through a mist extractor.

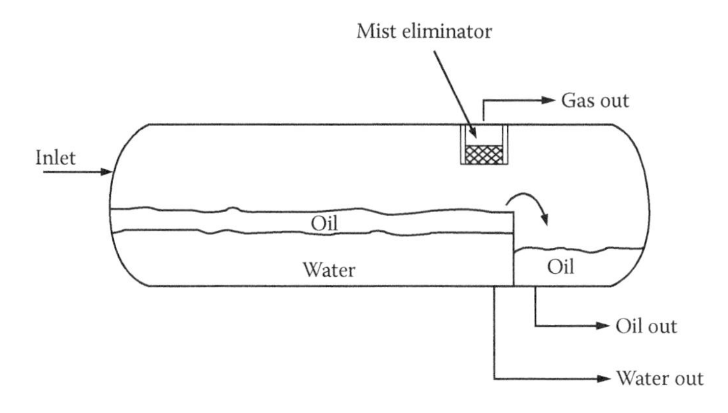

FIGURE 15.1
Schematic of a horizontal skimmer.

Figure 15. 2 shows a schematic of a vertical skimmer that is equipped with an inlet spreader and a water outlet collector, which work to even the distribution of the incoming and outgoing flow, respectively. As with the horizontal skimmers, the produced water enters the vessel below the oil–water interface. Water flows downward while the oil droplets rise upward to the oil pad. Vertical vessels are generally less efficient than horizontal ones, because of the counter-current flow pattern of water and oil. The oil is skimmed over the weir into the oil collection section, where it is withdrawn from the vessel. The water outlet is at the bottom of the vessel through the water collector. The liberated gas leaves at the top of the vessel through a mist extractor.

Vertical vessels are preferred over horizontal vessels when treating water containing sand or other solids. A sand drain at the bottom of the vertical vessel always provide a simpler and more effective means for cleaning the vessels as compared to the sand drains in horizontal vessels. Further, vertical vessels are better than horizontal vessels with regard to handling liquid surges. Surging in horizontal vessels tends to create internal waves, which result in a false indication of a high liquid level within the vessel and leads to false high-level shutdown.

Another type of skimmer is the API separator, which is basically a horizontal, rectangular cross-section tank. This type of skimmer is mostly used for treatment under atmospheric conditions.

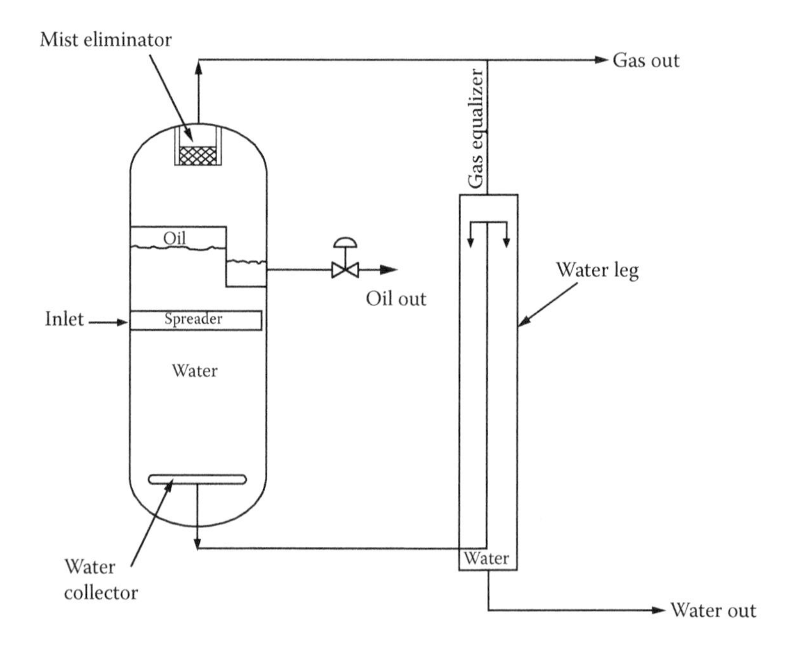

FIGURE 15.2
Schematic of a vertical skimmer.

15.4.4 Plate Coalescers

There are mainly two types of plate coalescer: the *parallel plate interceptors* (PPIs) and the *corrugated plate interceptors* (CPIs), as shown in Figures 15.3 and 15.4, respectively. Both types consist of a set of parallel plates that are spaced a short distance apart and are inclined by an angle of 45°.

The PPI was the first form of plate coalescers where a series of inclined parallel plates is installed inside an API separator. The water flow is split between the plates; therefore, the oil droplets need first to rise along the short distance between two consecutive plates where coalescence occurs. Due to gravity, the large oil droplets move upward along the bottom surface of the inclined plate and then vertically upward to the oil collection section, where oil is skimmed out of the tank. Sediments in the water move downward to the bottom of the tank, where they can be removed.

The CPI is the most commonly used plate interceptor in the industry. The CPI was an improvement over the PPI, where the surface of the parallel plates was made corrugated with the axis of the corrugations being parallel to the direction of water flow. The water to be treated flows downward through the

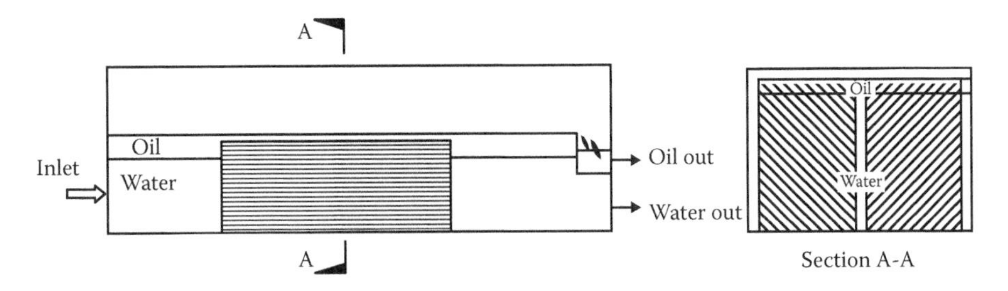

FIGURE 15.3
Parallel plate interceptor.

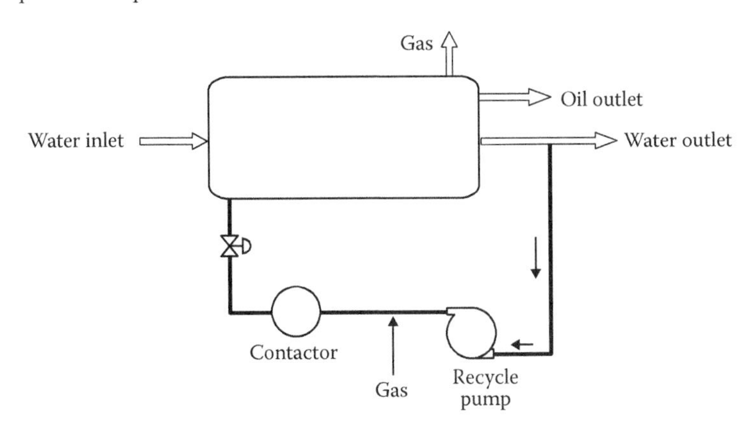

FIGURE 15.4
Corrugated plate interceptor.

CPI pack. The oil rises upward, counter to the water flow and accumulates at the corrugations. The accumulated oil flows along the axis of corrugations and upward to the oil–water interface.

Both PPIs and CPIs are normally used for water treatment under atmospheric conditions. The 45° inclination of the plates may present a problem when produced water contains appreciable amounts of sediments or sand, particularly oil-wet sand. The solids have the tendency to adhere to the surface of the plates at such an angle of inclination, which may cause clogging of the plates. To avoid such a problem and to enable treatment at pressures higher than atmospheric, modified CPI equipment known as *cross-flow devices* have been developed. In such equipment, the angle of inclination of the plates is made steeper than 45° (normally 60° to the horizontal) and the plate pack is placed inside a pressure vessel (vertical or horizontal) such that the water flow is perpendicular to the axis of corrugations in the plates. Vertical vessels are generally preferred for handling sediments and sand problems. Regular CPI units are less expensive and more efficient than cross-flow devices, but the latter should be used for treatment under pressure and for water containing large amounts of sand or sediments.

15.4.5 Serpentine-Pipe Packs

The serpentine-pipe (SP) pack is another device that is used to promote coalescence of the oil droplets and thus facilitates their separation by gravity. The coalescence concept for the SP pack is, however, different from the equipment described in Section 15.4.4. Water is forced to flow through a serpentine path that is properly sized to create turbulence that is sufficient to cause coalescence without causing shearing of oil droplets below a specified size. The SP packs are available in standard dimensions ranging from 2 to 8 in diameter for handling water flow ranging from 900 to 73,000 bbl/day (BPD). Such packs are designed to develop a drop size distribution curve with a maximum drop size of 1000 μm. By producing such a drop size distribution, gravity settling becomes very efficient. In fact, SP packs can result in about 50% additional oil removal as compared to gravity settling alone. The SP pack is normally placed inside any gravity settling vessel with the water inlet diameter being the same as the SP diameter. SP packs can be staged in series to allow successive coalescence and removal of oil as the water flows from one stage to the next.

15.4.6 Flotation Units

Flotation units utilize a completely different concept in removing oil droplets from water. In this type of treatment equipment, a large number of small gas bubbles are produced within the water. As the gas bubbles rise upward, they carry the oil droplets to the surface, where they accumulate and are then skimmed out of the unit. Flotation units are classified into two types based on the method by which the gas bubbles are produced. These are the *dissolved gas units* and the *dispersed gas units.*

15.4.6.1 Dissolved Gas Flotation Units

As shown in Figure 15.5, a portion of the treated water (between 20% and 50% of the effluent) is taken and saturated with natural gas in a contactor at a pressure between 20 and 40 psi. The amount of gas used in standard cubic feet (SCF) ranges from 0.2 to 0.5 SCF/bbl of water to be treated. The gas-saturated water is recycled back into the unit, which operates at a pressure lower than that of the gas–water contactor. Due to the reduction in pressure, the dissolved gas breaks out of solution as small bubbles. The gas bubbles carry the oil droplets with them as they move to the surface. The size and depth of the unit are determined to provide retention times between 10 and 40 min. The equipment manufacturer normally determines the detailed design parameters of the unit based on the specific operating conditions.

15.4.6.2 Dispersed Gas Flotation Units

In dispersed gas flotation units, the gas bubbles are created, introduced, and dispersed into the bulk of the water to be treated. This is basically done by two methods. In one method, the gas bubbles are created and dispersed in the water by inducing a vortex using a mechanical rotor driven by an electric motor. Figure 15.6 shows a schematic cross section of a unit utilizing this method and manufactured by Petrolite Corporation.

The vortex induced by the rotor creates vacuum within the vortex tube. Due to this vacuum, gas is withdrawn into the vortex and is dispersed in the water. The gas bubbles carry the oil droplets as froth to the surface, where the oil is skimmed and collected in the recovery channel for removal out of the unit.

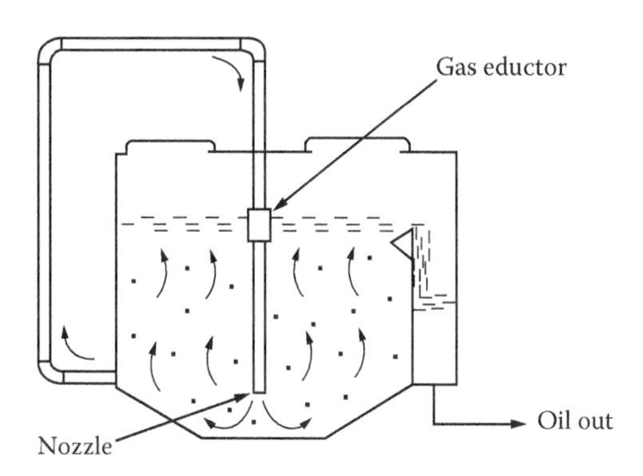

FIGURE 15.5
Dissolved gas flotation unit with eduction.

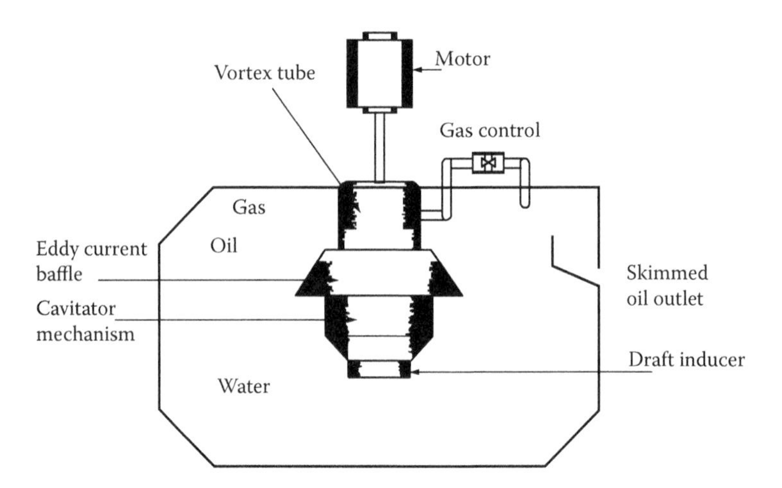

FIGURE 15.6
Schematic of dispersed gas flotation unit.

The other method of creating and dispersing the gas bubbles utilizes an inductor device as shown in Figure 15.7, where a portion of the treated water is recycled back to the unit using a pump. The recycled water flows through a venturi and, due to the reduction in pressure, sucks gas from the vapor space at the top of the unit. The gas is released through a nozzle near the

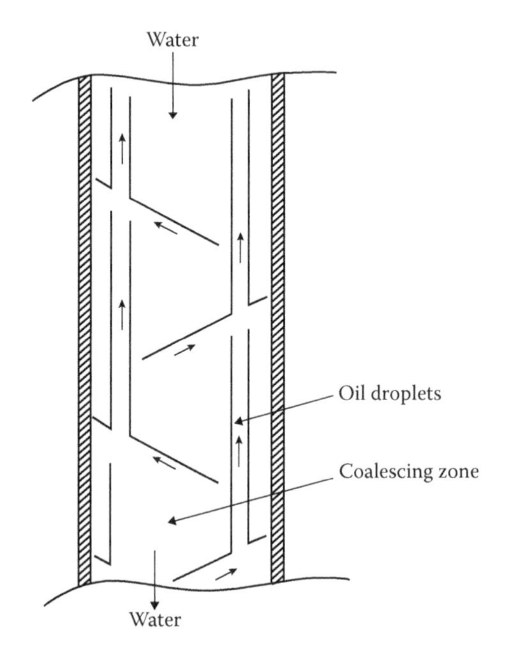

FIGURE 15.7
Dispersed gas flotation unit using an inductor device.

bottom in the form of small bubbles that carry the oil droplets to the surface as they rise. Finally, the oil is skimmed and collected in a chamber for removal from the unit.

Normally, a dispersed gas flotation unit consists of three or four of the cells described. The water to be treated moves from one cell to the next for further removal of oil. Typically, the oil removal efficiency of one cell is about 50%. Therefore, a three-cell unit will have an overall efficiency of 87%, whereas a four-cell unit will have an overall efficiency of 94%. Flotation unit manufacturers have patented design and produce standard units that are typically designed to handle produced water flow at a rate of about 5000 BPD. For higher flow rates, additional units are added in parallel. Flotation units are capable of removing oil droplets smaller than 30 μm.

15.5 Offshore Water Disposal Equipment

Produced water in offshore operations should not be dumped directly into the sea after treatment. In addition to the treated produced water, rainwater and equipment-washdown water represent other sources of oil-contaminated water that need to be disposed of properly. For this purpose, offshore production platforms should be equipped with some form of a disposal device that disposes of the water deep enough below the surface of the sea and away from the wave action to prevent sheens from occurring. The most common of these disposal devices are the *disposal piles, skim piles,* and *SP piles*; these are described in the following sections.

15.5.1 Disposal Piles

Disposal piles are the simplest form of offshore water disposal devices. The disposal pile is simply a large diameter open-ended pipe that is attached to the platform and extends to a specific minimum depth below the surface of the sea. The diameter of the pile is determined based on the total flow of water to be disposed, and the water and oil gravities. In shallow water, the disposal pile should extend down to near the seafloor. In deep water, however, the depth of the pile below the normal water level is determined such that a high level in the pile will be sensed and the shutdown signal measured before the oil in the pile comes within 10 ft of the bottom.

Disposal piles are used to collect treated produced water, deck drains, treated sand, and liquids from drip pans and dispose of them deep below the surface. Disposal piles are also useful as traps for oil in the event of equipment failure or upset operating conditions. The deck drainage, normally rainwater and washdown water, is saturated with oxygen and may contain sand and other solids. Therefore, it should not be treated in the same equipment as produced water to avoid corrosion and plugging problems. Disposal piles are particularly useful for disposal of the platform drainage.

15.5.2 Skim Piles

The skim pile is basically a disposal pile equipped with a series of inclined baffle plates and oil collection risers, as shown in Figure 15.8. The presence of these baffle plates serves two functions. They reduce the distance a given oil droplet has to rise to be separated from the water and create zones of no flow below each plate. The oil droplets rise to the zone of no flow between two successive plates, where coalescence and gravity separation occurs. The coalesced large oil droplets travel up the bottom side of the plate and into the oil collection riser to the surface of the pile where oil could be skimmed out.

Skim piles have two specific advantages over standard disposal piles. Skim piles are more efficient in separating oil from water. Skim piles also provide for some degree of cleaning sand that may be present in the water from oil.

15.5.3 SP Piles

In this type of device, the disposal pile is equipped with a number of equally separated SP packs and oil risers. As water flows through an SP pack,

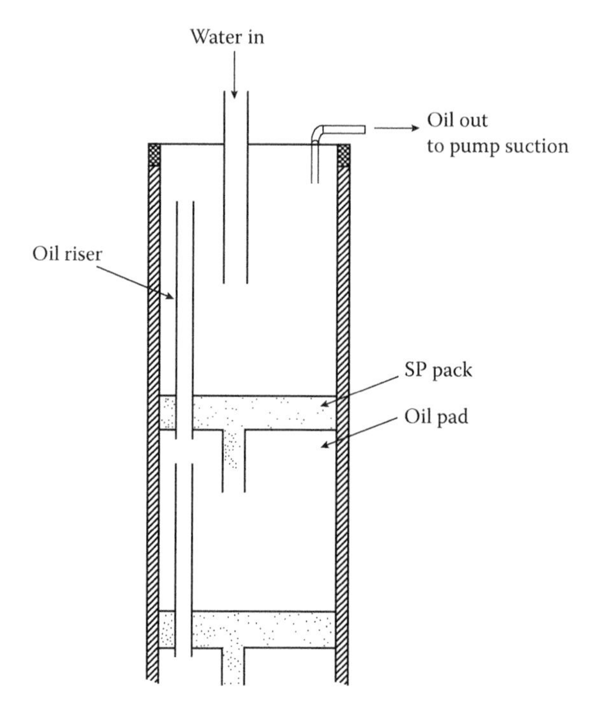

FIGURE 15.8
Schematic of a skim pile.

coalescence of oil droplets occurs due to the induced turbulence. As the water travels out of the SP pack to the next SP pack, the larger oil droplets rise to form an oil pad below the upper SP pack. This continues as the water goes from one SP pack to the next. The oil accumulated below the bottom SP pack rises to the oil pad above through the risers until it reaches the surface to be pumped out.

The SP packs are normally designed to develop oil droplets to a maximum size of 750 μm. The number of SP packs needed is determined from the desired overall efficiency of oil removal and the calculated efficiency of a single pack will be explained later.

15.6 Design of Water Treatment and Disposal Equipment

To properly design the produced water system, certain data and information must be available. In some cases, needed data could be obtained from actual measurements and analyses. In other cases, however, the engineer may have to assume reasonable values for missing or unattainable data. The following data and information are required for the design of the treatment system:

- Flow rate of the produced water, Q_{pw} (BPD)
- Specific gravity, γ_{pw}, and viscosity, μ_{pw}, of the produced water
- Concentration of oil in the produced water (in mg/L or ppm)
- Oil droplets size distribution in the produced water
- Specific gravity of oil at flowing conditions, γ_o
- Required effluent (treated water) quality (in mg/L or ppm)
- Concentration of soluble oil in the effluent
- Rainfall rate, Q_{rw}, and flow rate of washdown water, Q_{wd}

It should be noted that the equipment discussed in this chapter cannot separate the soluble oil. The treating equipment is designed to separate only dispersed oil. Therefore, the soluble oil concentration should be subtracted from the allowable oil concentration of the effluent to obtain the allowable dispersed oil concentration used for design purposes. This is used to determine the size of oil droplets that must be removed from the water, which greatly affects the size and selection of the treating equipment.

In the following sections the equations used to determine the size of the various treating equipment are first presented, then a guideline of equipment selection procedure is highlighted. Throughout the derivation of the equations, the following notations and units are used:

Q_{pw} = produced water flow rate (BPD)
Q_{rw} = deck's rainwater flow rate (BPD)
Q_{wd} = deck's washdown water flow rate (BPD)
μ_{pw} = produced water viscosity (cP)
μ = produced water viscosity (lb-s/ft^2)
d_m = smallest diameter of oil droplets to be removed (μm)
t_{rw} = retention time for water (min)
t_w = retention time for water (s)
t_o = settling time for oil droplets (s)
V_w = average water velocity through the skimmer (ft/s)
V_o = settling velocity of oil droplets (ft/s)
D = diameter of cylindrical skimmer (in)
L = effective length of skimmer (ft)
L_s = total (seam-to-seam) length of skimmer (ft)
H = height of water in the skimmer (ft)
W = width of rectangular cross-section skimmer (ft)
A = flow area for water (ft^2)
h = perpendicular distance between plates of CPI and PPI (in)
θ = angle of inclination of plate interceptors with the horizontal (deg)

15.6.1 Sizing Horizontal Cylindrical Skimmers

The dimensions of the skimmer must be sufficient to allow for separation of the smallest oil droplet that needs to be separated and for the desired retention time. Let D be the skimmer diameter (in inches) and L_s be the skimmer length (in feet). We shall assume that the skimmer is half full with water, which moves horizontally from the inlet to the outlet of the skimmer. Due to the presence of various internal components, the effective length for separation, L, will be less than L_s. To be on the conservative side, we shall also assume that the oil droplets will settle vertically upward from the bottom of the skimmer to the surface of the water; that is, the oil droplets will travel a distance equal to the radius of the skimmer to reach the surface of the water. The time it takes the oil droplet to rise to the surface, t_o, must equal the time it takes the water to move from the inlet to the outlet of the skimmer, t_w.

From Stokes' law, the terminal rise velocity of the oil droplet, u_o, is given by

$$u = 1.787 \times 10^{-6} \frac{(\gamma_w - \gamma_o) d_m}{\mu_{pw}} \frac{\text{ft}}{\text{s}} \tag{15.1}$$

where d_m is the diameter of the oil droplet (in μm). Therefore,

$$t_o = \frac{(D/2 \times 12)}{u_o} \text{ s}$$

$$t_o = 2.34 \times 10^4 \frac{D\mu_{pw}}{(\gamma_w - \gamma_o)d_m} \text{ s} \tag{15.2}$$

The average velocity of the water, u_w, is obtained by dividing the flow rate by the flow area; that is,

$$u_w = \frac{\left[5.61Q_{pw}/(24 \times 3600)\right]}{\left[0.5(\pi/4)(D/12)^2\right]^{-1}} \frac{\text{ft}}{\text{s}}$$

The water retention time is obtained by dividing the effective length, L, by the average velocity; therefore,

$$t_w = 41.9786 \frac{D^2 L}{Q_{pw}} \text{ s} \tag{15.3}$$

Equating Equations 15.2 and 15.3, we get

$$DL = 557.43 \frac{Q_{pw}\mu_{pw}}{(\gamma_w - \gamma_o)d_m} \text{ in ft} \tag{15.4}$$

For turbulence and short-circuiting, an efficiency factor of 1.8 is recommended; therefore, Equation 15.4 becomes

$$DL = 1000 \frac{Q_{pw}\mu_{pw}}{(\gamma_w - \gamma_o)d_m} \text{ in ft} \tag{15.5}$$

Equation 15.5 provides a relationship between the diameter, D, and effective length, L, of the skimmer, which allows all oil droplets of diameter d_m or larger to rise to the surface of the water. Therefore, any combination of D and L that satisfies Equation 15.5 is theoretically acceptable.

Another constraint on the dimensions of the skimmer is that the size should be large enough to provide the necessary retention time, t_{rw}. The retention time is obtained by dividing the volume of the skimmer occupied by water, V_{pw}, by the water flow rate, that is,

$$t_{rw} = \frac{V_{pw}}{Q_{pw}} \tag{15.6}$$

where

$$V_{pw} = 0.5\left(\frac{\pi}{4}\right)\left(\frac{D}{12}\right)^2 L \text{ ft}^3$$

$$Q_{pw}\left(\frac{ft^3}{\text{min}}\right) = Q_{pw}(\text{BPD})\left(\frac{5.61}{24 \times 60}\right)$$

Substituting in Equation 15.6 and rearranging, we obtain

$$D^2L = 1.43 t_{rw} Q_{pw} \text{ in}^2 \text{ ft} \tag{15.7}$$

Therefore, the combination of diameter and effective length that satisfies both settling and retention time constraints should be selected. This is obtained by plotting D versus L for both Equations 15.5 and 15.7 and determining the values of D and L that satisfy both equations. The actual length of the skimmer, L_s, is then determined from

$$L_s = 1.33 \text{ L} \tag{15.8}$$

15.6.2 Sizing Horizontal Rectangular Cross Section (API) Skimmer

The same procedure described in Section 9.6.1 is used to determine the width, W, effective length, L, and height of the water, H, of the skim tank. In this case, we have

$$t_o = \frac{H}{u_o} \tag{15.9}$$

$$t_w = \frac{L}{u_{pw}} \tag{15.10}$$

Equating Equations 15.9 and 15.10 with u_o given by Equation 15.1, and u_{pw} given by dividing the water flow rate [$5.61Q_{pw}/(24 \times 3600)$] by the flow area ($HW$), we obtain

$$WL = 36.5\frac{Q_{pw}\mu_{pw}}{(\gamma_w - \gamma_o)d_m}\ \text{ft}^2 \tag{15.11}$$

For this type of skimmer, the recommended efficiency factor for turbulence and short-circuiting is 1.9; therefore, Equation 15.11 becomes

$$WL = 70\frac{Q_{pw}\mu_{pw}}{(\gamma_w - \gamma_o)d_m}\ \text{ft}^2 \tag{15.12}$$

The retention time as obtained from Equation 15.10 should also be equal to the volume of the skimmer occupied by water (HWL) divided by the water flow rate; therefore,

$$t_{rw} = 15401\frac{HWL}{Q_{pw}}\ \text{s} \tag{15.13}$$

The height of the water, H, is usually limited to one-half the width, W; therefore, substituting 0.5 W for H in Equation 15.13 and dividing by 60, the retention time in minutes is given by

$$t_{rw} = 125\frac{W^2L}{Q_{pw}}\ \text{min} \tag{15.14}$$

Rearranging, we obtain

$$W^2L = 0.008t_{rw}Q_{pw}\ \text{ft}^3 \tag{15.15}$$

Any combination of W and L that satisfies both Equations 15.12 and 15.15 satisfies both the settling and retention time constraints.

15.6.3 Sizing Vertical Cylindrical Skimmer

In a vertical cylindrical skimmer, the oil droplets move vertically upward against the vertically downward flowing water. For settling to occur, the velocity of the oil droplet, u_o, must be at least equal to the average velocity of the water, u_{pw}, determined by dividing the water flow rate by the circular cross-sectional area of the skimmer; that is,

$$u_w = \frac{5.61Q_{pw}/(24 \times 3600)}{(\pi/4)(D/12)^2} \frac{\text{ft}}{\text{sec}}$$

Equating u_w to u_o expressed by Equation 15.1 and solving for the diameter of the skimmer, D, we obtain

$$D^2 = 6691 \frac{Q_{pw}\mu_w}{(\gamma_w - \gamma_o)d_m} \text{ in}^2 \tag{15.16}$$

Equation 15.16 is good for diameters up to 48 in. For larger diameters, where turbulence and short-circuiting occur, Equation 15.16 should be multiplied by a factor F that is greater than 1. The value of F depends largely on the design of the inlet and outlet flow spreaders, oil collector, and baffles.

The retention time is obtained by dividing the height of the water in the skimmer, H, by the average flow velocity of the water:

$$t_{rw} = H\left(\frac{5.61Q_{pw}/(24 \times 3600)}{(\pi/4)(D/12)^2}\right)^{-1} \text{ s} \tag{15.17}$$

Convert t_{rw} into minutes in Equation 15.17 and solve for the height, H:

$$H = 0.7\frac{t_{rw}Q_{pw}}{D^2} \text{ ft} \tag{15.18}$$

Equation 15.16 determines the minimum diameter of the skimmer that satisfies the settling constraint. Any diameter that is equal to or larger than that determined from Equation 15.16 is acceptable. Equation 15.18 determines the height of the water for the selected diameter to satisfy the retention time constraint. The total height (seam-to-seam length) of the skimmer, L_s, is obtained by adding 3 ft to H; that is,

$$L_s = H + 3 \text{ ft} \tag{15.19}$$

Example 15.1

A field separation and treatment plant produces 8000 BPD of water containing 2000 mg of oil per liter of water with a maximum oil droplet size of 500 μm. The water viscosity and specific gravity are 1.1 cP and 1.07, respectively, and the oil specific gravity is 0.87. It is required to treat the produced water to reduce the oil content to 800 mg/L.

Taking the retention time as 10 min, determine the dimensions of a (a) horizontal cylindrical skimmer, (b) horizontal, rectangular cross-section skimmer; and (c) vertical cylindrical skimmer to provide the required treatment.

Solution

The first step is to determine the size of the oil droplet that must be removed to achieve the required treated water quality. Because the relationship between the oil droplet size and concentration is approximately linear, the droplet diameter, d_m, that must be removed is obtained from

$$d_m = 500\left(\frac{800}{2000}\right) = 200\ \mu\text{m}$$

a. *Horizontal cylindrical skimmer.* The dimensions of the skimmer must satisfy both settling and retention time constraints, Equations 15.5 and 15.7, respectively. Using Equation 15.5,

$$DL = \frac{(1000)(8000)(1.1)}{(1.07-0.87)(200)^2} \tag{15.20}$$

$$DL = 1100$$

Using Equation 15.7,

$$D^2L = (1.43)(8000)(10) = 114{,}400 \tag{15.21}$$

Assuming various values for D and determining the corresponding values of L from Equations 15.20 and 15.21 the following table is constructed:

	Settling (Equation 15.20)		Retention Time (Equation 15.21)	
D (in)	L(ft)	L_s(ft)	L(ft)	L_s(ft)
60	18.33	24.44	31.78	42.26
72	15.28	20.36	22.07	29.35
84	13.10	17.46	16.21	21.56
96	11.46	15.27	12.41	16.51
108	10.19	13.58	9.81	13.05
120	9.17	12.22	7.94	10.57

Examination of the table shows that the retention time constraint governs the design up to a diameter of 96 in; then, the settling constraint governs. Suitable selections would be 84 in × 22 ft or 96 in × 17 ft.

b. *Horizontal rectangular cross-section skimmer.* Using Equation 15.12 (settling constraint),

$$WL = \frac{(70)(8000)(1.1)}{0.2(200)^2} = 77 \text{ ft}^2 \tag{15.22}$$

Using the retention time constraint, Equation 15.14

$$W^2L = (0.008)(8000)(10) = 640 \text{ ft}^3 \tag{15.23}$$

		L(ft)	**L(ft)**
W(ft)	H_w **= 0.5 W**	**Settling (Equation 15.22)**	**Retention (Equation 15.23)**
5	2.5	15.40	26.6
6	3.0	12.83	17.28
7	3.5	11.00	13.06

Again, the retention time constraint governs the design of the skimmer. The suitable selections are skimmers that are 6 ft wide by 18 ft long, and 7 ft wide by 14 ft long. The height of the water in the skimmer is limited to one-half of the width.

c. Vertical cylindrical skimmer. The settling constraint Equation 15.16 provides the minimum diameter for the skimmer:

$$D^2 = \frac{(6691)(8000)(1.1)}{0.2(200)^2} = 7360$$

$$D_{\min} = 85.79 \text{ in}$$

Because the diameter is greater than 48 in, the settling equation should be multiplied by a factor that is greater than 1 to account for turbulence and short-circuiting. Using a factor of 1.5 gives

$$D_{\min} = 105 \text{ in}$$

Taking a diameter of 108 in and substituting in Equation 15.18 (the retention-time constraint), the height of the water is determined:

$$H = \frac{(0.7)(8000)(10)}{(108)^2} = 4.8 \text{ ft}$$

The height of the skimmer (seam-to-seam length), L_s, is given by

$$L_s = H + 3 = 4.8 + 3 = 7.8 \text{ ft}$$

Therefore, the skimmer is 108 in diameter and 8 ft high.

Any of the three skimmers will provide the required water treatment. The final selection depends on cost, availability, and space limitation.

15.6.4 Sizing of Skimmers with Parallel Plate Interceptors

In this case, the oil droplets need only to rise to the underside of plates in a time, t_o, that is equal to the time it takes the water to travel through the effective length of the skimmer, t_w. We shall assume that the effective length in the settling process is 70% of the actual length of the PPI pack and that the plates occupy 10% of the flow area (HW).

$$t_o = \frac{(h/12\cos\theta)}{u_o} \tag{15.24}$$

where h is the perpendicular distance between plates of CPI and PPI. Substituting for u_o from Equation 15.1

$$t_o = 4.681\frac{h\mu_{\text{pw}}}{(\gamma_w - \gamma_o)d_m\cos\theta}\text{ s} \tag{15.25}$$

The retention time of water is given by

$$t_w = \frac{L}{u_w} = 0.7\frac{L_s}{u_w} \tag{15.26}$$

where

$$u_w = \frac{5.61Q_{\text{pw}}/(24\times 3600)}{0.9HW}\ \frac{\text{ft}}{\text{s}} \tag{15.27}$$

Therefore,

$$t_w = 9.703\times 10^3\frac{LHW}{Q_{\text{pw}}}\text{ s} \tag{15.28}$$

Equating Equations 15.25 and 15.28,

$$HWL = 4.824 \frac{Q_{pw} h \mu_{pw}}{(\gamma_w - \gamma_o) d_m \cos\theta} \tag{15.29}$$

For efficient settling of the oil droplets, it has been reported that the Reynolds number (Re) with the characteristic dimension being 4 times the hydraulic radius should not exceed 1600. This can be used to develop an equation for the minimum flow area (HW) as follows. The hydraulic radius, R, is obtained by dividing the area between plates by the wetted perimeter, that is,

$$R = \frac{(h/12)(W/\cos\theta)}{2((W/\cos\theta)+(h/12))}$$

$$R = \frac{(h/12)}{2(1+(h\cos\theta/12W))}$$

Considering that $(h \cos \theta/12) \ll W$, R may be approximated by

$$R = \frac{h}{24}$$

Therefore,

$$\text{Re} = \frac{\rho_w D u_w}{\mu g} = 1600$$

where g is the acceleration of gravity, ft/sec^2. Substituting for u_w from Equation 15.27, $D = 4R$, $\rho_w = 62.4\ \gamma_w$ lb/ft^3, and $\mu = 2.088 \times 10^{-5}\ \mu_{pw}$ and solving for HW, we obtain

$$HW = 6.982 \times 10^{-4} \frac{Q_{pw} h \gamma_w}{\mu_{pw}} \text{ ft}^2 \tag{15.30}$$

Equation 15.30 is first used to determine the minimum value of HW, then Equation 15.29 is used to determine L for any value of HW above the minimum determined.

15.6.5 Sizing of Skimmers with Corrugated Plate Interceptors

The corrugated plate interceptor packs are normally available in standard sizes with $H = 3.25$ ft, $W = 3.25$ ft, $L = 5.75$ ft, $h = 0.69$ in, and $\theta = 45°$. The size of the skimmer used with the CPI is determined by the number of CPI packs installed. It should be noted that the flow rate through the standard CPI pack should not exceed 20,000 BPD in order to stay within the Reynolds number limitation.

Using Equation 15.29 and the dimensions of standard CPI packs, the number of CPI packs needed, N, is determined using

$$N = 0.11\frac{Q_{\text{pw}}\mu_{\text{pw}}}{(\gamma_w - \gamma_o)d_m} \tag{15.31}$$

15.6.6 Sizing of Skimmers with Cross-Flow Devices

As described in Section 15.5, cross-flow devices require the installation of internal spreaders and collectors for uniform distribution of the water among the plates. A spreader efficiency must be introduced in sizing skimmers with cross-flow devices, because spreaders do not guarantee a 100% uniform distribution of the flow. Equation 15.29 could be used for sizing after dividing the right-hand side by 0.75; therefore,

$$HWL = 6.432\frac{Q_{\text{pw}}h\mu_{\text{pw}}}{(\gamma_w - \gamma_o)d_m \cos\theta} \tag{15.32}$$

15.6.7 Determining the Required Number of Serpentine-Pipe Packs

As described before in Section 5.5, the SP packs are used to coalesce the oil droplets and greatly increase their size to speed the separation process. Depending on the treatment requirements, more than one stage may be needed. In such a case, the stages will be arranged in series. The number of SP pack stages, n, can be determined as follows. Let C_i be the oil concentration in the influent water and C_o be the maximum allowed oil concentration in the effluent (treated) water. Therefore, the overall separation or treatment efficiency, E, is determined from

$$E = 1 - \frac{C_o}{C_i} \tag{15.33}$$

Alternatively, because the droplet size distribution is approximately linear, E could be determined from

$$E = 1 - \frac{d_o}{d_i} \tag{15.34}$$

where d_o is the largest oil droplet diameter in the effluent and d_i is the largest oil droplet diameter in the influent water.

Now, let d_m be the oil droplet diameter that can be separated by a skimmer of a certain size. The function of the SP pack is to coalesce such oil droplets to increase the diameter to d_{max} (typically, d_{max} for standard SP packs is approximately 1000 μm). The efficiency of the SP pack could then be determined from

$$E_{SP} = 1 - \frac{d_m}{d_{max}} \tag{15.35}$$

Therefore, for n SP packs staged in series, the overall efficiency could be expressed in terms of the efficiency of the SP pack by

$$E = 1 - (1 - E_{SP})^n \tag{15.36}$$

15.6.8 Sizing of Flotation Units

Flotation unit manufacturers have various standard units with specific dimensions and flow rate limitations. The role of the facilities engineer is, therefore, limited to determining whether flotation units are needed for the treatment process and the number of units needed. The need for flotation units depends on the required quality of the treated (effluent) water. Normally, if the maximum allowable oil droplet size in the effluent water is about 30 μm or less, flotation units should be considered. The efficiency of the flotation unit, which is typically between 87% and 94% depending on the number of cells, is then used to determine the quality (oil concentration, or maximum oil droplet size) of the influent water that could be treated by the flotation unit from Equation 15.33 or Equation 15.34. The quality of the influent water is then compared to the quality of the produced water to determine whether a primary treatment stage is needed.

Example 15.2

For the same conditions of Example 15.1, design the required treatment facilities such that the treated water does not contain more than 80 mg of oil per liter of water.

Solution

The maximum oil droplet size that can exist in the treated water is determined from

$$d_m = \frac{(80)(500)}{2000} = 20\ \mu m$$

This size is too small to be separated by skimmers of any reasonable size. Therefore, either flotation units or SP packs should be used. Let us examine both options.

USING FLOTATION UNITS

Flotation units are normally available as standard units for a flow rate of 5000 BPD. Two units must be used in parallel because 8000 BPD of water is to be treated. Assume that a four-cell unit is being used and that the efficiency of the unit is 90%. The quality of the water that could be treated by the flotation unit is, therefore, determined from

$$E = 1 - \left(\frac{d_{mo}}{d_{mi}}\right)$$

$$0.9 = 1 - \left(\frac{20}{d_{mi}}\right)$$

$$d_{mi} = 200\ \mu m$$

Since the diameter is smaller than the droplet size in the produced water, a primary treatment of the water is necessaary before it can be treated in the flotation unit. The primary treatment is required to remove all oil droplets of 200 μm and larger. This is the same as the requirement for Example 15.1. Therefore, any of the three skimmers designed in the previous example could be used. The treatment system could, therefore, consist of a horizontal cylindrical skimmer 96 in × 17 ft, or a vertical skimmer of 108 in diameter by 8 ft high, and two flotation units in parallel.

USING SP PACKS

The overall required efficiency is first calculated from

$$E = \frac{2000 - 80}{2000} = 0.96$$

Assume that we will use the vertical skimmer designed in Example 15.1. The droplet size that can be treated in such a skimmer is 200 μm.

Therefore, the efficiency of the SP pack (assuming it grows 1000 μm droplets) is

$$E_{SP} = 1 - \left(\frac{200}{1000}\right) = 0.8$$

From Equation 15.32,

$$0.96 = 1 - (1 - 0.8)^n$$

$$n = 2$$

Therefore, the treatment could be achieved using two 108 in diameter vertical skimmers with SP packs.

The selection of any of the aforementioned two options will depend on cost and other considerations such as the availability of gas for flotation units.

15.6.9 Sizing of Offshore Disposal Devices

15.6.9.1 Disposal Piles

As mentioned in Section 15.6.1, disposal piles are not designed for produced water treatment; they are simply used to dispose of the treated produced water. However, disposal piles are also used to dispose of the deck drains, normally rainwater or washdown water. Such water is usually contaminated by oil and, therefore, the disposal pile will be expected to provide the means to settle oil droplets of certain size. In most cases, the disposal pile will be designed to separate oil droplets 150 μm and larger. This, however, depends on local regulations.

Equation 15.16 developed for vertical skimmers is used to determine the diameter of the disposal pile; the equation is rewritten as follows:

$$D^2 = 6691 \frac{Q_w \mu_w}{(\gamma_w - \gamma_o)} \text{in}^2 \tag{15.37}$$

where Q_w is given by

$$Q_w = Q_{pw} + Q_{rw} + Q_{wd} \text{ BPD} \tag{15.38}$$

Normally, the larger of the rainwater and washdown water flow rates is used in Equation 15.38, as it is unlikely to have both at the same time. The water viscosity, μ_w, is taken as 1 cP.

The length of the pile submerged below the normal sea level is determined such that the high-level and shutdown alarms are sensed before the oil reaches 10 ft from the bottom. This could be determined from a hydrostatic pressure balance at the bottom of the pile. The normal tide and expected storms should be considered.

Let L_w be the normal water level, L_{st} be the design storm range, L_t be the average tide range, and assigning 4 ft for the alarm and shutdown levels, the length of the pile submerged, L, will be given by

$$L = 10 + \frac{(4 + L_{st} + L_t + L_w)\gamma_o}{\gamma_w - \gamma_o} \text{ ft} \tag{15.39}$$

In deep waters, however, the minimum length should be 50 ft. In shallow water, the length of the pile should extend as deep as possible.

15.6.9.2 Skim Piles

Due to the complexity of the flow regime in skim piles, no analytical equation exists for sizing such piles. However, an adequate and simple equation that is based on field experience, developed in this section, may be used for determining the size of skim pile.

Let L be the length of the submerged section of the pile as determined from Equation 15.39. The length of the baffle section, L_b, is then determined from

$$L_b = L - 15 \text{ ft} \tag{15.40}$$

If d is the diameter of the pile (in inches) and assuming that the baffles and oil collection risers occupy 25% of the volume, the volume occupied by water, v, is then given by

$$V = 0.75\left(\frac{\pi D^2}{4 \times 1144}\right) L_b \text{ ft}^3 \tag{15.41}$$

The retention time in the baffle section, t_r, is obtained by dividing the volume, V, by the water flow rate, Q_w, which is determined from Equation 15.38; therefore,

$$t_r = \frac{V}{5.61 Q_w / (24 \times 60)} \text{ min} \tag{15.42}$$

Experience has shown that a retention time of 20 min is sufficient in meeting disposal requirements. Therefore, combining Equations 15.41 and 15.42 assigning a value of 20 to t_r and solving for $D^2 L_b$ we get

$$D^2 L_b = 19.1\ Q_w \text{ in}^2 \text{ ft} \tag{15.43}$$

15.6.9.3 SP Piles

The procedure for determining the number of SP packs needed is similar to that used for skimmers. A pile diameter is first selected and the efficiency per SP stage is determined. The number of stages is then determined from Equation 15.36 for an overall efficiency of 90%.

Example 15.3

An offshore production platform having a deck area of 3000 ft² produces 4500 BPD of water. Primary and secondary treatment stages are used and result in a final water quality of 45 mg oil per liter of water, which meets the offshore disposal requirements. The average rainfall in the area is 1.5 in/h and the deck is equipped with a 50 gpm washdown hose.

Design the necessary disposal facility to handle the treated produced water and deck drain. Assume a water viscosity of 1.1 cP and a difference in specific gravity of 0.2.

Solution

The first step is to calculate the total amount of water to be disposed:

$$\text{Rainwater} = 1.5\frac{\text{in}}{\text{h}} \times \frac{1\ \text{ft}}{12\ \text{in}} \times 3000\ \text{ft}^2 \times 24\frac{\text{h}}{\text{day}}$$

$$\times \frac{1}{5.61}\frac{\text{bbl}}{\text{ft}^3} = 1604\ \text{BPD}$$

$$\text{Washdown water} = 50\ \text{gpm} \times (60 \times 24)\frac{\text{min}}{\text{day}} \times \frac{1\ \text{bbl}}{48\ \text{gal}}$$

$$= 1500\ \text{BPD}$$

A consideration is given in the design for the fact that rain water rate is larger than the wash-down rate. Therefore,

$$Q_w = 4500 + 1604 = 6104\ \text{BPD}$$

We first check the simplest disposal device (i.e., the disposal pile). Using Equation 15.37 with $d_m = 150$ μm,

$$D^2 = \frac{(6691)(6104)(1.1)}{0.2(150)^2} = 9983.57$$

$$D = 99.9\ \text{in}$$

This is too large. Therefore, we check the skim pile. Using Equation 15.43,

$$D^2 L_b = (19.1)(6104) = 116{,}586 \text{ in}^2 \text{ ft}$$

Assume various values for D and calculate the corresponding values of L_b:

D (in)	36	42	48
L_b (ft)	89.9	66.1	50.6

Therefore, we can use a 48 in diameter pile with a 50 ft baffle section. The total submerged length is, therefore Equation 15.40,

$$L = L_b + 15, \text{ or}$$

$$L = 50 + 15 = 65 \text{ ft}$$

which is greater than the minimum required length of 50 ft.

The skim pile is an adequate choice. The SP pile could also be used where the treated produced water should be introduced below the SP packs because it does not need any treatment. The skim pile is, however, simpler and less expensive.

REVIEW QUESTIONS

1. Discuss the reasons for treating produced water that will be
   a. Disposed into the sea
   b. Disposed into an underground formation
   c. Used as injection water for improved oil recovery
   d. Used for generating steam for thermal recovery operations
2. Discuss the advantages and disadvantages of treating produced water under pressure versus atmospheric treatment.
3. Describe the functions of the following equipment in promoting efficient produced water treatment:
   a. Parallel plates interceptors
   b. Corrugated plate interceptors
   c. Serpentine-pipe coalescers
4. Describe the basis of operation of flotation units and discuss the reasons for using such units.
5. Describe the function and basis of operation of the various types of offshore disposal piles.

6. Discuss the effects of the various parameters in the settling equation (Equation 15.1) on the efficiency of separating oil droplets from produced water. Discuss possible means of controlling these parameters to improve the separation (treatment) process.
7. Design the water treatment and disposal facilities for an offshore production and treatment platform for the conditions listed. Perform the design calculations for all types of treatment and disposal equipment discussed in this chapter, and then determine the best combination of facilities giving reasons for your decisions.

Produced water rate:	12,000 BPD
Oil content:	2000 mg oil/L water
Maximum oil droplet size:	500 μm
Required treated water quality:	70 mg/L
Dissolved oil:	25 mg/L
Water specific gravity:	1.074
Oil specific gravity:	0.89
Water viscosity:	1.03 cP
Treating temperature:	70°F
Platform area:	4500 ft^2
Average annual rainfall rate:	1.7 in/h
Available washdown hose:	50 gpm

16

Field Storage Tanks, Vapor Recovery System (VRS), and Tank Blanketing

16.1 Field Storage Tanks

Production, refining, and distribution of petroleum products require many different types and sizes of storage tanks. Small bolted or welded tanks might be ideal for production fields while larger, welded storage tanks are used in distribution terminals and refineries. There are many factors that should be considered in the selection of storage tanks in oil field operations. Field operating conditions, storage capacities, and specific designs are most important. Storage tanks are often cylindrical in shape, perpendicular to the ground with flat bottoms, and with a fixed or floating roof. Atmospheric storage tanks, both fixed roof and floating roof tanks are used to store liquid hydrocarbons in the field.

Storage tanks are needed in order to receive and collect oil produced by wells before pumping to the pipelines and to allow for measuring oil properties, sampling, and gauging.

The design of storage tanks for crude oil and other hydrocarbon products is a function of the following factors:

- The vapor pressure of the materials to be stored
- The storage temperature and pressure
- Toxicity of the petroleum material

16.1.1 Tank Classification

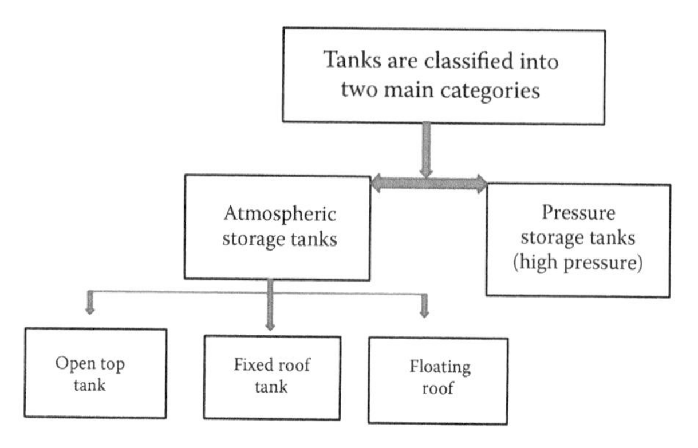

According to the National Fire Protection Association (NFPA 30-2008, 2011), *atmospheric storage tanks* are defined as those tanks that are designed to operate at pressures between atmospheric and 6.9 kPa gauge. Such tanks are built in two basic designs: the cone-roof design where the roof remains fixed, and the floating-roof design where the roof floats on top of the liquid and rises and falls with the liquid level. *Pressure storage tanks*, on the other hand, are used to store liquefied gases such as liquid hydrogen (LH) or a compressed gas such as compressed natural gas. They can be referred to as *high-pressure tanks*. Storage tanks can also be classified as aboveground storage tanks (AST) and underground tanks (UST).

There are usually many environmental regulations and others that apply to the design and operation of each category depending on the nature of the fluid contained within.

16.1.2 Types of Tanks

There are four basic types of tanks that are commonly used to store crude oil and its products.

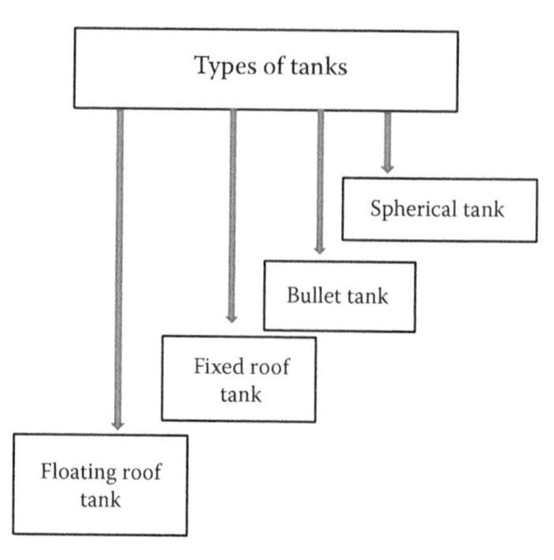

16.1.2.1 Floating Roof Tanks

Floating roof tanks are advantageous, compared to fixed roof tanks, as they prevent vapor emissions (that are highly combustible). The floating roof rises and falls with the liquid level inside the tank, thereby decreasing the vapor space above the liquid level. This will eliminate the chances of fire or an internal tank explosion. Floating roofs are considered a safety requirement as well as a pollution prevention measure for many industries including petroleum refining. In external floating roof storage tanks, the roof is made to rest on the stored liquid and is free to move with the level of the liquid. These tanks reduce evaporation losses and control breathing losses while filling. They are preferred for storage of petroleum products with a true vapor pressure of 10.3 to 76.5 kPa absolute.

Another alternative to external floating roofs is an internal roof that combines the concept of conical fixed roof tanks that lie on top of pontoons. They too are affected by the withdrawal and storage losses that are mitigated using similar means.

16.1.2.2 Fixed Roof Tanks

Fixed roof tanks consist of a cylindrical shell with a permanently welded roof that can be flat, conical, or dome shaped. Such tanks are used to store materials with a true vapor pressure of less than 10.3 kPa absolute. Horizontal cylinders and spheres are generally used for full pressure storage of hydrocarbon or chemical products. The atmospheric or low-pressure storage tanks are widely used from the production fields to the refinery. The most common shape used is the vertical, cylindrical storage tank. Gross capacities can range from 100 bbl to over 1.5 MM bbl in a single storage tank. Corresponding tank sizes range from approximately 10 ft to over 412 ft diameter, for some of the largest floating roof tanks ever constructed.

During the process of storing crude oil, light hydrocarbons such as natural gas liquids, volatile organic compounds, hazardous air pollutants, and some inert gases vaporize and collect between the liquid level and the fixed roof tanks. As the liquid level in the tank varies, these gases slowly release out to the atmosphere. Installing vapor recovery units is the answer to avoid these losses, as explained later.

16.1.2.3 Bullet Tanks

It is a fact that all storage tanks and vessels are built with a bullet-shaped or cylindrical shell for some good reasons. In general, storage tanks and vessels are pressurized to some degree and vary in shapes as well. For storing hydrocarbons, it is shaped like a bullet or cylinder or sometimes spherical because it distributes pressure more evenly compared to vessels with edges. The ratio of volume by surface area is high in case of a cylinder. The highest is for a sphere. But there are problems in manufacturing and transporting spherical vessels. Low surface area means lesser material cost in manufacturing these bullet-shaped vessels. In addition, because of its shape, there are fewer stress concentration points in bullet-shaped vessels.

16.1.2.4 Spherical Tanks (Storage Spheres)

Spherical tanks are preferred for storage of high-pressure fluids. A sphere is a very strong structure, because it has even distribution of stresses on its surface both internally and externally. Spheres, however, are much more costly to manufacture than cylindrical or rectangular vessels. Storage spheres need ancillary equipment similar to tank storage (e.g., access manholes, safety valves, access ladders). An advantage of spherical storage vessels is that they have a smaller surface area per unit volume than any other shape of vessel. This means that the quantity of heat transferred from warmer surroundings to the liquid in the sphere will be less than that for cylindrical or rectangular storage vessels.

16.1.3 Materials of Construction of Storage Tanks

Although the earliest storage tanks used by the petroleum industry were constructed from various types of wood, they are currently fabricated from steel or optional nonmetallic materials. Steel and concrete remain the most popular choices for tanks. However, glass-reinforced plastic, thermoplastic, and polyethylene tanks are increasing in popularity. They offer lower build costs and greater chemical resistance, especially for storage of specialty chemicals. However, they suffer some problems. The temperature limits of plastic tanks are approximately 40°F to 150°F. Some operators prohibit the use of plastic tanks in hydrocarbon service, because plastic tanks are considered to degrade more quickly than metal tanks when exposed to fire.

Before the development and perfection of welding processes, petroleum storage tanks used either bolted or riveted construction techniques. The tanks would be designed and supplied as segmental elements for final assembly on site. Field-welded storage tanks easily meet industry needs for increased storage capacity whether at a remote production site, at the refinery, or at the marketing terminal.

16.1.4 Additional Comments

Since most liquids can spill, evaporate, or seep through even the smallest opening, special consideration must be made for their safe and secure handling. This usually involves building a containment dike around the tank so that any leakage may be safely contained.

Metal tanks in contact with soil and containing petroleum products must be protected from corrosion to prevent escape of the product into the environment.

Cathodic protection is the most effective and common corrosion control techniques for steel in contact with the soil.

Production facilities generally rely on either open-top tanks or fixed roof tanks operating at or slightly above atmospheric pressure.

16.2 Vapor Recovery System

16.2.1 Introduction

In production operations, underground crude oil contains many lighter hydrocarbons in solution. When oil is brought to the surface it experiences drastic pressure drop by going through the gas–oil separator plant (GOSP).

The evolution of hydrocarbon vapors is dependent on many factors:

- The product's physical characteristics
- The operating pressure of upstream equipment
- Tank storage condition

During storage, light hydrocarbons dissolved in the crude oil or in the condensate, including methane, other volatile organic compounds (VOCs), and hazardous air pollutants (HAPs), vaporize or flash out. These vapors collect in the space between the liquid and the fixed roof tank. As the liquid level in the tank fluctuates, these vapors are vented to the atmosphere, or flared. Alternatively, a vapor recovery compressor (or blower) may be installed to direct vapors vented from storage to downstream compressors for sales or injection. A significant economic savings is obtained by installing vapor recovery units (VRUs) on the storage tanks. These units are capable of capturing about 95% of the vapors. Losses of dissolved light are identified as:

- Flash losses in the GOSP
- Working losses due to change in the fluid level inside the tank during pumping, filling, or emptying
- Standing or breathing losses that occur with daily and seasonal temperature changes

Vacuum relief valves are needed to keep a vacuum from occurring because of tank breathing and pumping operations. If a vacuum develops, the tank roof will collapse.

16.2.2 Economic and Environmental Benefits

VOC and HAP emissions to the atmosphere cause pollution of the air we breathe. These emissions can be controlled by either destruction or by recovery using VRUs. VRUs are designed to comply with U.S. Environmental Protection Agency (EPA) standards, provide economic profits to the oil and gas producers, and eliminate stock vapors from the atmosphere. Waste gas is the lost product, hence a lost revenue. Gases flashed from crude oil or condensate and captured by VRUs can be sold at profit or used at an oil field facility. Options for utilizing the recovered gases are to be used as a fuel for oil field operations, to be collected to natural gas gathering stations and sold, or to be used as a stripping agent.

In order to estimate the economic return when installing a VRU, one should follow the following procedure.

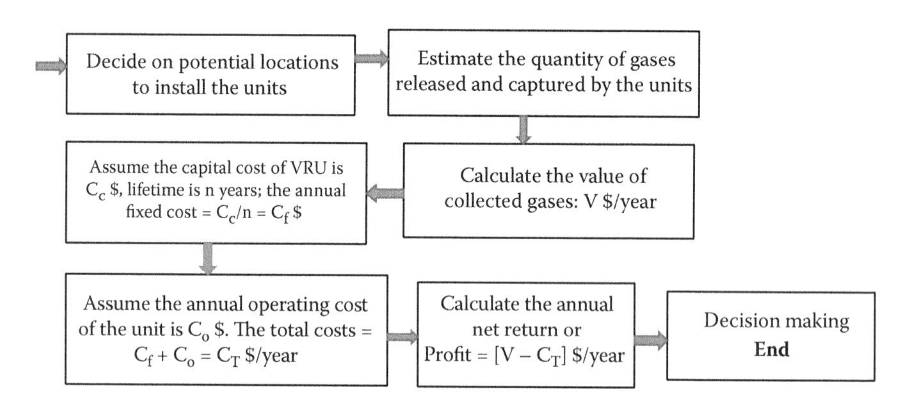

16.2.3 Basic Vapor Recovery System Modules

Vapor recovery systems are designed to recover VOC and HAP emissions from hydrocarbon sources. They are used in the chemical process and petroleum industries to recover escaped vapors for either reuse or destruction (usually by combustion).

Vapor recovery systems can capture target compounds in a gas stream by using single or multiple module that utilize one of the following processes:

- Condensation system—A gas stream is chilled to a temperature where the target pollutants or compounds condense out of the stream. The cooling liquids used depend primarily on the compounds being targeted and the emission limits required.
- Activated carbon system—Activated carbon's ability is used to adsorb certain molecules from gaseous mixtures. The adsorption process continues until the carbon is either fully saturated and removed, or regenerated in situ for reuse.

- Lean oil absorption system—Usually is carried out under pressure. An absorbent oil or liquid (solvent) of higher molecular weight than that of the contaminated vapor stream is used. Usually it is introduced, countercurrent in a packed column to make close contact with the feed vapor stream. Absorbed components are separated and recovered from the rich solvent, which is recycled back to the column as *lean* solvent.
- Jet ejector system—Venture-jet ejectors are used to suck in target gases. By utilizing a high-pressure motive gas, the system entrains target compounds out of low-pressure streams.

The following should be considered in choosing a vapor recovery module:

- Some modern vapor recovery systems implement multiple techniques in combination for complex or specialized gas streams.
- The type of vapor recovery system used depends largely on the type of compounds being captured.
- The most common vapor pollutants are hydrocarbons, which can be removed in most system types.
- Operating and maintenance costs are important when selecting a system. These costs tend to be high, especially when utilizing high-pressure compressors or low-temperature condensers.

Basically, when we talk about a vapor recovery system, what we are shooting at is to hook our storage tanks (or the source of gases) to some kind of a *breather* system with the following functions. During the day, with temperature rise, excess vapors caused by evaporation of the hydrocarbons from a storage tank are released and collected by the VRU. At night, when the vapors cool and condensation occurs, leading to partial vacuum, vapors from the VRU will be admitted into the tanks. While pumping in and pumping out liquids to and from storage tanks, vapors could be vented (collected and drawn in respectively), by such a breather system (main component of a VRU). Figures 16.1 and 16.2 illustrate this concept.

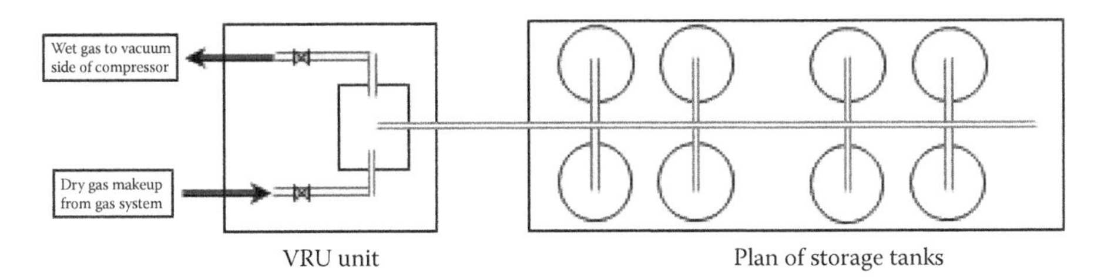

FIGURE 16.1
Vapor regulating system connected to storage tanks.

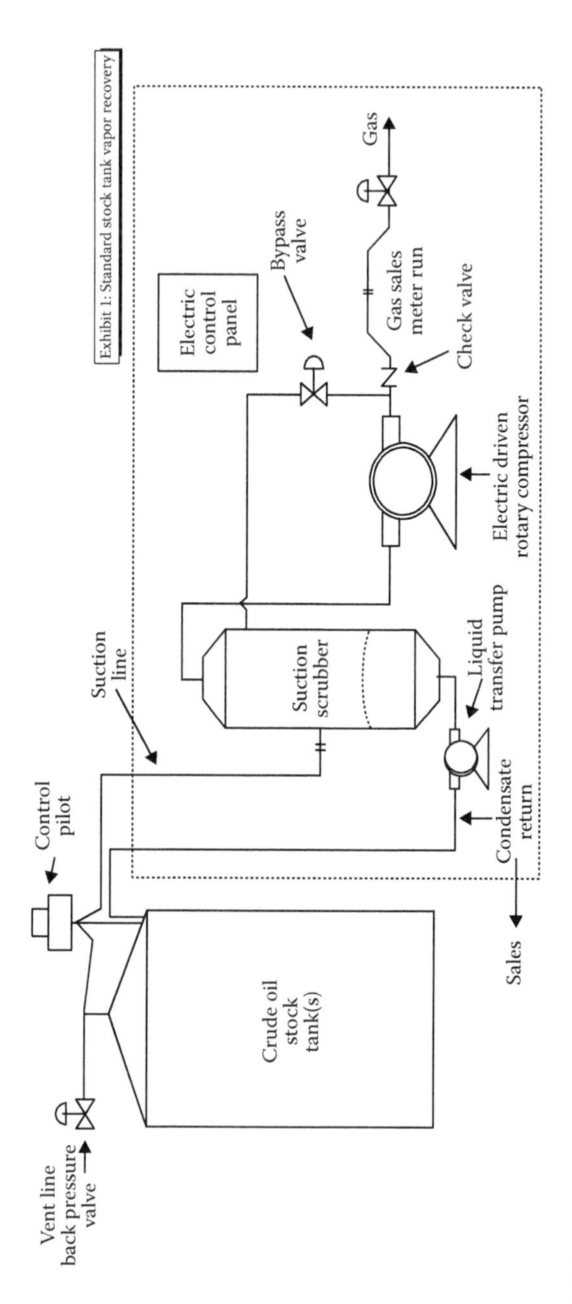

FIGURE 16.2
Basic components of a VRU.

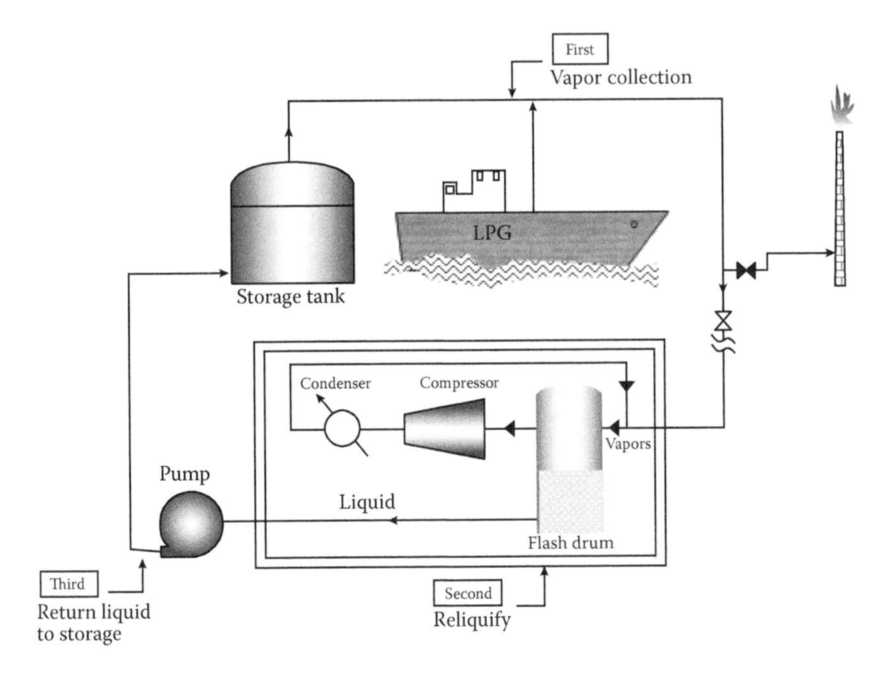

FIGURE 16.3
Main functions of a vapor recoverey system.

The closed cycle comprising vapor capturing or collection, followed by condensation and pumping back to storage tank, is demonstrated in Figure 16.3.

16.3 Tank Blanketing

16.3.1 Background

Tank blanketing or padding is the process of applying a cover of gas to the empty space in a storage container, usually a liquid. Its purpose is either to protect or contain the stored product, or prevent it from harming personnel, equipment, or the environment. Thus the gas phase maintained above the liquid will protect the liquid against air contamination and reduce the hazard of detonation. The gas source is located outside the vessel. In most cases the blanketing gas is nitrogen, although other gases may be used. In 1970, Appalachian Controls Environmental (ACE) was the world's first company to introduce a tank blanketing valve.

16.3.2 Benefits

A few of the benefits of blanketing include a longer life of the product in the container, reduced hazards, and longer equipment life cycles. Blanketing may prevent liquid from vaporizing into the atmosphere. It can maintain the atmosphere above a flammable or combustible liquid to reduce ignition potential. It can make up the volume caused by cooling of the tank contents, preventing vacuum and the ingress of atmospheric air. Blanketing can simply prevent oxidation or contamination of the product by reducing its exposure to atmospheric air. It can also reduce the moisture content. Gas such as nitrogen is supplied in a very pure and dry state.

When considering the application of blanketing for combustible products, the greatest benefit is process safety. Since fuels require oxygen to combust, reduced oxygen content in the vapor space lowers the risk of combustion hazards.

16.3.3 How Blanketing Functions

The most common gas used in blanketing is nitrogen. Nitrogen is widely used due to its inert properties, as well as its availability and relatively low cost. There must be a way of allowing the blanketing gas into the system and a way to vent the gas should the pressure get too high. The blanketing system must be capable of pressurizing the vapor space and accurately maintaining that pressure.

When the pressure inside the container drops below a set point, a valve opens and allows the blanketing gas to enter. Once the pressure reaches the set point, the valve closes. As a safety feature, many systems include a pressure vent that opens when the pressure inside exceeds a maximum pressure set point. This helps to prevent the container from rupturing due to high pressure. Since most blanketing gas sources will provide gas at a much higher than desired pressure, a blanketing system will also use a pressure reducing valve to decrease the inlet pressure to the tank. Blanketing systems usually operate slightly higher than atmospheric pressure (a few inches of water column above atmospheric). Higher pressures than this are generally not used as they often yield only marginal increases in results while wasting large amounts of expensive blanketing gas.

Another approach employs a simple, direct-operated pressure-reducing valve (PRV) to blanket the tank. However, these devices are best suited to a continuous flow rate. When used to blanket a tank, they must throttle over a wide flow range. Ranging from a shut-off to a full flow condition significantly varies the controlled pressure due to droop and lockup conditions.

17

Oil Field Chemicals (OFC)

There is an increasing demand for various specialty oil field chemicals (OFC) since there is a need to tap more oil from the existing reserves. OFC are classified into three main groups, according to their applications as follows:

1. Drilling fluids
2. Cementing and stimulation
3. Oil production chemicals

In this chapter, we will limit our discussion to the third group, oil production chemicals, which are used at all stages, in the oil field from wellhead to finished products including oil production at the well bore, gas–oil separation, and oil and gas treatment all the way through to the delivery end points of both quality oil and gas products.

In general, the usage of most production chemicals is directly related to the volume of associated water produced and is thus linked to the age of the field. Chemicals in this group include:

- Corrosion and scale inhibitors
- Biocides
- Demulsifiers
- Surfactants
- Solvents used in gas treatment and processing

For the selection of OFC, the traditional practice has always been based on their known individual usage and performance. However, the selection and the evaluation of OFC are guided by some technical parameters as explained in this chapter.

The effective mitigation of corrosion using corrosion inhibitors in oil field operations is addressed in this chapter through a brief elaboration on some corrosion fundamentals.

17.1 Introduction

Crude oil naturally occurs in the form of water-in-oil emulsion during its production. Composition of oil emulsions depend on the water-to-oil ratio, the natural emulsifier systems contained in the oil, and the origin of the emulsion. The naturally occurring emulsifiers in the crude oil have a very complex chemical nature. They differ from one oil well to another and they also depend upon the age of the oil wells; hence to overcome their effect petroleum emulsion–demulsifiers must be selectively developed.

Chemicals of various types, on the other hand, are used in every stage of drilling, completing, and producing oil and gas wells. Many services performed in the oil field rely on specialty fluids or chemicals that fulfill specific functions within the oil/gas operations.

Achieving greater oil field efficiency and productivity depends on well-site operations including surface processing of the produced oil/gas mixture. These operations should be cost effective in order to maximize the recovery of the terminal quality products, while minimizing the impact on the environment. Pivotal to these operations are specialty oil field chemicals (OFC) that are used in these operations.

OFC are used for improving the field operations and the output of drilling fluid, well stimulation, production chemicals, cementing, work over and completion, and enhance oil recovery (EOR) applications. The demand for specialty oil field chemicals is influenced by various factors such as increasing production of crude oil, location of reservoirs and depth of drilling, oil reservoirs availability, government policies, and trade-offs among various chemical compounds.

17.2 Evaluation and Selection

About half of the worldwide market continues to be accounted for by the highly mature fields in North America. A very high percentage of the chemical use in this region is associated with gas fields.

In oil field production, the greatest part of crude oils occurs as oil–water emulsion. These oil emulsions are stable mixtures of immiscible components formed as a result of the intimate mixing of the oil with the formation water highly stabilized with naturally occurring emulsifiers (stabilizers). To meet the market demand for quality crude oil, OFC are used extensively in oil/gas production operations.

Extensive testing is to be performed on the chemicals before they are used. This involves two basic steps. The first is a preliminary performance series of tests, where the formulations are evaluated for any undesirable side effects that they might have on the production process. The second series of tests involve evaluation of candidates for their performance in the field.

However, before the final selection of OFC, it is desirable to consider the effect of the following parameters:

- The interdependence of emulsion, corrosion, mineral scale deposits, and paraffin deposits.
- The corrosion rate may be enhanced when demulsifiers are injected since the pipe walls may change from oil-wet to water-wet.
- Paraffin dispersants cause crystals to repeal each other, whereas demulsifiers tend to increase coalescence of molecules.
- Filming corrosion inhibitors reduce corrosion by producing a protective scale at the metal surface. This scale may act as seed for scale and paraffin crystals formation.
- Corrosion may increase deposit of paraffin and scale since it roughens metallic walls.

To summarize, it is very important to have a good understanding of all the technical aspects of each case and to evaluate carefully the net results when applying OFC.

17.3 Classification of Chemicals

Oil field chemicals are generally classified based on type and application. Those on the basis of type include corrosion inhibitor and scale inhibitor, biocides, de-emulsifiers, pour-point depressants (PPD), surfactants, and polymers. Chemicals based on applications include drilling fluid, well stimulation, cementing, and production chemicals.

17.4 Corrosion Inhibitors and Scale Inhibitors

Corrosion is an unfortunate reality in nature and has a costly effect on equipment, in particular in oil field operations. Corrosion is the deterioration of a metal as a result of chemical reactions between it and the surrounding environment. It could be simply stated that corrosion of a metal means the loss of electrons by the metal. If we say, for example, that iron metal is corroded, we simply say that the Fe (atom) loses electrons to become Fe^{++} (an ion); that is, Fe (atom), when it corrodes, it is converted to → Fe^{++}+ ee.

A corrosion inhibitor is a substance when added in small concentration to an environment reduces the corrosion rate of a metal exposed to the environment. In the oil industry, corrosion inhibitors are designed to protect against the following: water, both fresh and brine; biological deposits; carbon dioxide (anaerobic corrosion); and hydrogen sulfide and other organic acids associated with oil operations in all oil-field types. Two types of corrosion inhibitors are generally applied:

1. Anodic inhibitors—They usually act by forming a protective oxide film on the surface of the metal, causing a large anodic shift of the corrosion potential. This shift causes the metallic surface to move into the passivation region. These inhibitors are called *passivators*. Chromates, nitrates, and molybdates are examples of anodic inhibitors.
2. Cathodic inhibitors—They act by either slowing the cathodic reaction itself or selectively precipitating on the cathodic areas in order to limit the diffusion of reducing species to the surface.

Corrosion inhibitor chemicals are injected at the level of wellhead, gas separation columns, pipelines header, and other locations as per customer, where it is pumped along with gas/oil, oil and gas into the pipeline, at specified locations, so as to provide a thin protective film on the metal surface that inhibits the reaction of corroding compounds such as acids, oxygen, and saline water with the metal surface, thereby protecting the oil field assets from the corrosion attack.

Corrosion inhibitor products are a combination of esters of fatty alcohols and amines, imidazolines, their salt, quaternary ammonium compounds, and amides mixed in the right proportion and formulated per the local requirement of each oil and gas field.

A scale inhibitor, on the other hand, is a chemical treatment used to control or prevent scale deposition in the production conduit or completion system or other applications. These are the oil field chemicals similar to corrosion inhibitors and are injected at the level of wellhead and other locations in the oil and gas processing steps so as to avoid the scale formation in the oil field installations. Scale inhibitors mainly protect the crude carrying pipelines, well casing, crude-processing facilities, and protect water-handling equipment from heavy scaling due to the scale formation of salts present in the crude and produce water. Scale inhibitors are usually a single or a combination of polyphosphonates, phosphoric acids, acrylic acid, and AMPS (2-acrylamido-2-methylpropane sulfonic acid, an acrylic monomer). They are mainly useful for scale prevention and protection of oil field assets.

17.5 Biocides

Biocidal products are defined as active substances and preparations containing one or more active substances put up in the form in which they are supplied to the user, intended to destroy, render harmless, prevent the action of, or otherwise exert a controlling effect on any harmful organism by chemical or biological means.

Bactericides are especially used in water injection and produce water treatment where the impurities are separated, and bacteria multiplication is avoided so that the water can be reused without any harm and does not degrade the crude quality.

These products are formulated through a combination of quaternary compounds, fatty acid ethoxylates, and tetrakis hydroxymethyl phosphonium sulfate (THPS) based compounds.

17.6 De-Emulsifiers

A large amount of petroleum, which is produced from petroleum-bearing formations, is contaminated by water or aqueous solutions of sodium chloride or other salts in emulsified form. Such water-containing systems occur predominantly in the form of water-in-oil emulsions. The natural-occurring emulsifiers in the crude oil have a very complex chemical nature, they differ from one oil well to another and they also depend upon the age of the oil wells, hence to overcome their effect, petroleum emulsion demulsifiers must be selectively developed. In practice, the water is separated by adding to the water–petroleum system very small amounts of emulsion breaking substances named de-emulsifiers.

A large number of de-emulsifiers of varying compositions for the indicated purpose have been proposed. The reason that de-emulsifiers of widely diverging chemical composition have been suggested is primarily due to the fact that petroleum, dependent on its origin, has different composition and de-emulsifiers that may be suitable for breaking water-in-oil emulsions of crude oil from one location may be unsuitable for accomplishing this result if the crude oil emanates from a different district or source. This means that the activity of prior art de-emulsifiers is specific to the nature of the respective petroleum or crude oil composition.

De-emulsifiers are used for treatment of crude upstream for separation of impurities such as water, heavy salts, and wax. The sludge is biodegradable.

Water, contained in petroleum in emulsified form, is separated by adding to the petroleum a conventional de-emulsifier and silica in fine particle form. The silica is advantageously in physical mixture with the de-emulsifier. The amount of silica, calculated on the amount of de-emulsifier, is about between 0.1% to 10% per weight.

The essential blocks are block polymer of castor, block polymer of fatty alcohol, propoxylate fatty amine, block polymer of fatty acid, propoxylate tall oil, propoxylate polyamine, linear alcohol ethoxylates, and a mixture of amine and alcohol ethoxylates. The combination of these basic products to make a demulsifier depends on the properties of crude and field conditions, and hence the product is customized to each customer and their specific field in a scientific manner.

Alkyl sulfates and alkyl aryl sulfonates, as well as petroleum sulfonates in the form of amino salts, have been proposed for de-emulsification purposes.

The demulsifier concentrations generally range from less than 5 ppm (approximately 1 gal/5000 bbl) to more than 200 ppm (approximately 8 gal/1000 bbl). The most common range is between 10 and 50 ppm. Whatever the demulsifier dosage and range, it may be possible to reduce and optimize the demulsifier usage by evaluating various components in the treatment program.

17.7 Surfactants

Surfactants are compounds that lower the surface tension (or interfacial tension) between two liquids or between a liquid and a solid. Surfactants may act as detergents, wetting agents, foaming agents, and dispersants. They are used for corrosion control, foaming, scale control, paraffin dispersants, asphaltene control, and demulsification.

Lauryl alcohol ethoxylates are a special class of nonionic surfactants. It is manufactured by mixing ethylene oxide with fatty alcohols that have alkyl carbon atoms. They have about 12 to 14 lauryl alcohol atoms and are commercially designated by the name LAE. They have high calcium ion tolerance, very high solubility for oily substances, and are easily biodegradable. They are also hygroscopic, hydrophilic, and lipophilic in nature.

17.8 Desalting Chemicals

Desalting chemicals are blends of amines and resins that are used to reduce salt level in crude oil both in upstream and downstream operations. Desalting oil field chemicals are blended with de-emulsifiers for upstream crude treatment. They are also used in refineries to cut down the salt content in crude oil.

17.9 Pour Point Depressants

The pour point of a fuel or oil is the lowest temperature at which it will pour when cooled under defined conditions. In general, the pour point is indicative of the amount of wax in oil. At low temperatures, the wax tends to separate, trapping a substantial amount of oil, inhibiting oil flow. It is also defined as the temperature at which it becomes semisolid and loses its flow characteristics. In crude oil, a high pour point is generally associated with high paraffin content.

Polymethacrylates are used as viscosity index improvers and pour point depressants. They impart exceptional shear stability and low temperature performance in treated oils. Pour point depressants are essentials to be added for crude oil flow lines all over oil field assets.

17.10 Chemicals Used in Gas Treatment and Processing

17.10.1 Sour Gas Treatment

Gas sweetening and acid gas removal refers to a group of processes that use aqueous solutions of various alkylamines (commonly referred to simply as amines) to remove hydrogen sulfide (H_2S) and carbon dioxide (CO_2) from gases. It is commonly used in petrochemical plants, natural gas processing plants, and other industries.

A number of solvent amines are used in gas treating: diethanolamine (DEA), monoethanolamine (MEA), methyldiethanolamine (MDEA), and diglycolamine (DGA). The most commonly used amines in industrial plants are the alkanolamines DEA, MEA, and MDEA.

17.10.2 Gas Dehydration

A number of chemicals are used depending on the method selected for dehydration. One of the most common methods is absorption using liquid desiccants such as glycols and methanol. Three types of glycols are used: ethylene glycol (EG), diethylene glycol (DEG) and triethylene glycol (TEG). Another method is adsorption using solid desiccants such as alumina and silica gel. Methanol is used to inhibit hydrate formation in the natural gas network.

18

Piping and Pumps

Pipelines along with pumps are needed as an efficient means of transporting crude oil, hydrocarbon products, natural gas, and other important fossil fuels, quickly, safely, and smoothly. Pipelines need to be constantly and reliably operated and monitored in order to ensure maximum operating efficiency, safe transportation, and minimal downtimes, and to maintain environmental and quality standards. Powerful pumps, on the other hand, are needed for oil transport of crude oil within the oil field and for the delivery of oil to terminal points.

In this chapter, the role of pipelines and pumps in oil field operations is highlighted. This includes the following:

- Gathering systems in the oil field
- Crude oil delivery network
- Sizing of pipeline and selection of wall thickness
- Economic balance in piping and optimum pipe diameter
- Other aspects of piping
- Classification and types of pumps
- How to select a pump
- Calculation of the horsepower (HP) for a pump

18.1 Pipelines

18.1.1 Introduction

Pipelines are the second most important form of oil and gas transportation. Their uses are more complex than uses of tankers, which by their nature only move crude oil or products and gas from or to a rather limited number of points on the oceans or navigable rivers. Pipelines, however, are used for gathering systems in oil fields, for moving the crude oil to refineries or marine terminals, and often for moving refined products from refineries to local distribution points.

Market demand growth can, of course, outstrip a pipeline's basic ability to handle the demanded volumes. The first way to solve this problem is to increase the speed with which the oil passes along the line by adding pumping stations. But since pipeline friction increases geometrically with the speed of flow, at some point it becomes economical to add more pipes. This process is called *looping,* and it consists of laying another pipeline along-side the existing one. In summary, pipelines serve a vital function in the transportation of both oil and natural gas.

18.1.2 Gathering Systems in Oil Fields

Producing oil fields commonly have a number of small diameter gathering lines that gather crude oil from the wells and move it to central gathering facilities called oil batteries. In general, there are four types of pipelines that are in common usage:

1. Oil field gathering pipelines; their function in the oil field is of great impact on production operations.
2. Larger diameter feeder pipelines transport the crude oil from the oil field to loading ports and nearby refineries.
3. Long-distance pipelines, which naturally shorten the alternative sea route.
4. Pipelines that transport oil from ports of discharge to inland refineries, located in industrial areas remote from a seaport. These are called transmission pipelines.

Pipelines used to carry crude oil and petroleum products differ a great deal in size, ranging from 2 in to as much as 36 in diameter. In some cases, even 48 in piping is used.

As far as the design of an oil gathering system, flowlines and trunklines make a combination of different schemes, as shown in Figures 18.1 and 18.2. These designs are described as follows:

- Individual flow lines—Through which oil wells are connected to a central gas–oil separator plant (GOSP).
- Trunklines and short flow lines—Well effluents in this scheme are directed into a large trunkline via short flowlines. Crude oil is directed to the GOSP by a trunkline.
- Trunkline and branches—Here major trunklines that are 30–50 miles long and 24–30 in diameter gather crude oil from branch trunklines, which in turn collect oil from short flowlines connected to the well. The branch trunklines are smaller in diameter than the main trunklines, 16–20 in, and much shorter.
- Wellhead separation—Where the oil exiting different wells is delivered to the GOSP through very short gathering lines. The separated oil is then transferred from the GOSP using trunklines.

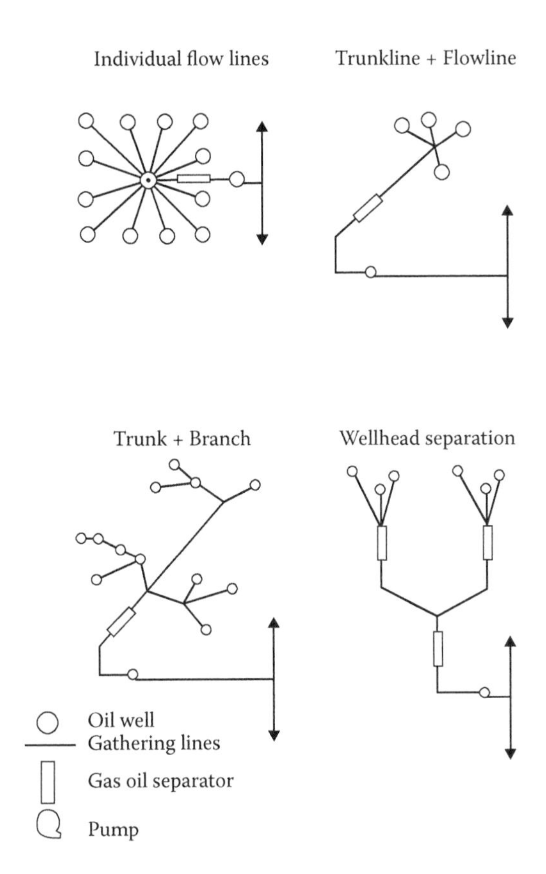

FIGURE 18.1
Oil field gathering system.

18.1.3 Crude Oil Delivery Network

Figures 18.3 and 18.4 illustrate how crude oil is transported from a wellhead through a delivery network.

18.1.4 Sizing of Pipelines

In the design of a pipe, one should be aware of two fundamental concepts: (1) the diameter of a pipe is a function of the flow rate of the fluid: $D = f(Q)$; and (2) the thickness of a pipe is a function of the working pressure inside the pipe: $t = f(p)$.

By sizing, we mean to determine the pipe diameter first. An engineer in charge must specify the diameter of pipe that will be used in a given piping system. Normally, the economic factor must be considered in determining the optimum pipe diameter.

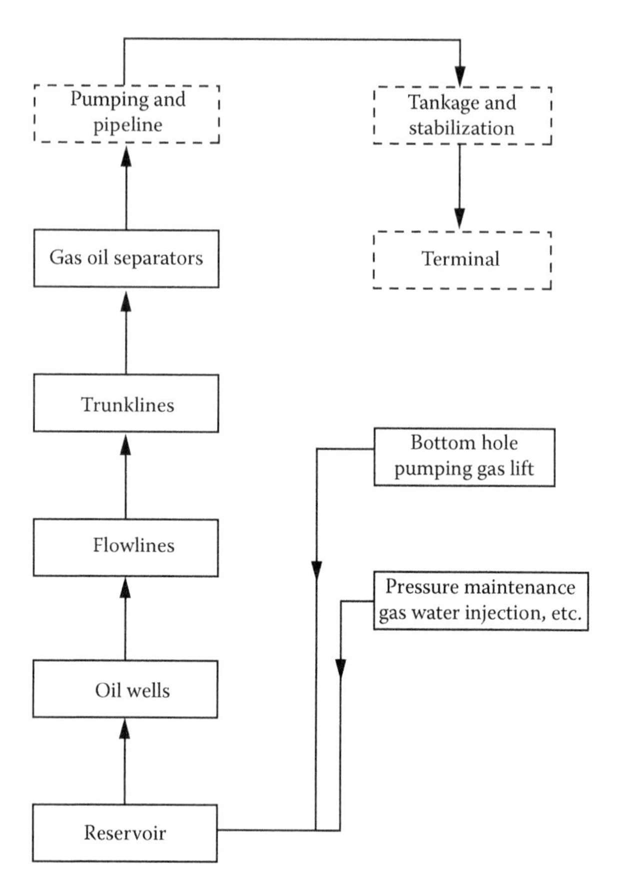

FIGURE 18.2
Diagrammatic sketch for a pipeline gathering system.

To calculate pipe diameter for noncompressible fluids, one can apply the well known equation:

$$Q = u.\ \mathrm{A(cross\ section\ area\ of\ pipe)}$$

$$= u.(\pi/4)d^2$$

Pipe diameter, d, is readily calculated from this equation for a specified flow rate, Q (bbl/hr) and for an assumed fluid velocity, u (ft/sec).

Example 18.1

Calculate the diameter of a pipeline handling 10,000 bbl of oil per hour, assuming that the velocity of flow is about 5 ft/sec.

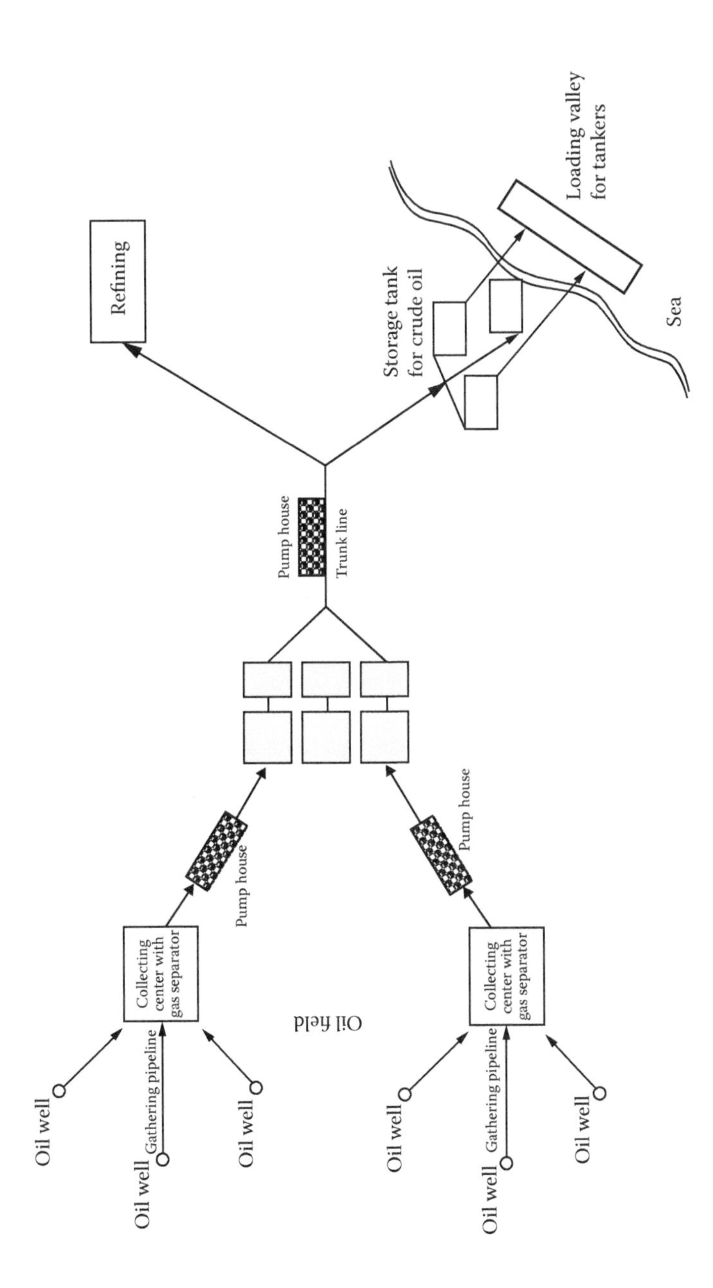

FIGURE 18.3
Network for the delivery of crude oil from an oil field.

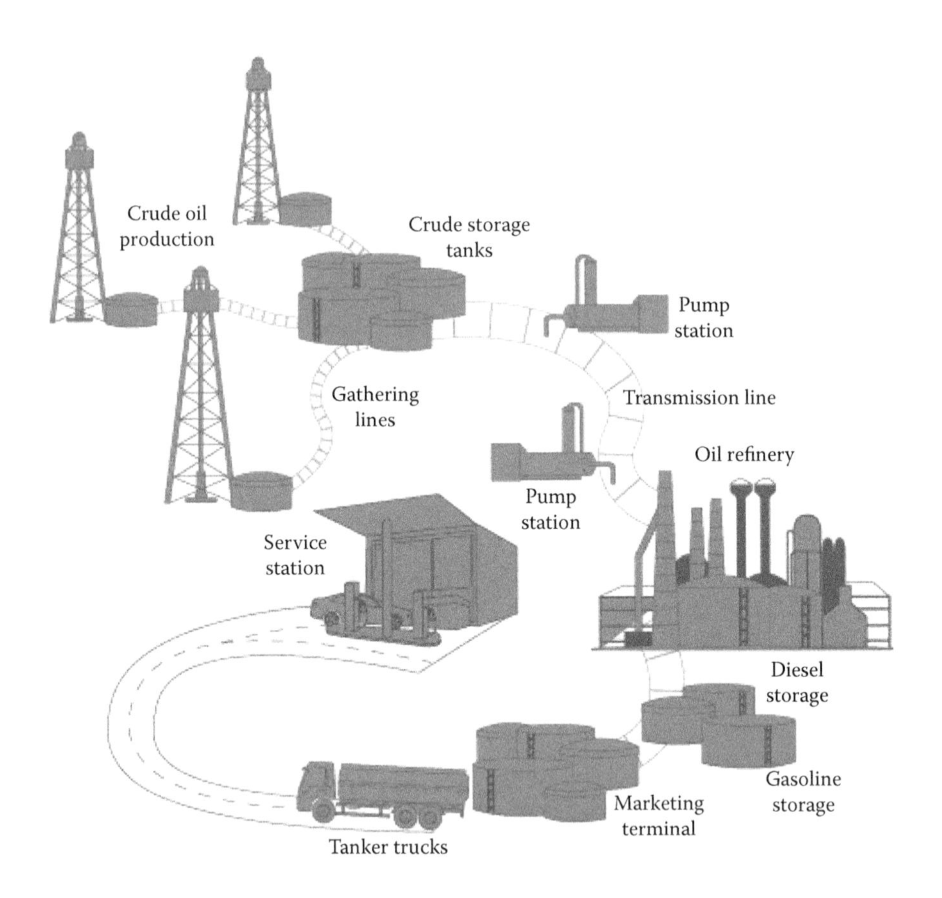

FIGURE 18.4
Network from wellheads to terminal points. (From Canadian Energy Pipeline Association, http://www.cepa.com/about-pipelines/types-of-pipelines/liquids-pipelines.)

Solution

$$Q = u.(\pi/4)d^2$$

$$[10{,}000\ \text{bbl/hr}\ (4.2\ \text{ft}^3/\text{bbl})]/3600\ \text{sec/hr} = 5\ \text{ft/sec}\ [(3.1416/4)\ d^2]\ \text{ft}^2$$

$$\text{Solving for } d = 1.724\ \text{ft}$$

18.1.4.1 Economic Balance in Piping and Optimum Pipe Diameter

When pumping a specified quantity of oil over a given distance, two alternatives exist and a decision has to be made as to whether to use a large diameter pipe with a small pressure drop or whether to use a smaller diameter pipe with a greater pressure drop. The first alternative involves a higher capital cost with lower running costs; the second, a lower capital cost with higher running costs specifically because of the need for more pumps. So, it

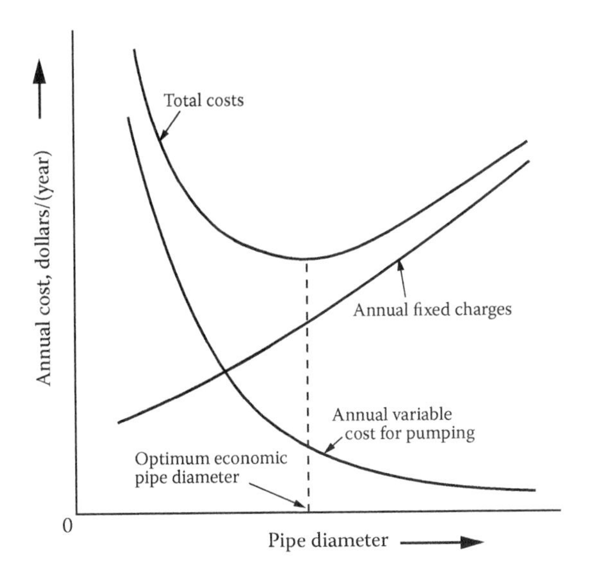

FIGURE 18.5
Optimum economic pipe diameter.

is necessary to arrive at an economic balance between the two alternatives. Unfortunately, there are no hard and fast rules or formulas to use; every case is different.

Costs of actual pumping equipment undoubtedly must be considered, but the area in which the pipes will *run* is also important. For instance, to obtain the same pumping effort in the desert as opposed to a populated area could involve much higher costs in the form of providing outside services and even creating a small, self-contained township.

In the flow of oil in pipes, the fixed charges are the cost of the pipe, all fittings, and installation. All these fixed costs can be related to pipe size to give an approximate mathematical expression for the sum of the fixed charges.

In the same way, direct costs, or variable costs comprising mostly the costs of power for pressure drop plus costs of minor items such as repairs and maintenance, can be related to pipe size. For a given flow, the power cost decreases as the pipe size increases. Thus direct costs decrease with pipe size. And total costs, which include fixed charges, reach a minimum at some optimum pipe size. The ultimate solution leading to the optimum diameter is found from the graph shown in Figure 18.5.

18.1.4.2 Stepwise Procedure to Calculate the Wall Thickness

1. Determine D_i, guided by the allowable pressure drop in a pipeline (ΔP).
2. Select a material of construction; S (tensile strength) is determined.

3. Knowing our operating pressure, Schedule Number = 1000 P/S is calculated.
4. If severe corrosion is anticipated in your pipe system, choose a larger Schedule Number.
5. Pick up a Nominal Pipe Size (NPS) with the specified Schedule Number that gives D_i for our flow equal or slightly greater than D_i obtained before.
6. As a final check, use the following equation to calculate the safe working pressure:

$$\text{Schedule Number} = 1000\, P/S = 2000\, (t_m/D_{av})$$

Solving this relationship to obtain:

$$P_s = 2\, S_s\, (t_m/D_{av})$$

where P is the operating pressure; P_s is the safe working pressure; S_s is the safe working fiber stress; S is the tensile strength, the greatest longitudinal stress a material can bear without tearing apart; t_m is the minimum thickness of pipe; and D_{av} is the average diameter of D_i and D_o.

There are many factors that affect the pipe wall thickness, including the maximum and working pressures, maximum and working temperatures, chemical properties of the fluid, the fluid velocity, the pipe material and grade, and the safety factor or code design application.

18.1.4.3 Relationship between Pipe Diameter and Pressure Drop

Two scenarios can be followed: either to assume a value for the velocity, u, and calculate ΔP; or consider an allowable value for the pressure drop, ΔP, and calculate the corresponding u.

The *first scenario* is illustrated by the following block diagram:

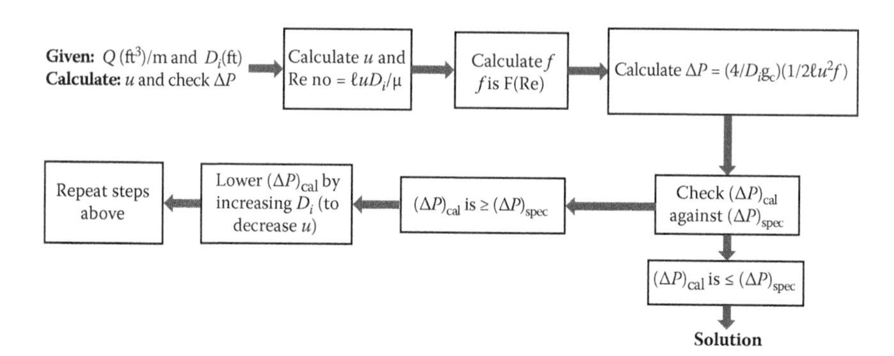

To determine the fluid velocity in a pipe, the rule of thumb economic velocity for turbulent flow is used, as reported by Peters and Timmerhaus (4th ed.):

Type of Fluid	Reasonable Assumed *u* (ft/s)
Water or fluid similar to water	3–10
Low pressure steam (25 psig)	50–100
High pressure steam (>100 psig)	100–200

Economics of scale are the major element in pipeline economies. From a theoretical point of view, doubling the pipeline diameter will tend to increase the amount delivered by more than fourfold in a given period of time—other factors remaining constant. This implies that total cost might double while the cost per unit delivered would decline. Crude oil moves at a speed of 5 kilometers per hour via pipeline, roughly a walking speed.

18.1.5 Other Aspects of Piping

Pipelines used for oil can be converted to natural gas, and vice versa, if the basic economic or strategic considerations make it appropriate. Similarly, if supply or demand conditions change, the direction of flow through pipelines can be reversed simply by turning around the pumping stations along the pipeline routes.

Oil pipelines are made from steel or plastic tubes with an inner diameter typically from 4 to 48 inches. Most pipelines are typically buried at a depth of about 3 to 6 feet. To protect pipes from impact, abrasion, and corrosion, a variety of methods are used. These can include wood lagging (wood slats), concrete coating, rockshield, and high-density polyethylene.

The major cost components of pipeline construction are material, labor, right-of-way (ROW) damages, and other miscellaneous costs. In most cases, material and labor account for more than 65% of construction cost. The investment distribution of constructing pipelines for both crude and oil product are almost similar.

The oil is kept in motion by pump stations along the pipeline, and usually flows at speed of about 4 to 20 ft/s. Multiproduct pipelines are used to transport two or more different products in sequence in the same pipeline. Usually in multiproduct pipelines there is no physical separation between the different products. Some mixing of adjacent products occurs, producing interface, also known in the industry as *transmix*.

Crude oil contains varying amounts of paraffin wax and in colder climates wax buildup may occur within a pipeline. Often these pipelines are inspected and cleaned using pipeline inspection gauges, *pigs*, also known as *scrapers* or *Go-devils*. Smart pigs (also known as intelligent or intelligence pigs) are used to detect anomalies in the pipe such as dents, metal loss caused by corrosion, cracking, or other mechanical damage.

A comprehensive field instrumentation system designed to provide operation-relevant data directly from all facilities along a pipeline is important to install. This is needed for the operation control system to run the facility with optimum productivity and efficiency, through a Supervisory Control and Data Acquisition (SCADA) system.

18.2 Pumps

18.2.1 Introduction

A fluid moves through a pipe or a conduit by increasing the pressure of the fluid using a pump that supplies the driving force for flow. In doing so, power must be provided to the pump. There are six basic means that cause the transfer of fluid flow: gravity, displacement, centrifugal force, electromagnetic force, transfer of momentum, and mechanical impulse. Excluding gravity, *centrifugal force* is the means most commonly used today in pumping fluids.

Centrifugal force is applied by using a *centrifugal pump* or *compressor*, of which the basic function of each is the same: To produce kinetic energy (KE) by the action of centrifugal force, and then converting this KE into pressure energy (PE) by efficient reduction of the velocity of the flowing fluid.

Fluid flow in pipes applying centrifugal devices have in general the following basic advantages and features:

- Fluid discharge is relatively free from pulsations.
- No limitation on throughput capacity of the operating pump.
- Discharge pressure is a function of the fluid density, i.e., $P = f(\ell_f)$.
- To provide efficient performance in a simple way with low first cost.

18.2.2 Classification and Types of Pumps

Pumps can be classified into three major groups according to the method they use to move the fluid: direct lift, displacement, and gravity pumps. Pumps can also be classified by their method of displacement as positive displacement pumps, impulse pumps, velocity pumps, and gravity pumps.

Pumps operate by a reciprocating or rotary mechanism. Mechanical pumps may be submerged in the fluid they are pumping or placed external to the fluid. A concise summary for the comparison between different types of pumps is in Table 18.1.

Pumps are used for many different applications. Understanding which pump type one needs for his application is very important. For the oil and gas industry, some basic features are listed next:

TABLE 18.1

Comparison between Types of Pumps

Type of Pump	Features
Centrifugal	Most common, high capacity, discharge lines can be shut off (safe) to handle liquids with solids
Reciprocating	Low capacity and high head, can handle viscous fluids, used to discharge bitumen (asphalt) in vacuum distillation columns
Rotary positive-displacement	Combination of rotary motion and positive displacement, used in gas pumps, screw pumps, and metering pumps
Air displacement	Nonmechanical, air-lift type, used for *acid eggs* and jet pumps

- Pumps should handle the fluids with low shear and least damage to droplet sizes causing no emulsions for the effective separation of water from oil.
- Pumps should be self-priming and experience no gas locking.
- The requirement of having low net positive suction head (NPSH) is an advantage. This is advantageous for vessel-emptying applications such as closed drain drums or flare knockout drums or any applications encountering high-vapor pressure liquids.
- Pumps should handle multiphase fluids.

18.2.3 How to Select a Pump

The following guide to pump types should prove to be helpful for better understanding the advantages and specifications for each pump type (PumpScout.com, n.d.).

- API process pumps—Designed to meet the 610 standard set by the American Petroleum Institute (API).
- Boiler feed pumps—Built to control the amount of water that enters a boiler. They are centrifugal pumps, and most are multistage.
- Chemical pumps—Built to handle abrasive and corrosive industrial materials. They can be either centrifugal or positive displacement type.
- Circulator pumps—Used to circulate fluid through a closed or looped system. They are usually centrifugal pumps, but a few use positive displacement technologies.
- Dewatering pumps—A dewatering process involves using a centrifugal pump (submersible or vertical turbine) to remove water from a construction site, pond, mine shaft, or any other area.
- Fire pumps—A type of centrifugal pump used for firefighting. They are generally horizontal split case, end suction, or vertical turbine.

- High-pressure pumps—Used in many applications including water blast, hydro-mining, and jet cutting. There can be a wide variety of pump types including positive displacement pumps, rotary pumps and reciprocating pumps, or centrifugal pumps.
- Industrial pumps—Used in industrial applications such as slurry, wastewater, industrial chemicals, oil and gas, etc. There are dozens of different industrial pumps both in positive displacement and centrifugal pump types.
- Marine pumps—Built to pump sea water. They are often used in large salt water tanks to continuously circulate water so it stays fresh.
- Mixed flow pumps—Incorporates the features of both axial flow pumps and radial flow pumps. Axial flow pumps operate on a vertical plane and radial flow pumps operate on a horizontal plane to the flow direction of water.
- Mud pumps—Built to transfer heavy sludge or mud. Some larger versions are used to raise the pressure. They are sometimes used on oil rigs to pressurize and circulate fluid.
- Petrochemical pumps—Made to transfer petroleum products that are often very viscous and corrosive. They can be magnetic drive pumps, diaphragm pumps, piston pumps, and others.
- Pneumatic pumps—Use compressed air to pressurize liquid through the piping system.
- Pressure pumps—Used to create either high or low pressure. They can be metering pumps and sometimes booster pumps.
- Process pumps—There are many types of centrifugal pumps or positive displacement pumps used in process applications. The type of pump and construction details varies depending on the application in which these pumps are used.
- Slurry pumps—A heavy duty pump that is made to handle thick, abrasive slurries. They are made of durable materials and capable of handling abrasive fluids for long periods of time.
- Solar pumps—Powered by the sun. They can be diaphragm pump (DP) or centrifugal pumps.
- Water pumps—A type of equipment used to move water through a piping system. They rely upon principles of displacement, gravity, suction, and vacuums to move water. They can be both positive displacement or centrifugal pumps.
- Well pumps—Designed to draw water to the surface from an underground water source. Depending on the well depth and configuration, the pumps can be jet pumps, centrifugal pumps, or submersible pumps.

In conclusion, the final selection of a pump for a particular operation is influenced by many factors, including the following:

- Pump capacity (size) which is a function of the flow rate to be pumped
- Fluid properties, both physical and chemical
- Operating conditions
- Type of power supply
- Type of flow distribution

18.2.4 Calculation of the Horsepower for a Pump

The following method is recommended to calculate the horsepower (HP) for a pump as a function of the flow rate and the total equivalent head, or the gauge pressure. It is much simpler to apply than using the mechanical energy equation:

$$\text{HP (Hydraulic Horsepower)} = [H.\ell.Q]/3960 = [P.Q]/1714$$

$$\text{Brake HP (actual)} = \text{HP (Hydraulic)}/\alpha$$

where H is head (in feet), ℓ is specific gravity, Q is flow rate (gpm), P is gauge pressure (in pounds per square inch), and α is pump efficiency (60% is used for centrifugal pumps).

Example 18.2

Find the HP for a pump that is handling 500 gpm of oil, against a 1000 ft pipeline with 19 ft equivalent to pipe fittings and valves. Assume friction losses account for 20% of the total head, and the specific gravity of the oil is 0.8.

Solution

$$\text{Total equivalent } H = 1000 + 19 + 0.2\ (1019) = 1223 \text{ ft}$$

$$\begin{aligned} \text{HP} &= [(1223)(500)(0.8)]/3960 \\ &= 123.5 \end{aligned}$$

$$\begin{aligned} \text{The Brake H.B. (Horsepower)} &= 123.5/0.6 \\ &= 206 \end{aligned}$$

Appendix A: Conversion Tables

Unit	Multiplied By	Approximate Conversion Factor	Equals	Unit
Volume				
Barrels of oil (bbl)	×	42	=	US gallons (gal)
Barrels of oil (bbl)	×	34.97	=	Imperial gallons (UK gal)
Barrels of oil (bbl)	×	0.136	=	Tonnes of oil equivalent (toe)
Barrels of oil (bbl)	×	0.1589873	=	Cubic meters (m^3)
Barrels of oil equivalent (boe)	×	5,658.53	=	Cubic feet (ft^3) of natural gas
Tonnes of oil equivalent (toe)	×	7.33	=	Barrels of oil equivalent (boe)
Cubic yards (y^3)	×	0.764555	=	Cubic meters (m^3)
Cubic feet (ft^3)	×	0.02831685	=	Cubic meters (m^3)
Cubic feet (ft^3) of natural gas	×	0.0001767	=	Barrels of oil equivalent (boe)
US gallons (gal)	×	0.0238095	=	Barrels (bbl)
US gallons (gal)	×	3.785412	=	Liters (l)
US gallons (gal)	×	0.8326394	=	Imperial gallons (UK gal)
Imperial gallons (UK gal)	×	1.201	=	US gallons (gal)
Imperial gallons (UK gal)	×	4.545	=	Liters (l)
Mass Weight				
Short tons	×	2000	=	Pounds (lb)
Short tons	×	0.9071847	=	Metric tonnes (t)
Long tons	×	1.016047	=	Metric tonnes (t)
Long tons	×	2240	=	Pounds (lb)
Metric tonnes (t)	×	1000	=	Kilograms (kg)
Metric tonnes (t)	×	0.9842	=	Long tons
Metric tonnes (t)	×	1.102	=	Short tons
Pounds (lb)	×	0.45359237	=	Kilograms (kg)
Kilograms (kg)	×	2.2046	=	Pounds (lb)
Length				
Miles (mi)	×	1.609344	=	Kilometer (km)
Yards (yd)	×	0.9144	=	Meters (m)
Feet (ft)	×	0.3048	=	Meters (m)
Inches (in)	×	2.54	=	Centimeters (cm)
Kilometer (km)	×	0.62137	=	Miles (mi)

(Continued)

Unit	Multiplied By	Approximate Conversion Factor	Equals	Unit
Area				
Acres	×	0.40469	=	Hectares (ha)
Square miles (mi^2)	×	2.589988	=	Square kilometers (km^2)
Square yards (yd^2)	×	0.8361274	=	Square meters (m^2)
Square feet (ft^2)	×	0.09290304	=	Square meters (m^2)
Square inches (in^2)	×	6.4516	=	Square centimeters (cm^2)
Energy				
British thermal units (Btus)	×	1,055.05585262	=	Joules (J)
Calories (cal)	×	4.1868	=	Joules (J)
Kilowatt hours (kWh)	×	3.6	=	Megajoules (MJ)
Therms	×	100,000	=	British thermal units (Btus)
Tonnes of oil equivalent	×	10,000,000	=	Kilocalories (kcal)
Tonnes of oil equivalent	×	396.83	=	Therms
Tonnes of oil equivalent	×	41.868	=	Gigajoules (GJ)
Tonnes of oil equivalent	×	11,630	=	Kilowatt hours (kWh)
Cubic feet (ft^3) of natural gas	×	1025	=	British thermal units (Btus)

Approximate Heat Content of Petroleum Products Million Btu (MMBtu per Barrel)

Energy Source	MMBtu/bbl	Energy Source	MMBtu/bbl
Crude oil	5.800	Natural gasoline	4.620
Natural gas plant liquids	3.735	Pentanes plus	4.620
Asphalt	6.636	Petrochemical feedstocks	
Aviation gasoline	5.048	Naphtha < 401°F	5.248
Butane	4.326	Other oils ≥ 401°F	5.825
Butane–propane (60/40) mixture	4.130	Still gas	6.000
Distillate fuel oil	5.825	Petroleum coke	6.024
Ethane	3.082	Plant condensate	5.418
Ethane–propane (70/30) mixture	3.308	Propane	3.836
Isobutane	3.974	Residual fuel oil	6.287
Jet fuel, kerosene-type	5.670	Road oil	6.636
Jet fuel, naphtha-type	5.355	Special naphthas	5.248
Kerosene	5.670	Still gas	6.000
Lubricants	6.065	Unfinished oils	5.825
Motor gasoline, conventional 5	5.253	Unfractionated stream	5.418
Motor gasoline, oxygenated or reformulated	5.150	Waxes	5.537
Motor gasoline	3.539	Miscellaneous	5.796

Appendix B: Values for Gas Constant

B1.1 Ideal Gas

Although no gas is truly ideal, many gasses follow the ideal gas law very closely at sufficiently low pressures. The ideal gas law was originally determined empirically and is simply

$$PV = nRT$$

where

P = absolute pressure (not gauge pressure)
V = volume
n = amount of substance (usually in moles)
R = ideal gas constant
T = absolute temperature (not °F or °C)

Values of R	Units ($V\,P\,T^{-1}\,n^{-1}$)
8.3144621(75)	$J\,K^{-1}\,mol^{-1}$
8.31446	$VC\,K^{-1}\,mol^{-1}$
5.189×10^{19}	$eV\,K^{-1}\,mol^{-1}$
0.08205736(14)	$L\,atm\,K^{-1}\,mol^{-1}$
1.9872041(18)	$cal\,K^{-1}\,mol^{-1}$
$1.9872041(18) \times 10^{-3}$	$kcal\,K^{-1}\,mol^{-1}$
$8.3144621(75) \times 10^{7}$	$erg\,K^{-1}\,mol^{-1}$
$8.3144621(75) \times 10^{-3}$	$amu\,(km/s)^2\,K^{-1}$
8.3144621(75)	$L\,kPa\,K^{-1}\,mol^{-1}$
$8.3144621(75) \times 10^{3}$	$cm^3\,kPa\,K^{-1}\,mol^{-1}$
8.3144621(75)	$m^3\,Pa\,K^{-1}\,mol^{-1}$
8.3144621(75)	$cm^3\,MPa\,K^{-1}\,mol^{-1}$
$8.3144621(75) \times 10^{-5}$	$m^3\,bar\,K^{-1}\,mol^{-1}$
8.205736×10^{-5}	$m^3\,atm\,K^{-1}\,mol^{-1}$
82.05736	$cm^3\,atm\,K^{-1}\,mol^{-1}$
84.78402×10^{-6}	$m^3\,kgf/cm^2\,K^{-1}\,mol^{-1}$
$8.3144621(75) \times 10^{-2}$	$L\,bar\,K^{-1}\,mol^{-1}$
$62.36367(11) \times 10^{-3}$	$m^3\,mmHg\,K^{-1}\,mol^{-1}$
62.36367(11)	$L\,mmHg\,K^{-1}\,mol^{-1}$
62.36367(11)	$L\,Torr\,K^{-1}\,mol^{-1}$

(*Continued*)

Values of R	**Units ($V\ P\ T^{-1}\ n^{-1}$)**
6.132440(10)	ft lbf K^{-1} g-mol^{-1}
1545.34896(3)	ft lbf R^{-1} lb-mol^{-1}
10.73159(2)	ft^3 psi R^{-1} lb-mol^{-1}
0.7302413(12)	ft^3 atm R^{-1} lb-mol^{-1}
1.31443	ft^3 atm K^{-1} lb-mol^{-1}
998.9701(17)	ft^3 mmHg K^{-1} lb-mol^{-1}
1.986	Btu lb-mol^{-1} R^{-1}

References

Abdel-Aal, H. K. *Surface Petroleum Operations*. Jeddah, Saudi Arabia: Saudi Publishing House, 1998.

Abdel-Aal, H. K., and R. Schmelzlee. *Petroleum Economics and Engineering, An Introduction*. New York: Marcel Dekker, 1976.

Abdel-Aal, H. K., and A. A. Shaikh. Desalting of oil using multiple orifice mixers: An empirical correlation for the water of dilution. Presented at the Third Iranian Congress of Chemical Engineering, 1977.

Abdel-Aal, H. K., and M. A. Shalabi. Noncatalytic partial oxidation of sour natural gas versus catalytic steam reforming of sweet natural gas. *Industrial & Engineering Chemistry Research* 35 (1996): 1787.

Abdel-Aal, H. K., and M. A. Alsahlawi, eds. *Petroleum Economics and Engineering*, 3rd ed. Boca Raton, FL: Taylor & Francis/CRC Press, 2014.

Abdel-Aal, H. K., A. Bakr, and M. A. Al-Sahlawi, eds. *Petroleum Economics and Engineering*, 2nd ed. New York: Marcel Dekker, 1992.

Abdel-Aal, H. K., M. Aggour, and M. A. Fahim. *Petroleum and Gas Field Processing*. New York: Marcel Dekker, 2003.

Al-Ghamdi, A., and S. Kokal. Investigation of causes of tight emulsions in gas oil separation plants. SPE Proceedings, Middle East Oil Show, Bahrain, June 9–12, 2003.

Allen, T. O., and A. P. Roberts. *Production Operations*, vols. 1 and 2. Tulsa, OK: Oil and Gas Consultant International, 1993.

Al-Tahini, A. Crude oil emulsions. Co-op Report, Department of Chemical Engineering, KFUPM, Dhahran, Saudi Arabia, 1996.

American Petroleum Institute. *Primer of Oil and Gas Production*. Washington, DC: American Petroleum Institute, 1976.

American Petroleum Institute. Industry sectors. http://www.api.org/aboutoilgas/sectors/, 2007. Retrieved May 12, 2008.

American Petroleum Institute. Robust Summary of Information on Crude oil, Case No. 800205-9, 2011.

American Society of Civil Engineers. ASCE Subject Headings: Corrosion, Oil pipelines, 2009.

Arnold, K., and M. Stewart. *Surface Production Operations*, vol. 2. Houston, TX: Gulf Publishing, 1988.

Arnold, K., and M. Stewart. *Design of Oil Handling Systems and Facilities*, vol. 1. Houston, TX: Gulf Publishing, 1989a.

Arnold, K., and M. Stewart. *Surface Production Operations*, vol. 2. Houston, TX: Gulf Publishing, 1989b.

Arnold, K., and M. Stewart. *Surface Production Operations*, vol. 1. Houston, TX: Gulf Publishing, 1998a.

Arnold, K., and M. Stewart. *Surface Production Operations: Design of Oil-Handling Systems and Facilities*, vol. 1, 2nd ed. Richardson, TX: Gulf Publishing, 1998b.

Basseler, O. U. *De-Emulsification of Enhanced Oil Recovery Produced Fluids*. St. Louis, MO: Tretolite Div., Petrolite Corp., 1983.

Berger, B. D., and K. E. Anderson. *Modern Petroleum: A Basic Primer of the Industry*, 2nd ed. Tulsa, OK: Penn Well Books, 1981.
Borchardt, J. K., and T. F. Yen. *Oil-Field Chemistry*, vol. 396. Washington, DC: American Chemical Society, 1989.
Bourgoyne, A. T., Jr., K. K. Millheim, M. E. Chenevert, and F. S. Young, Jr. *Applied Drilling Engineering*, 2nd ed. Society of Petroleum Engineers, 1991.
Bradley, H. B. *Petroleum Engineering Handbook*. Richardson, TX: Society of Petroleum Engineers, 1987.
Campbell, J. M. *Gas Conditioning and Processing*, vol. 2. Norman, OK: Campbell Petroleum Series, 1976.
Campbell, J. M. *Gas Conditioning and Processing*, vol. 1. Norman, OK: Campbell Petroleum Series, 1978.
Canadian Energy Pipeline Association. http://www.cepa.com/about-pipelines/types-of-pipelines/liquids-pipelines.
Carios, E., L. Vega, R. Pardo, and J. Ibarra. Experimental study of a poor boy downhole gas separator under continuous gas-liquid flow. Presented at the SPE Artificial Lift Conference-Americas, Cartagena, Columbia, May 21–22, 2013. SPE-165033-MS.
Chilingarian, G. V., and C. M. Beeson. *Surface Operations in Petroleum Production*. New York: Elsevier Publishing, 1969.
Chilingarian, G. V., J. O. Robertson, Jr., and S. Kumar. *Surface Operations in Petroleum Production*, I. Amsterdam: Elsevier Science, 1987.
Craft, B. C., and M. Hawkins. *Applied Reservoir Engineering*, 2nd ed. Englewood Cliffs, NJ: Prentice-Hall, 1991.
Craft, B. C., W. R. Holden, and E. D. Graves. *Well Design: Drilling and Production*. Englewood Cliffs, NJ: Prentice-Hall, 1962.
Crude Oil Quality Group. Crude oil quality programs: What is involved and why is quality important? http://www.coqa-inc.org/docs/default-source/meeting-presentations/20010927Conoco.pdf, September 27, 2001. Retrieved August 17, 2015.
Crude Oil Quality Group. Crude oil contaminants and adverse chemical components and their effects on refinery operations. Presented at the General Session Houston, TX, May 27, 2004. http://www.coqa-inc.org/docs/default-source/meeting-presentations/components-paper.pdf?sfvrsn=2. Retrieved August 17, 2015.
Devold, H. *Oil and Gas Production Handbook*. SRH Publisher, 2012.
Doctor, V. H., and H. D. Mustafa. Crude oil cleaning with solvent. *Hydrocarbon Asia* 1 (2003): 62–64.
Donohue, D. A. T., and R. W. Taylor. *Petroleum Technology*. Boston: IHRDC, 1986.
Economides, M. J., and L. Kappos. UNESCO—EOLSS sample chapters exergy, energy system analysis and optimization—Vol. II, *Petroleum Pipeline Network Optimization*. Oxford: EOLSS Publishers, 2002.
Energy Intelligence Group. U.S. Energy Information Administration. *International Crude Oil Market Handbook*. 2011. http://eia.gov/todayinenergy/detail.cfm?id=7110. Retrieved February 2, 2014.
Fahim, M. A., T. A. Alsahhaf, and A. A. Elkilani. *Fundamentals of Petroleum Refining*. Amsterdam: Elsevier, 2010.
Fink, J. *Petroleum Engineer's Guide to Oil Field Chemicals and Fluids*. Waltham, MA: Elsevier, 2012.

Gas Processors Suppliers Association. *Engineering Data Book*, 15th ed. Tulsa, OK: Gas Processors Suppliers Association, 1987.

Green, D. W., and G. P. Willhite. *Enhanced Oil Recovery*. Richardson, TX: Society of Petroleum Engineers, 1998.

Halliburton. "Water Production and Disposal" in *Coalbed Methane: Principles and Practice*, pp. 421–459. http://www.halliburton.com/public/pe/contents/Books_and_Catalogs/web/CBM/H06263_Chap_09.pdf, 2008. Retrieved August 17, 2015.

Hearts, J. R., P. H. Nelson, and F. L. Paillet. *Well Logging for Physical Properties*. New York: Wiley, 2000.

Holland, C. D. *Fundamentals of Multicomponent Distillation*. New York: McGraw–Hill, 1981.

Inkpen, A., and M. H. Moffett. *The Global Oil & Gas Industry: Management, Strategy and Finance*. Tulsa, OK: Penn Well Corporation, 2011.

Kidnay, A. J., and W. Parish. *Fundamentals of Natural Gas Processing*. CRC Press, 2006.

Kima, E. A. Oil field chemicals synergy upstream Arab heavy crude handling facilities. SPE 154099, Society of Petroleum Engineering, 2012.

Kister, H. Z. *Distillation Operations*. New York: McGraw-Hill, 1988.

Kokal, S., and M. Wingrove. Emulsion separation index: From laboratory to field case studies. SPE Proceedings, Annual Technical Conference and Exhibition, Dallas, October 1–4, 2000.

Lee, W. J. *Well Testing*. Richardson, TX: Society of Petroleum Engineers, 1982.

Linga, H., F. A. Al-Qahtani, and S. N. Al-Qahtani. New mixer optimizes crude desalting plant (SPE 124823). Paper presented at the 2009 SPE Annual Technical Conference and Exhibition, New Orleans, LA, 2009.

Link, P. K. *Basic Petroleum Geology*, 3rd ed. Tulsa, OK: Oil and Gas Consultant International, 2001.

"List of oil fields," *Wikipedia*, last modified July 19, 2015. https://en.wikipedia.org/wiki/List_of_oil_fields.

Lunsford, K. M., and J. A. Bullin. Optimization of amine sweetening units. Proceedings of the 1996 AIChE National Meeting, New York, 1996.

Maddox, R. N., J. H. Erbar, and A. Shariat. *PDS Documentation*. Stillwater, OK: CPC, Inc., 1976.

Manning, F. S., and R. E. Thomson. *Oilfield Processing of Petroleum*. Tulsa, OK: Penn Well Publishing, 1991.

Manning, F. S., and R. E. Thompson. *Oilfield Processing, vol. 2: Crude Oil*. Tulsa, OK: Penn Well, 1995.

Masseron, J. *Petroleum Economics*. Paris: Editions Technip, 1990.

McCain, W. D. Jr. *The Properties of Petroleum Fluids*, 2nd ed. Tulsa, OK: Penn Well, 1990.

McKetta, J. J., ed. *Petroleum Processing Handbook*. New York: Marcel Dekker, 1992.

Mennon, V. B., and D. T. Wassam. De-emulsification. In *Encyclopedia of Emulsion Technology*, edited by P. Becher. New York: Marcel Dekker, 1984.

Merchant, P., and S. M. Lacy. Water-based demulsifier formulation and its use in dewatering and desalting crude hydrocarbon oils, US Patent 455123 gA, 1985.

Merichem Company. *Economic Treatment of Whole Crude Oil (Proprietary and Confidential Information)*. Crude Oil Quality Association, March 1, 2012.

Meyers, R. A., ed. *Handbook of Petroleum Refining Processes*. New York: McGraw-Hill, 1996.

Meyers, R. A. *Handbook of Petroleum Refining Processes*, 3rd ed. New York: McGraw-Hill, 2003.

Moins, G. Stabilization process comparison helps selection. *Oil and Gas Journal* 78 (1980): 163–173.

Najafi, M., and B. Ma, eds. *ICPTT 2009: Advances and Experiences with Pipelines and Trenchless Technology for Water, Sewer, Gas, and Oil Applications*. Reston, VA: American Society of Civil Engineers, 2009.

Nalco Chemical Co. *Theories of Emulsion Breaking*, vol. 3. Sugarland, TX: Technology Series CTS, 1983.

Nelson, W. L. *Petroleum Refinery Engineering*, 4th ed. New York: McGraw-Hill, 1958.

NFPA 30-2008, Basic Requirements for Storage Tanks Society of Fire Protection Engineers, New York, 2011.

Perry, R. H., and D. Green. *Perry's Chemical Engineers' Handbook*, 50th ed. New York: McGraw-Hill, 1984.

Perry, R. H., and D. Green. *Perry's Chemical Engineers' Handbook*. New York: McGraw-Hill, 1999.

Peters, M., K. Timmerhaus, and R. West. *Plant Design and Economics for Chemical Engineers*, 4th ed. New York: McGraw-Hill Education, 2003.

PumpScout.com. Pump types guide a comprehensive guide to all industrial pump types. http://www.pumpscout.com/articles-scout-guide/pump-types-guide-aid100.html, n.d. Retrieved August 17, 2015.

Robbins, W. K., and C. S. Hsu. Petroleum, composition. In *Kirk-Othmer Encyclopedia of Chemical Technology* [online]. John Wiley & Sons, 2000.

Rudd, D. F., G. J. Powers, and J. J. Sivola. *Process Synthesis*. Englewood Cliffs, NJ: Prentice-Hall, 1973.

Schlumberger. Oilfield glossary: Production facilities. http://www.glossary.oilfield.slb.com/en/Disciplines/Production-Facilities.aspx, n.d. Retrieved April 5, 2014.

Sivalls, C. R. *Glycol Dehydration Design Manual*. Odessa, TX: Sivalls Inc., 1976.

Sivalls, C. R. *Oil and Gas Separation Design Manual*. Odessa, TX: Sivalls Inc. Bradley, B. H., Petroleum Engineering Handbook, SPE, Richardson.

Society of Petroleum Engineers. *Production Facilities*. SPE Reprint Series No. 25. Richardson, TX: Society of Petroleum Engineers, 1989.

Szilas, A. P. *Production and Transportation of Oil and Gas*. Amsterdam: Elsevier Science, 1975.

Tennyson, R. N., and R. P. Schaff. Guidelines can help choose proper process for gas treating plants. *Oil and Gas Journal* 10 (1977): 78–86.

Thro, M. E., and K. E. Arnold. Water droplet size determination for improved oil treater sizing. SPE 69th Annual Technical Conference and Exhibition, 1994.

Types by transport pipeline by function, Wikipedia, last modified July 31, 2015. https://en.wikipedia.org/wiki/Pipeline_transport.

U.S. Energy Information Administration. *Annual Energy Outlook 2015 with Projections to 2040*. http://www.eia.gov/forecasts/aeo/pdf/0383(2015).pdf, 2015. Retrieved March 1, 2015.

Valle-Riestra, J. R. *Project Evaluation in the Chemical Process Industries*, 1st ed. New York: McGraw-Hill, 1983.

Van Vector, S. A. How oil from the north went south. Presentation at the Electrical Policy Research Group, University of Cambridge, March 3, 2008.

Ventura, California, Copyright 1979, American Institute of Mining, Metallurgical, and Petroleum Engineers, Inc.

Vonday, D. Spherical process vessels. *Oil and Gas Journal* 121–122, April 8, 1957.

Wang, X., and M. Economides. *Advanced Natural Gas Engineering*. Houston, TX: Gulf Publishing Company, 2009.

Whinery, K. F., and J. M. Campbell. A method for determining optimum second stage pressure in 3-stage separation. *Journal of Petroleum Technology* 4 (1958): 53–54.

Wolf, Jr., J. R., T. C. McLean. Husky Oil Company. Oilfield Engineering and Consulting, Inc. SPE California Regional Meeting April 18–20, 1979.

Yang, Z., H. Qu, and C. Liu. Corrosion analysis of gathering pipeline in oil field. In *ICPTT 2009: Advances and Experiences with Pipelines and Trenchless Technology for Water, Sewer, Gas, and Oil Applications*, edited by M. Najafi and B. Ma, 1552–1560. Reston, VA: American Society of Civil Engineers, 2009.

Yocum, B. T. Mathematical models for design and optimization. Proceedings of the Second AIME Regional Technical Symposium, Dhahran, Saudi Arabia, March 1969.

Index

Page numbers ending in "f" refer to figures. Page numbers ending in "t" refer to tables.

H